C0-APH-380

23. Colloquium
der Gesellschaft für Biologische Chemie
13.–15. April 1972 in Mosbach/Baden

Protein-Protein Interactions

Edited by R. Jaenicke and E. Helmreich

With 234 Figures

Springer-Verlag Berlin · Heidelberg · New York 1972

Professor Dr. R. JAENICKE, Biochemie II, Fachbereich Biologie, Universität Regensburg, D-8400 Regensburg, Universitätsstr. 31

Professor Dr. E. HELMREICH, Direktor des Physiologisch-Chemischen Institutes der Universität Würzburg, D-8700 Würzburg, Koellikerstr. 2

ISBN 3-540-05992-X Springer-Verlag Berlin · Heidelberg · New York
ISBN 0-387-05992-X Springer-Verlag New York · Heidelberg · Berlin

Druck: Carl Ritter & Co., Wiesbaden

Preface

The present volume contains the proceedings of the 23rd Mosbach Colloquium on "Protein-Protein Interactions". It includes the paper by Prof. S. LIFSON who unfortunately was unable to attend.

Discussions were included whenever possible in order to make the complete proceedings accessible to the future reader.

The colloquium was generously supported by the firms who are corporate members of the Gesellschaft für Biologische Chemie.

Dr. W. B. GRATZER and Dr. L. JAENICKE, as well as some of the invited speakers, were a tremendous help in preparing the program and we wish to thank them for their helpful suggestions and advice. Our appreciation extends to Prof. E. AUHAGEN and Prof. H. GIBIAN, and their secretaries, who shared the burden of organizing the colloquium.

In addition, we wish to express our gratitude to Springer-Verlag, and especially to Dr. H. MAYER-KAUPP for editorial help and for the rapid publication of this volume.

We hope that the reader, like the audience in the Mosbach market hall, will gain an impression of the present state of protein research and of the way in which intermolecular interactions hold specific implications for the structure and function of proteins.

Regensburg and Würzburg, autumn 1972

R. JAENICKE
E. HELMREICH

Contents

Multienzyme Complexes

Antigen-Antibody Interactions

Self-Assembly

Intercellular Interactions

Introduction

R. Jaenicke

Biochemie II, Fachbereich Biologie, Universität Regensburg, Regensburg, Germany

The label "protein-protein interactions", the subject of this year's Mosbach Colloquium, could be put on a broad variety of problems in biophysics and biochemistry. Therefore the selection of topics in our program needs some explanation.

The complex molecular organization of living organisms is based on the association of different classes of bio-macromolecules, i.e. nucleic acids, proteins, polysaccharides, and lipids. The ribosome, simple viruses, enzyme complexes, and contractile systems may be considered different stages in a hierarchy of supra-molecular assemblies. The next level would be organelles which build up the cell, such as the cell nucleus, mitochondria, chloroplasts, etc., and finally the cell itself and the assembly of cells to differentiated tissues.

There is active research in all these areas. In fact, the amount of available information is so great, one could very well dwell on one specific associating system for the whole of this symposium, and in this way illustrate the significance of intermolecular interactions in protein molecules.

Instead, an attempt has been made to sum up some of the more important facets of the theme in order to give a bird's-eye view of the whole field. The multiplicity of the phenomena will become apparent in the introductory lectures which cover the structure-function relationship in allosteric enzymes as well as the general aspects of intermolecular forces and energy calculations in proteins and simple protein models. Then follows a number of examples which give an idea of the extent to which X-ray crystallography, electron microscopy and labelling techniques have been able to elucidate specific interactions in proteins with quaternary structure and in certain complexes of distinct proteins. This discussion of methods and their potentialities is supplemented by the consideration of thermodynamic and kinetic aspects of cooperativity in assembly processes.

On the next level of molecular organization structural and functional aspects are discussed for a number of heterologous protein systems, such as the glycolytic chain or stable multienzyme complexes, as well as the contractile proteins of muscle, and complexes of protein antigen and antibody. The remaining contributions on the self-assembly of TMV protein and cell-cell interactions give an insight into the process of morphogenesis, the final step in the hierarchy of molecular and supramolecular assemblies.

Before we start the program I would like to thank all the speakers. The fact that we have this extensive program full of contributions by distinguished speakers shows that almost all our invitations were accepted. Let us take this as a good omen for a successful Colloquium.

Molecular Forces

S. Lifson

Department of Chemical Physics
The Weizmann Institute of Science, Rehovot, Israel

It seems appropriate that a conference on protein-protein interactions should include an introductory discussion on the subject of molecular forces. The nature of molecular forces is universal and our understanding of such forces in small molecules, particularly those which have the same groupings of atoms as the side groups and the backbone of proteins, promotes our understanding of the forces within and between protein molecules. The more we understand molecular forces in proteins, the better will be our understanding of the structure of protein molecules and of their complexes, their interaction with substrates and other molecules, and ultimately the detailed quantitative relation between molecular structure and biological function in living systems.

The nature of molecular forces is in principle well understood. Molecular and atomic forces are all essentially Coulomb, i.e. electrostatic forces, and their dynamic behavior is determined by the Schroedinger equation, i.e. by the laws of quantum mechanics. The solution of the Schroedinger equation for various molecular systems has thus become one of the major trends in the study of molecular interactions and constitutes a large part of theoretical quantum chemistry. There are, however, serious limitations in the application of quantum chemistry to the study of large molecules. It is impossible to obtain exact solutions to their Schroedinger equations, and the larger the molecular system the more unfavorable is the effect of the approximations used, until one is not sure whether one is studying the nature of the molecules or of the approximations. Another trend which has been followed in the study of molecular forces is based upon a phenomenological approach. Phenomenologically, we distinguish a great number of forces which may be studied separately and which together determine the physical and chemical properties of molecular systems. This approach is the subject of the present discussion.

Intermolecular Forces

Interactions between molecules are both attractive and repulsive and fall into three categories:

a) *Electrostatic interactions.* When molecules carry net electrical charges there is obviously a Coulombic interaction between them which is repulsive for equal signs and attractive for opposite signs of the charges and decreases with inverse distance. Similarly, polar molecules, which have permanent electrostatic dipole moments, exert a dipole interaction on each other. Such interactions are well represented by the classical electrostatic theory.

b) *Dispersion forces.* Even electro-neutral, non-polar molecules exert on each other forces which are functions of the intermolecular distance as well as of the electronic structure. The theory of the attractive interaction between atoms and molecules is due to London, and such forces are often called London forces, or dispersion forces.

While the mathematical derivation of the theory of the dispersion force is complex, the qualitative explanation of the nature of this force and its origin is simple and easy to grasp. When two molecules approach each other, the electron cloud distribution of each molecule is somewhat perturbed by the influence of the electrostatic forces exerted by both the positively charged nuclei and negatively charged electrons of the other molecule. The mutual perturbation of the electron charge distributions reduces the total energy of the interacting molecules and this creates an attractive force. London showed that this interaction energy is related to the polarizabilities of the interacting molecules and decreases approximately with the inverse 6th power of the intermolecular distance, if the distance is large compared with the molecular diameters, and if the molecules are spherical. An important property of the London forces is their additivity, that is to say, that the energy of a system of many molecules is the sum of the interaction energies of all pairs of molecules. Furthermore, the force between two large polyatomic molecules may be attributed to the interatomic forces between all the atoms of one molecule and all the atoms of the other molecule. This concept of additivity has been extended to include *intra-*

molecular interaction between atoms in polyatomic molecules. According to this concept, or rather this assumption, atoms of the same molecule which are not chemically bonded to each other interact with each other as if they belong to different molecules. One should keep in mind, however, that this assumption of the additivity of interatomic nonbonded interactions in polyatomic molecules is neither as theoretically well-founded nor as practically reliable as the original London theory of attraction between two spherically symmetric, polyelectronic atoms.

c) *Repulsive forces.* When molecules or atoms approach each other so closely that their electronic clouds start to overlap, there is a very strong repulsive force between them which increases steeply with decreasing distance. This is the force which ensures that two different entities cannot simultaneously occupy the same space. Quantum mechanical theory has a qualitative explanation for this repulsive force, as resulting from the Pauli exclusion principle. The analytical description of the variation of the force with distance is approximately represented by an exponential decreasing function $e^{-\alpha(r-r_0)}$ where r is the atomic distance, and α and r_0 are constant parameters, chosen to fit the experimental data.

Attractive and repulsive forces act simultaneously and cannot be isolated. The equilibrium distance between molecules in crystals, for example, is determined by the balance between these forces, which is obtained when the sum of all forces acting on each molecule is zero, namely, when the potential of the total force is minimum. Lennard-Jones proposed that the total interatomic potential of non-bonded interaction between non-polar molecules be written in the form

$$V_{LJ}(r) = A/r^{12} - B/r^6 \tag{1}$$

which is commonly known as the Lennard-Jones (LJ) potential, or the 12-6 potential. The first term in this potential represents the repulsion and the second the attraction.

The 12th-power dependence on distance of the first term is, according to Lennard-Jones, an empirical choice mainly for computational convenience, and devoid of theoretical significance. Various other powers have sometimes been used. Recent studies have shown [1] that for alkane molecules 9th-power dependence seems preferable to 12th-power for atom-atom repulsions (see last section). The parameter B has been the subject of extensive theoretical studies relating the strength of attraction to atomic polarizabilities. However, the limited success of these efforts, and the complexity of the problem of interatomic attractions in polyatomic molecules, make it impractical to determine B on a priori theoretical grounds. Thus, it is common practice to fit the parameters A and B empirically so as to obtain best agreement between theory and experiment.

Another common representation of intermolecular or interatomic interactions is the so-called Buckingham or exp-6 potential

$$V_B(r) = Ae^{-\alpha r} - B/r^6 \tag{2}$$

where the repulsive term can also be written as $e^{-\alpha(r-r_0)}$. A, α and B are again adaptable parameters. This potential may perhaps be better than the LJ potential in representing the repulsive force, but requires one more adaptable parameter for each atom which, being in the exponent, is more difficult to fit by the least-squares method (see below). Thus the choice between exp-6 and LJ potentials is still an open question.

d) *The hydrogen bond.* All three types of intermolecular interaction play an important role in determining the structure of protein molecules and their mutual interactions. However, special importance is accorded a particular kind of intermolecular interaction, the hydrogen bond. This bond seems to be intermediate between an intermolecular interaction and a chemical bond. It is stronger than the former and weaker than the latter. Like electrostatic dipole interactions, it varies with distance and orientation but, like the chemical bond, it may perhaps require quantum mechanical theory to explain its nature. Various analytical formulae in the literature try to describe the hydrogen bond, but these are rather tentative and we should admit that our present knowledge of the dependence of the hydrogen bond on distances and orientations can be described by qualitative considerations only.

Interactions of the kind described by equation 1 or 2, namely interactions between non-polar molecules including both the attractive and repulsive terms, are often referred to as Van der Waals interactions. This term is used mainly with reference to the general qualitative character of such interactions, rather than as a quantitative description.

Intramolecular Forces

A polyatomic molecule in its equilibrium state has a well-defined geometric structure or conformation. This structure is determined by the forces which bind atoms together. These forces are of quantum-mechanical origin and are basically well understood. It is useful to distinguish phenomenologically between the bonding forces which determine the strength and length of the chemical bond between two atoms, the bending forces which determine the angles between adjacent chemical bonds, and the torsional forces which control the rotations around chemical bonds.

a) *The torsional potential.* When a bond connects two polyatomic groups in a molecule, for example the $C-C$ bond in CH_3-CH_3, it can serve as an axis of rotation of the two groups relative to each other. Such a rotation is called internal rotation. The internal rotation angle is also termed the torsional or dihedral angle. The internal rotation around single bonds is almost free from energetic hindrances. The torsional potential is thus the softest among the molecular potentials. The potential barrier to rotation amounts to about 3 kcal for rotation around the bond in ethane and is somewhat higher for other tetrahedral carbon atoms. It is even lower in cases where one of the carbon forms an adjacent double bond. Due to Pitzer, much is known about the torsional potential of ethane and some of its derivatives. Far less is known about other cases, e.g. the torsional potentials of the $N-C^{\alpha}$ and $C^{\alpha}-C^{1}$ of the polypeptide backbone. It is usual to describe this potential by the formula

$$\frac{1}{2} K\phi \, (1 + \cos n\phi) \tag{3}$$

where ϕ is the angle of rotation, and n is the number of maxima and minima in one full rotation, determined by the symmetry properties of the bond. In general, $K\phi$ is again an adaptable parameter determined by experiment, though elaborate quantum-mechanical *ab initio* calculations have succeeded in reproducing the potential barrier of ethane.

b) *The bending potential.* The angles between bonds connected to a given atom are essentially determined by the tendency of such bonds to be as far apart as possible. In other words, the interaction between the bond orbitals is a repulsive one, tending to reduce the energy as the bond angle increases.

Thus, if only two bonds are connected to a single atom, as in BeH_2, they will be colinear. Similarly, three bonds connected to an atom are always coplanar, with bond angles of around 120°. Each bond angle separately would tend to increase, but since it is impossible for one of the three bond angles in the plane to increase without decreasing the others, the minimum of the total bending potential energy is reached when the three bond angles are about equal. (The angles would be exactly equal for identical bonds, as one would expect to observe in CH_3^+.) Similarly, four bonds connected to an atom form a tetrahedral structure[1], with all six angles between the four bonds equal to the tetrahedral angle 109,47° when the four bonds are identical, as is the case in CH_4. Deviations from the tetrahedral angle occur when the four bonds are not equal, as is the case in most tetrahedral carbon atoms. The deviations are usually small, 2 to 3°, unless the molecule is strained by other forces, as is the case in a number of ring molecules, like cyclobutane, cyclodecane and other cycloalkanes. There, the closure of the ring imposes significant deviations from the tetrahedral angles. In the same way, when five bonds are connected to an atom, as in pentavalent phosphorus compounds, three bonds are coplanar (equatorial) and two are perpendicular to the plane of the others (axial). This is the highest symmetry available for such a structure, and the fact that it is an energy minimum point can be derived from symmetry considerations.

Deviations from these rules occur when lone pairs of electrons form bond-like orbitals. For example, the two OH bonds in the water molecule form an angle of 105°, but there are two lone pairs of electrons which are connected to the oxygen atom and form lobes which behave like chemical bonds in the sense that they repel each other and the adjacent OH bonds. Thus, the two OH bonds and the two lone pairs form a tetrahedral structure around the oxygen atom. Similarly, the three NH bonds and the one lone pair of NH_3 form a tetrahedral structure, so that NH_3 is not planar.

---

[1] Tetrahedral structure means that the central atom is considered to be at the centre of a tetrahedron while the 4 atoms bonded to it are located at the vertices.

Thus, the qualitative nature of the intramolecular bending energy may be considered to be well understood. However, we are still unable to describe this energy quantitatively. Therefore, the common empirical representation of the bending potential of a polyatomic molecule is given in terms of the deviation of the a bond angles θ from the "equilibrium value" or "reference value" θ_0, and is usually assumed to be quadratic, i.e. of the form

$$\tfrac{1}{2}\,K_\theta(\theta - \theta_0)^2 \tag{4}$$

where K_θ is the "bending-force constant", a term taken from the theory of vibrational spectra, and θ_0 is the reference angle. The reference angle is assumed to be 120° for coplanar trivalent structures and 109.47° for tetrahedral structures; alternatively, the reference angle may be used as an adaptable parameter, like K_θ, where the values are adapted to give the best fit with the experimental data.

By definition, this empirical potential is good for small deviations from the equilibrium angle. For larger deviations, like those mentioned above for some cyclodecane molecules, it is not as reliable. Unfortunately, various efforts to replace it by some better functions have not so far been particularly successful. It is, however, worth bearing these limitations in mind.

c) *The stretching potential.* The molecular energy as a function of bond lengths represents the energy of formation of the chemical bond. It has a minimum at the equilibrium bond length and is adequately represented by the Morse potential

$$De^{-2\alpha(b-b_0)} - 2\,De^{-\alpha(b-b_0)} \tag{5}$$

which can also be written

$$D[(e^{-\alpha(b-b_0)} - 1)^2 - 1]\ ;$$

D is the dissociation energy of the bond, b_0 is the equilibrium bond length, and α is chosen so that the second derivative of the Morse function will give the force constant for the bond-stretching mode of vibration as determined by infrared or Raman spectroscopy.

Molecular Forces and Observable Properties of Molecular Systems

All observable properties of molecular systems are related in one way or another to the molecular forces. However, the relation is in many cases so complex that a theoretical derivation of such properties from molecular forces may be a formidable task. An example relevant to this conference is the study of the properties of water, so important for the study of proteins in aqueous solutions. By the theory of statistical mechanics, all macroscopic properties of water can be derived as averages over all its microscopic states, provided the energy of the microscopic states is known. However, the problem of averaging is itself so difficult for condensed systems with partial short-range order, like water, that even with a perfect knowledge of the forces acting between water molecules, a good theory of water properties would still offer an exceptional challenge.

Fortunately, there are also some observable properties which are much easier to relate theoretically to molecular forces. Experimental study of such properties serves as the major source of our accumulated knowledge of molecular forces. Conversely, such properties can be theoretically predicted with confidence, to the extent that our knowledge of molecular forces is satisfactory. We shall now review some of these properties and their relations to molecular forces or molecular energy.

a) Structure of Molecules and of Molecular Crystals

Interaction of molecular systems with electromagnetic radiation and with electron beams is a good source of experimental information on the detailed geometric structure of molecules and molecular crystals. The major tools in this field are X-ray diffraction, electron diffraction, and microwave spectroscopy. These give us the equilibrium geometric conformations of molecules in the solid or the gaseous state, and the detailed spacing and orientation of molecules in crystals.

The derivation of these observable properties from molecular energy is direct and simple: at equilibrium the total force (i.e. the sum of all forces) acting on each atom must be equal to zero; in other words, the total energy must be minimal.

Assume that the energy of the molecule is known to us, as a sum of all potential energies of stretching, bending and torsion of all bonds, and of repulsive and attractive interactions between non-bonded atoms, as described in the previous section. If we denote the total energy by V, and the geometric coordinates on which it depends by r_α, the condition that the molecule or crystal is at equilibrium is given simply by the requirement that

$$(\partial V/\partial r_\alpha)_{\mathbf{r}=\mathbf{r}_{eq}} = 0 \tag{6}$$

where $\mathbf{r}$ is the vector whose components are r_α, and $\mathbf{r}_{eq}$ is the equilibrium value of $\mathbf{r}$. This set of equations can be solved if $V(\mathbf{r})$ is known. Even if $V(\mathbf{r})$ is a complicated function of $\mathbf{r}$, and if the number of independent variables r_α is large, fast computers and elaborate programs can perform a numerical solution of these equations, yielding both the molecular equilibrium structure $\mathbf{r}_{eq}$ and the energy $V(\mathbf{r}_{eq})$.

However, the inverse process is much more difficult. The only way to obtain information on $V(\mathbf{r})$ from the experimental $\mathbf{r}_{eq}$ is by trial and error: one makes various assumptions on the nature of molecular potentials and checks how well the calculated molecular or crystal structures based on these assumptions fit the experimental facts.

A simple, though cumbersome, way of fitting molecular potentials to experimental data is to guess both the analytic form of the functions and the numerical value of their constant parameters. A more sophisticated way, particularly suited to the use of computers, is the fitting of the parameters by the method of least squares. This mathematical algorithm calculates, for a given set of energy functions, the best values of their energy parameters, in the sense that they give the best fit to the given set of experimental data[2].

It is not the purpose of this lecture to review the achievements in the field of theoretical conformational analysis. Suffice it to say that much of the existing information on the various energy potentials has been derived from the study of molecular conformations and of crystal structures; conversely, the use of these potentials has often led to a better understanding of the structural properties of molecular systems.

---

[2] As a measure of the fit of calculated and measured properties one takes the sum of squares of the differences between the calculated and observed values. The best values are then those which make the sum of squares least, or minimum.

We must recognize the limitations of structure analysis as a source of information about energy functions. Although it tells us the location of the minimum of the total energy, it supplies us with insufficient information about the behavior of the energy away from the minimum. It therefore leaves us with considerable freedom of choice, which makes it difficult to decide upon the best choice. Furthermore, even for a given set of potential functions, the best set of constant parameters is often not uniquely defined. For example, it is known that the parameters of a non-bonded potential are interdependent, or partly "redundant". This is to say that one can continuously change one parameter, and still get the same fit to experimental data, provided that one changes concurrently some other parameter in an appropriate manner. Thus, apparently different sets of parameters may actually be the same to within their inherent redundancy. This property is not always recognized, and may be a source of confusion in the literature.

Conformational analysis has been applied to proteins and polyaminoacids by LIQUORI, RAMACHANDRAN, SCHERAGA, FLORY and their coworkers. Discussion of this work could not be included in the present discussion of molecular forces, but a number of reviews are available [2, 3]. The abovementioned limitations of theoretical conformation analysis of simpler organic molecules obviously holds for such complex structures as proteins. Therefore the study of proteins could greatly benefit from advances in the conformation analysis of simpler model compounds. The impatience with the slow progress made in the latter field shown by many of those interested in the molecular biology of proteins is understandable, in view of the enormous intrinsic importance of the theoretical study of biopolymers and the many interesting results obtained even with our present knowledge of energy functions.

b) Vibrational Spectra of Molecules

Vibrational spectra, measured by infrared (IR) and Raman spectroscopy, are a source of another kind of information on the molecular energy $V(\mathbf{r})$. Whereas equilibrium conformations are determined by the energy V reaching a minimum, IR and Raman spectra are linked to the oscillatory motions of the molecules around equilibrium, and the associated increase in the potential energy. To put it in mathematical terms, while static equilibrium is determined by the *vanishing* of $\partial V/\partial r_\alpha$, the dynamic motion around equilibrium is determined by the *non*-vanishing of the matrix of the second derivatives of the energy, **F**, whose elements are $F_{\alpha\beta} = \partial^2 V/\partial r_\alpha \partial r_\beta$. These elements are commonly termed *force constants* in spectroscopy. They represent the harmonic forces which pull the molecule back to equilibrium when the equilibrium is disturbed. There are known mathematical procedures, which need not concern us here, for deriving the normal modes of harmonic vibrations from **F**.

Thus, if V(**r**) is known, a straightforward calculation can give us its second derivatives so that we can calculate the vibrational spectra of molecules, the experimental counterparts of which are the IR and Raman spectra of gases. A similar procedure holds for crystals, where the unit cell plays the role of a single molecule and the energy V also includes the sum of interactions of the unit cell with all the surrounding unit cells. The vibrational spectrum then includes the molecular vibrations modified by intermolecular interactions, as well as vibrations of the crystal lattice as a whole. All these can be calculated from V(**r**).

The inverse procedure, namely the determination of V(**r**), is much more difficult, since the changes in V(**r**) for very small deviations from equilibrium do not hold sufficient information to determine V(**r**). The theoretical normal mode analysis of molecular spectra is therefore limited to deriving the force constants from experiment and to finding correlations between the spectra of homologous series of molecules for which the force constants are *transferable*, i.e. are expected to be the same.

A significantly deeper insight into the nature of the potential energy functions is obtained by combining conformational analysis and vibrational analysis, a procedure which becomes feasible with the advent of fast computers with big memories. It is obviously more difficult to choose the appropriate functions and fit the appropriate parameters so as to obtain good agreement with both conformational and vibrational data. However, once such agreement is obtained, these energy functions give a significantly more faithful description of the molecular energy surface. Furthermore, the combined treatment considers the force "constants" as functions of the equilibrium conformation. Consequently, it makes it possible to correlate spectra of molecules of significantly different conformation, for example, the cycloalkane ring molecules mentioned earlier. Such molecules must have different force constants, since they have different geometries, different strains of bond angles, torsional angles and non-bonded interactions. However, these force constants can be derived from the same energy functions, by calculating the second derivatives of the total energy at the equilibrium conformation $\mathbf{r}_{eq}$, which is different for each molecule.

c) Thermodynamic Properties of Molecular Gases and Crystals

Rich and interesting sources of information about molecular forces are the various thermodynamic functions of gases and crystals. The energy of polyatomic molecules in the gas phase is composed of the energies of translations, rotations, vibrations and the ground-state electronic energy. The last term is just the value of V at equilibrium, i.e. $V(\mathbf{r}_{eq})$. The energy of vibration can be derived by calculating the vibrational modes of the molecule as described in the preceding section. According to the quantum mechanical principle of uncertainty, a polyatomic molecule must have a vibrational energy, even at absolute zero temperature, of $\frac{1}{2}h\nu_\alpha$ for each vibrational mode ν_α; at finite temperatures the vibrational energy increases as higher energy levels are being occupied in a statistical manner, according to the Boltzman distribution law. Statistical mechanics provides us with simple detailed formulae by which we can calculate the separate contribution of the vibrational modes to the free energy, enthalpy, inner energy, and entropy of gas molecules. Similarly, the calculation of the rotational contribution to the thermodynamic energy functions is easily derived, as it requires at most the moments of inertia of the molecule which are simple functions of $\mathbf{r}_{eq}$. The translational contribution is even simpler, being independent of molecular structure. Thus, by combining the results of sections a. and b., we can calculate from $V(\mathbf{r})$ all the thermodynamic functions of molecular gases and derive the many observable properties related to them.

A similar situation, though somewhat more complex, from the point of view of computer calculations, exists with respect to the thermodynamic functions of crystals which can be derived from the energy potential V(**r**) of the unit cell. A simple and very useful example is the heat of sublimation of molecular crystals, which is the difference between the enthalpies of a molecule (or one mole of molecules) in the crystalline and the gaseous states. It is a direct measure of the work done in separating the molecules packed in the crystal, and is instrumental for the evaluation of the intermolecular potential.

The Consistent Force Field

As we have seen, it is possible to calculate a great many observable properties of a molecular system when its potential energy functions are assumed to be known. Moreover, assuming that these functions are the same for given classes of molecules, it is possible to calculate all the observable properties for whole classes, or families, of molecules, thus subjecting the assumed energy potentials to the test of a large and diverse set of experimental facts. Empirical functions chosen by comparison with a small and limited set of data would in general have a poor chance of standing up to such a test. It is possible to improve the agreement of the calculated and measured quantities by least-squares fitting of the constant parameters of the potential function to the experimental data. Nevertheless, even the best set of parameters would not give sufficiently good agreement if the functional dependence of the energy on the intra- and inter-molecular coordinates is not appropriate. The next logical step, then, is to examine the weaknesses of the various energy potentials in the light of the discrepancies between calculated and measured properties, and to look for a better energy potential to replace a poor one.

This approach to the study of molecular forces has been adopted in our laboratory in recent years, and a set of energy functions derived by such a procedure has been called the Consistent Force Field (CFF) [4]. The following examples may illustrate the method and its potential power.

The first is related to the LJ potential. When this potential was used to calculate *intra*molecular interactions related to properties of alkane molecules in the gaseous phase, the least-squares procedure produced a set of "best" parameters for the repulsive and attractive forces for H....H and C....C interactions. When the study was subsequently extended to *inter*molecular interactions related to crystal properties like the unit cell dimensions and heats of sublimation of some alkanes, a quite different set of parameters was derived by the least squares procedure. Analysis of the results indicated that the repulsive energy is not adequately represented by the A/r^{12} potential, and a less steep potential is needed. Essaying the exponents 10,9,8 in succession, it appeared that 9 is indeed the best: first, it gave a better overall agreement with experiment, and second, the least-squares-fitting of the parameters for intra- and intermolecular data was very much the same. Thus we could conclude that the 9th power gave the best description of the repulsion potential among the repulsive potentials tested.

The second example is related to the nature of the hydrogen bond. As we noted in the section on intermolecular forces, it is still not known to what extent the hydrogen-bond potential is similar to that of a chemical bond, and to what extent it is the result of electrostatic interactions between the partial charges located at the different atoms which have a share in the hydrogen bond. We have recently started to examine this question by the CFF method [5]. We assumed the hydrogen bond potential to be composed of *both* the Morse potential and the electrostatic interactions, and let the least-squares procedure determine the best parameters. Our preliminary results indicate that the electrostatic interactions are predominant, and may well be sufficient to account for the experimental facts associated with the hydrogen bond.

References

1. Warshel, A., Lifson, S.: J. chem. Phys. **53**, 582 (1970).
2. Scheraga, H. A.: Chem. Rev. **71**, 195 (1971).
3. Ramachandran, G. N., Sasisekharan, V.: Advanc. in Protein Chem. **23**, 283 (1968).
4. Lifson, S., Warshel, A.: J. chem. Phys. **49**, 5116 (1968).
5. Hagler, A., Huler, E., Lifson, S.: Unpublished.

Structure, Function and Dynamics of a Regulatory Enzyme — Aspartate Transcarbamylase*

H. K. Schachman

Department of Molecular Biology and The Virus Laboratory
Wendell M. Stanley Hall
University of California, Berkeley, California 94720, USA

With 17 Figures

Introduction

During the past 10 years there have been revolutionary advances in our knowledge of the structure, function and dynamics of proteins composed of subunits. Problems scarcely imagined a decade ago either have been solved already or are being pursued vigorously. New experimental techniques and theoretical approaches are being exploited in attempts to understand how subunit interactions endow oligomeric proteins with the unique properties needed for their biological functions.

Despite the great progress in our understanding of protein structure and function and the powerful armament of tools available for studying proteins, we still face formidable challenges. Are the polypeptide chains in an oligomeric protein identical ? How are the individual polypeptide chains assembled in an oligomeric enzyme ? What forces hold them together and how strong are these forces ? Are these chains organized in discrete subunits and do the subunits have different functions ? Do the subunits change their conformations on assembly into the oligomers ? Are there local changes in the secondary and tertiary structures of the folded polypeptide chains upon interactions with ligands such as substrates or inhibitors ? Upon the addition of these ligands are there

* This work was supported by Public Health Service Research Grant GM 12159 from the National Institute of General Medical Sciences and by National Science Foundation Research Grant GB 4810.

gross changes in the quaternary structures of the oligomeric enzymes ? Are these conformational changes in the tertiary and quaternary structures coordinated and how are they linked ? How are these conformational changes in quaternary structure mediated ? What is the role of the bonding domains between subunits ? Do the subunits in each protein molecule isomerize in unison when a ligand is bound to only one subunit [1]; i.e., is the transition from one state to another concerted as in a two-state model ? Alternatively, do some subunits within an oligomer change their conformation while others in the same molecule remain unchanged [2]; i.e., are there intermediate states ? These represent some of the questions now confronting the protein chemist. Answers to most of them are available only for hemoglobin [3]. For the oligomeric enzymes our knowledge is still too scanty to permit answers. Moreover there is reason to expect that the answers will differ because of the diversity of proteins in terms of both their structures and their functions. Hence generalizations must be treated with caution.

This communication summarizes our recent studies on the enzyme, aspartate transcarbamylase (ATCase). The work presented here was performed largely by J. A. COHLBERG, M. W. KIRSCHNER, G. M. NAGEL, V. P. PIGIET, and Y. YANG and their contributions are cited in the various sections aimed at providing answers to some of the questions posed above.

In its role in regulating the biosynthesis of pyrimidines in *Escherichia coli*, ATCase not only catalyzes a specific chemical reaction but also responds to cellular levels of relatively unrelated metabolites [4, 5, 6]. This enzyme exhibits an unusual sigmoidal response to varying concentrations of the substrates, carbamyl phosphate and aspartate, and is subject to feedback inhibition by cytidine triphosphate. These two closely linked effects, termed homotropic and heterotropic interactions [1], have their origin apparently in subunit interactions within an elegantly constructed macromolecule comprising two distinct types of subunits, one for catalysis and the other for regulation [7]. ATCase is now known to contain 12 polypeptide chains, 6 for catalysis and 6 for regulation [8—12], and 6 zinc ions [6, 12—14]. Although the arrangement of the catalytic polypeptide chains within subunits had been clarified [10], there has been uncertainty, until recently [15], with regard to the structure of the regulatory subunit isolated from

ATCase. Moreover, there was little evidence concerning the structure of the regulatory subunits within the intact ATCase molecules. Results presented here show that the polypeptide chains in the Zn-regulatory subunit are strongly associated to dimers and that these regulatory dimers are integral structural entities of ATCase molecules [15, 16]. These findings, together with those from earlier studies [6, 10] and recent electron micrographs [17], provide the basis for a model of the quaternary structure of ATCase. The functional roles of the various subunits are examined through the use of ATCase-like molecules composed of native and chemically modified polypeptide chains.

Organization of Polypeptide Chains within Subunits in ATCase

Early studies by GERHART and PARDEE [5] showed that ATCase became insensitive to inhibition by CTP and lost its cooperativity when the enzyme was heated briefly to about 60 °C or was reacted with mercurials. Subsequently it was shown that the addition of p-mercuribenzoate led to the dissociation of the enzyme into two types of subunits which were readily separated and purified by sucrose gradient centrifugation [7]. The larger component with a molecular weight of 1×10^5 and a sedimentation coefficient of 5.8 S possessed the entire catalytic activity of the native enzyme. This catalytic subunit no longer exhibited the characteristic homotropic and heterotropic behavior of native ATCase. The second component, termed the regulatory subunit, had a molecular weight of 3×10^4 and a sedimentation coefficient of 2.8 S and retained the ability to bind the inhibitor, CTP. Upon removal of the mercurial the regulatory subunits were capable of combining with the catalytic subunits to give reconstituted ATCase molecules having properties similar to the native enzyme. The regulatory subunits thus appeared to be essential in the native enzyme in mediating the feedback inhibition by CTP and the cooperative effects responsible for the sigmoidal kinetic behavior.

The characterization of the subunits was greatly facilitated by the development of a separation procedure involving chromatography on DEAE-Sephadex A-50 [18]. Recently KIRSCHNER [19] devised an improved method which produced both subunits in excellent yields (greater than 90%) in about 3 h. ATCase was dis-

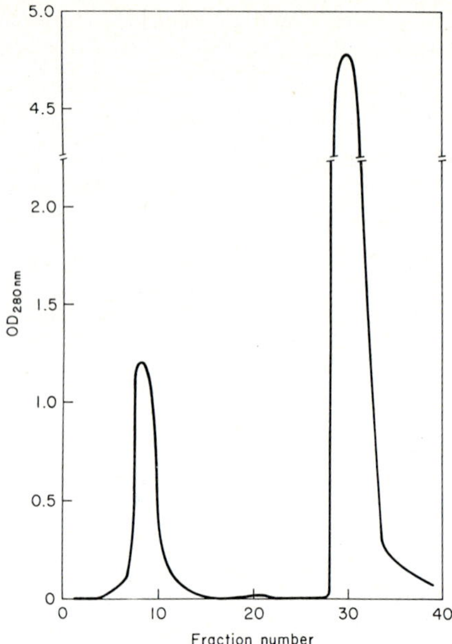

Fig. 1. Purification of regulatory and catalytic subunits. A sample of 100 mg of ATCase in 4 ml of solution was dialyzed against 500 ml of 0.01 M Tris-HCl, pH 8.25, containing 0.1 M KCl. Neohydrin (6.3 mg) was dissolved in 0.5 ml of 0.01 M Tris-HCl at pH 9.3 and the sample of protein was added quickly to the neohydrin and mixed by tipping the tube. After 10 min the sample was added to a DEAE-cellulose column, 0.5 cm × 20 cm, previously equilibrated with 0.01 M Tris-HCl, pH 8.25, containing 0.1 M KCl. The protein was eluted with the same buffer at a flow rate of 25 ml/h. At fraction 18 the concentration of KCl in the buffer was increased to 0.5 M. Absorbance at 280 nm was measured in 1 cm path-length cuvettes. Each fraction contained approximately 3 ml. The yield of the regulatory subunit was 93% of the theoretical and that for the catalytic subunit was 91% of the theoretical yield (from KIRSCHNER [19])

sociated by treatment with neohydrin, 1-(3-chloromercuri-2-methoxypropyl)-urea, and the reaction mixture was fractionated on a DEAE-cellulose column. As seen in Fig. 1 the regulatory subunit was eluted at low ionic strength as a sharp peak followed by the catalytic subunit which emerged after the ionic strength was raised to 0.5.

Early experimental evidence from the subunit composition [7], from molecular weight determinations of ATCase, the subunits and the polypeptide chains [7, 20], from amino-terminal analyses [20, 21], from ligand binding studies [22], and from x-ray diffraction analysis [23] were all consistent with the proposed tetrameric model for ATCase [24]. Subsequent investigations [8—13, 15], however, indicated that ATCase is a hexamer composed of six pairs of regulatory and catalytic polypeptide chains. Supporting evidence

Table 1. *Subunit composition of* ATCase[a]

	ATCase	Catalytic subunit	Regulatory subunit
Mol. wt. $\times 10^{-5}$	3.07 ± 0.03	1.03 ± 0.02	0.337 ± 0.004
Percent by weight	—	68	32
Wt. in daltons per ATCase molecule $\times 10^{-5}$	3.07	2.09	0.98
Subunits/ATCase molecule	—	2.0	2.9
Mol. wt. of polypeptide chains $\times 10^{-4}$	—	3.2 ± 0.1	1.72 ± 0.04
Polypeptide chains/subunit	—	3.2	2.0
Polypeptide chains/ATCase molecule	—	6.4	5.8

[a] These results are taken from Cohlberg et al. [16].

for this model is shown in Table 1 which summarizes Cohlberg's data for the structure of ATCase in terms of its constituent subunits and polypeptide chains [15, 16]. The molecular weights for ATCase, 3.07×10^5, and the catalytic subunit, 1.03×10^5, are in excellent agreement with those reported earlier [7, 12]. In contrast, the value, 3.37×10^4, for the Zn-regulatory subunit is higher than the previously reported values [7, 12]. Taken in conjunction with the composition of ATCase in terms of weight percent of the two types of subunits, these figures show that an ATCase molecule contains two catalytic and three regulatory subunits. Previous evidence regarding the number of regulatory subunits per ATCase molecule has been inconclusive because the molecular weight values for the regulatory subunit were intermediate between the values expected for a single polypeptide chain [20] and a dimer of such chains. Sedimentation equilibrium studies of the subunits in guanidine hydrochloride gave molecular weights for the catalytic polypeptide chains, 3.2×10^4, and for the regulatory chains, 1.72×10^4, which are in good agreement with those reported elsewhere [8, 12]. It follows that the catalytic subunits contain three polypeptide chains and that the regulatory subunits comprise two chains; hence there are six catalytic chains and six regulatory chains per ATCase molecule.

Although the isolated catalytic and Zn-regulatory subunits were found to exist as trimers and dimers, respectively, these observations do not constitute evidence that the trimers and dimers are structural entities within the native enzyme molecules. The respective polypeptide chains could have been spatially separated from one another in the native enzyme and association of them to form the distinct oligomers could have occurred subsequent to the disruption of the ATCase molecules upon treatment with mercurials. For proof of the existence of these discrete subunits as structural entities within the intact ATCase molecules we turn to other approaches.

Catalytic Polypeptide Chains and Subunits

In order to determine the organization of the catalytic polypeptide chains at both the intra- and inter-subunit levels of ATCase we studied the hybridization from native and chemically modified catalytic chains [10]. The approach is shown schematically in Fig. 2. Native catalytic subunits isolated from ATCase [18] were treated with succinic anhydride to give a relatively homogeneous, inactive electrophoretic variant with about 50% of the lysyl residues converted to succinyl-lysyl groups. This derivative had a sedimentation coefficient of 5.8 S, similar to that of the unmodified catalytic subunit. Upon treatment of the succinylated catalytic subunit, C_S, with guanidine hydrochloride (2 M) dissociation occurred to give material with a sedimentation coefficient of 1.9 S. Subsequent removal of the denaturant by dialysis gave a component with the same sedimentation coefficient, 5.8 S, as the unmodified catalytic subunit. Similar treatment of the native catalytic subunit, C_N, showed that it could be dissociated into unfolded polypeptide chains and then reconstituted to give active subunits having the same electrophoretic mobility and sedimentation coefficient as C_N. Hence hybridization experiments were conducted, as outlined in the lower part of Fig. 2. The resulting hybrid set contained four members, as seen in Fig. 3. These different species were readily detected by cellulose acetate electrophoresis [10] and PIGIET subsequently separated and purified them by DEAE-Sephadex chromatography [25]. The number, relative mobilities and enzyme activities of the intermediate bands in the hybrid set correspond to those expected from the hybridization of two molecules each composed of three similar polypeptide chains. Thus the four members of the hybrid set can be correlated with the structures: C_{nnn}, C_{nns}, C_{nss}, and C_{sss} where n and s refer to native and succinylated

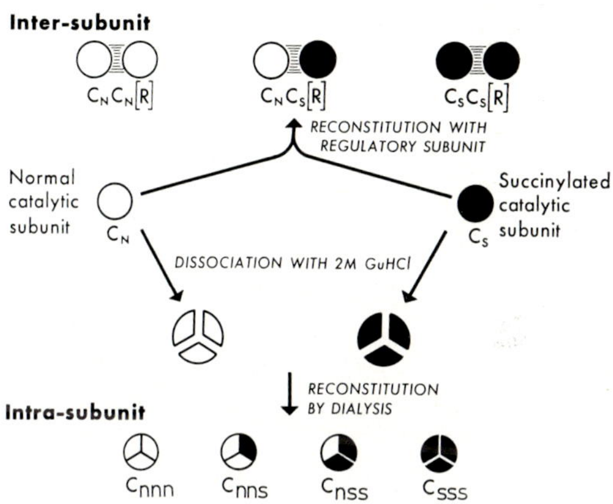

Fig. 2. Hybridization scheme for the catalytic subunits. Intra- and inter-subunit hybrids were prepared from native and succinylated catalytic subunits. Intra-subunit hybridization was effected by dissociating a mixture of the two preparations with 2 M guanidine hydrochloride followed by reconstitution through dialysis. Inter-subunit hybridization was accomplished by mixing the native and succinylated catalytic subunits and adding native regulatory subunits to permit reconstitution of ATCase molecules. The various types of hybrids at both levels are shown with lower-case letters corresponding to the types of polypeptide chains, n for native and s for succinylated. Upper-case letters correspond to subunits, with N corresponding to native and S corresponding to succinylated subunits

polypeptide chains, respectively. When C_N and C_S were incubated for long periods of time in the absence of a denaturing agent no hybrids were detected. This finding, coupled with the observation that the catalytic subunit at a concentration as low as one μg/ml had a sedimentation coefficient of 5.8 S (M. Springer, Y. Yang, and H. K. Schachman, unpublished results), shows that the catalytic subunits exist as stable trimers which show essentially no dissociation into single polypeptide chains.

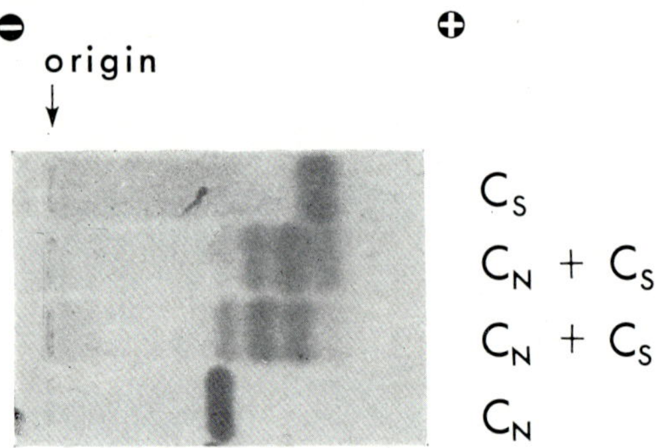

Fig. 3. Intra-subunit hybridization of native, C_N, and succinylated, C_S, catalytic subunits. All samples were dissociated in 2 M guanidine hydrochloride and then reconstituted by dialysis. The two hybrid sets, $C_N + C_S$, were mixtures of native and succinylated catalytic subunits at ratios of 0.6 and 0.9 for the upper and lower samples, respectively. The zone electrophoresis experiments were performed on 14.6 cm cellulose polyacetate strips (from Meighen et al. [10])

Hybridization was also achieved at the inter-subunit level as shown in Fig. 2. For this experiment C_N and C_S were mixed and regulatory subunits were then added to permit reconstitution to ATCase-like molecules. As seen in Fig. 4, a three-membered hybrid set was produced [10]. These three components were readily separated by chromatography on DEAE-Sephadex [10, 25]. From the number, position, expected elution order, and the relative specific activities Pigiet was able to identify the components in the hybrid set as $(C_N)_2$ [R], C_N C_S [R], and $(C_S)_2$ [R]. We can conclude, therefore, that there are two catalytic subunits in each ATCase molecule.

The characteristics of both the intra- and the inter-subunit hybridization reactions and the stability and properties of the

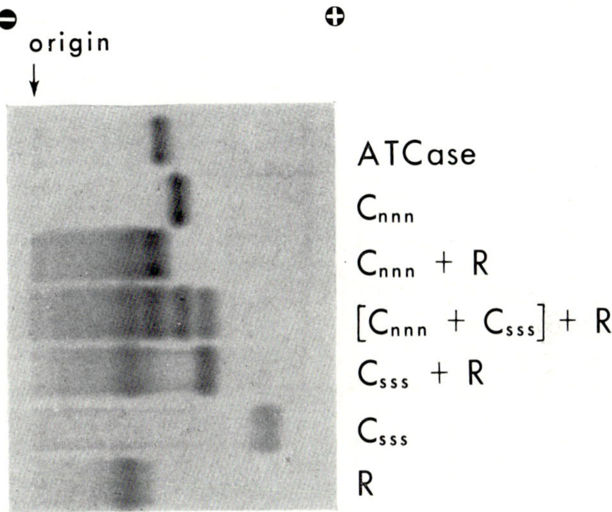

Fig. 4. Inter-subunit hybridization of the native (C_{nnn}) and succinylated (C_{sss}) catalytic subunits in ATCase. ATCase was reconstituted by the addition of excess regulatory subunits (R) to a solution containing native and/or succinylated catalytic subunits. The electrophoresis patterns for the various samples are indicated by the appropriate labels. In all reconstitution experiments excess regulatory subunits were added; thus in the third, fourth and fifth patterns a band corresponding to the excess regulatory subunits was observed. The fourth pattern (from the top) shows the inter-subunit hybrid set composed of three members (from Meighen et al. [10])

various hybrids, C_{nns}, C_{nss}, and C_N C_S [R] indicate that each of the catalytic subunits exists as a tightly folded trimer within the ATCase molecules. Had there been dissociation of the trimers into single polypeptide chains during the reconstitution process, a total of seven electrophoretic species would have been observed instead of three. These hybridization experiments thus provide conclusive evidence that the six catalytic chains exist in the form of two trimers within ATCase molecules [10].

Regulatory Chains and Subunits

Investigations of the structure of the isolated regulatory subunits and their arrangement within intact ATCase molecules were aided considerably by the recent development [14] of an improved procedure for preparing these subunits (see Fig. 1 for their separation). The methods used previously in this laboratory yielded subunits containing variable amounts of residual mercuric ions and exhibiting irreproducible absorption spectra and differing capacities to form ATCase when mixed with catalytic subunits. Exposure of these preparations to zinc chloride yielded homogeneous Zn-regulatory subunits containing 1.0 zinc ion per polypeptide chain. These latter preparations had a reproducible and low extinction coefficient in the ultraviolet region of the spectrum and showed complete competence in forming ATCase by reconstitution with catalytic subunits [14, 25]. Moreover, this reconstituted ATCase, unlike that formed with mercury-containing regulatory subunits, had the spectrum and enzymic behavior characteristic of the native enzyme [14, 25].

Sedimentation velocity and equilibrium studies conducted by Cohlberg on the mercury-containing regulatory subunits gave varying results depending upon the protein concentration and upon the amount of mercuric ions in the preparation [15]. The sedimentation coefficient of these preparations decreased markedly at protein concentrations below 1 mg/ml. Such behavior is characteristic of a rapidly reversible associating-dissociating system such as a monomer-dimer equilibrium [26]. Additional evidence that the mercury-containing regulatory subunits existed as an associating-dissociating system came from sedimentation equilibrium experiments [15]. The plots of ln c *vs.* r^2, one of which is presented in Fig. 5 A, showed upward curvature as expected for an interacting system. Curve fitting of the experimental data showed that the concentration distribution could be satisfactorily fit by assuming that the bulk of the protein existed in a monomer-dimer equilibrium with a small amount of aggregated species [15]. In contrast, Cohlberg found that the sedimentation coefficient of the Zn-regulatory subunits decreased slightly and linearly with increasing concentration; this behavior is characteristic of non-interacting globular proteins. Moreover, as shown in Fig. 5 B, the plots of ln c *vs.* r^2 were linear, corresponding to a molecular weight of 3.38×10^4 [16]. This finding, taken in conjunction with the amino-acid sequence of the regulatory chains [8], demonstrates that the Zn-regulatory subunits exist as dimers. Although no dissociation of the Zn-

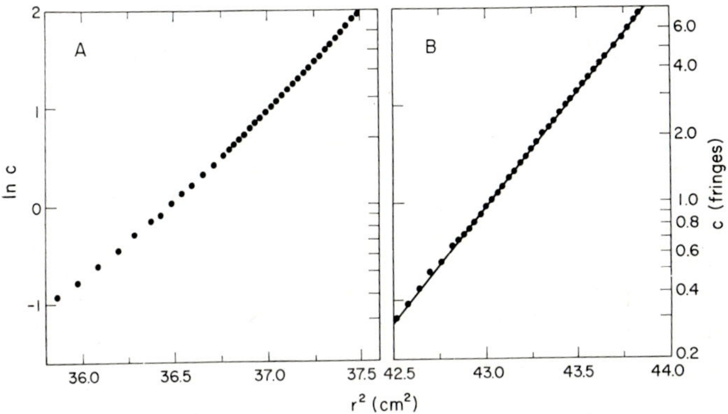

Fig. 5. Sedimentation equilibrium of the mercury-containing and Zn-regulatory subunits. Plot of the natural logarithm of the concentration in fringes vs. the square of the radial distance in cm² for experiments on solutions of mercury-containing regulatory subunit (A) and Zn-regulatory subunit (B) at initial concentrations of 1.5 mg/ml and 2.5 mg/ml, respectively. The temperatures of the solutions were 15.1 °C and 4.3 °C, respectively, and the rotor speed was 34,000 rpm. For both solutions the buffer was 0.04 M potassium phosphate at pH 7.0 containing 10 mM MET and 0.2 mM EDTA in (A) and 10 mM MET and 0.2 mM Zn Cl_2 in (B). The concentration scale, in fringes, is shown on the right (from Cohlberg [15])

regulatory subunit was detected in the sedimentation experiments and these subunits appeared as stable dimers at pH values from 7 to 10, it should be recognized that no special efforts were made to look for dissociation in very dilute solutions.

Since the physical chemical studies indicated that the regulatory subunits could exist as a monomer-dimer equilibrium with zinc ions stabilizing the dimers, it seemed of interest to determine the nature of the combining units in reconstitution experiments with native catalytic subunits and a mixture of native and chemically modified regulatory subunits. Preliminary attempts to prepare a suitable electrophoretic variant of the isolated regulatory subunit by treating it with succinic anhydride led to a derivative which did not form ATCase-like molecules upon the addition of native catalytic subunits [10]. Hence Nagel performed the succinylation on the

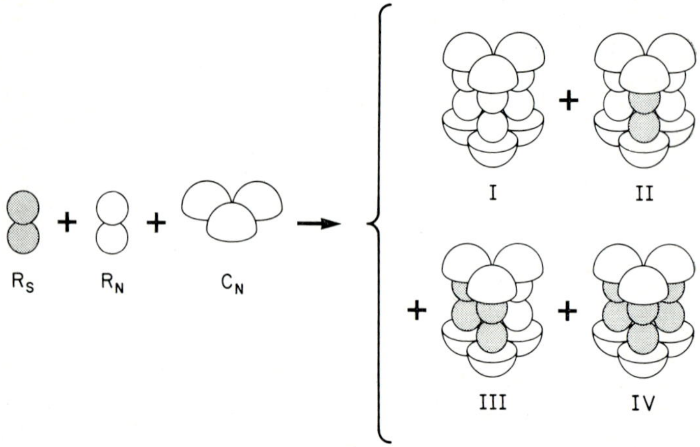

Fig. 6. Hybridization scheme for the formation of ATCase molecules containing native catalytic subunits, C_N, and both native, R_N, and succinylated, R_S, regulatory subunits. The four hybrids to be expected if there were three regulatory combining units in an ATCase molecule are shown on the right. This pattern is based upon the assumption that the native and modified regulatory subunits are preserved as dimers which function as recombining units. Had these dimers dissociated to form appreciable amounts of monomer, then a 7-membered hybrid set would be obtained (from Nagel et al. [27])

native ATCase molecules [27] which were then treated with a mercurial to cause dissociation into modified regulatory and catalytic subunits. These were separated by chromatography on DEAE-cellulose to give purified succinylated regulatory subunits, R_S, which were capable of combining in excellent yield with native catalytic subunits, C_N. With these modified regulatory subunits and native regulatory subunits, R_N, Nagel was able to prepare a hybrid set of reconstituted ATCase-like molecules upon the addition of catalytic subunits. As shown schematically in Fig. 6, a four-membered hybrid set would be obtained if there were three regulatory combining units in the reconstitution process. Rapid mixing of R_S, R_N, and C_N did, in fact, give a hybrid set of four equally spaced electrophoretic species on cellulose acetate membranes [28]. This result indicates that there are three regulatory combining

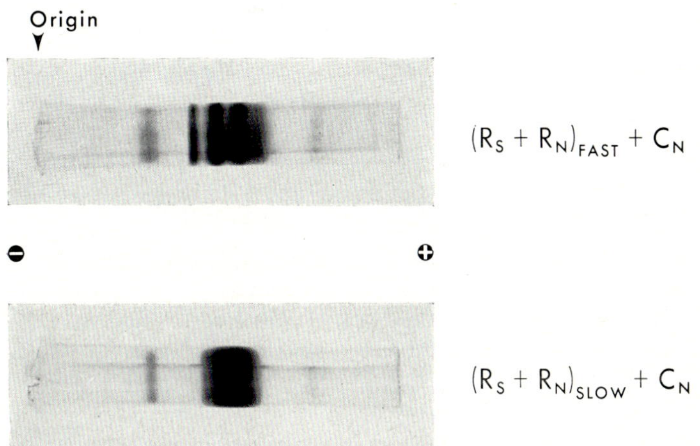

Fig. 7. Hybrid set of ATCase containing native catalytic subunits and a mixture of native and succinylated regulatory subunits. Both electrophoresis patterns were obtained on polyacrylamide gels. The upper pattern was obtained with a solution prepared by rapid mixing of the three components. Both R_N and R_S were added to a tube containing C_N by means of a motor-driven arrangement which insured mixing within 5 sec. The four members of the hybrid set are readily seen. When, however, R_S and R_N were mixed for one minute before C_N was added, the pattern shown below was obtained. No resolution of the putative seven bands could be obtained (from NAGEL and SCHACHMAN [28])

units in the reconstitution process and that the six regulatory polypeptide chains in ATCase must exist as three dimers. When R_S and R_N were mixed for about 1 min prior to the addition of C_N, NAGEL observed a complex hybrid set of unresolved bands. Fig. 7 shows the electrophoretic patterns obtained on polyacrylamide gels for the hybrid sets obtained upon rapid and slow mixing. Apparently the prior mixing of R_N and R_S led to hybridization of the regulatory dimers which, upon subsequent addition of catalytic subunits, yielded a complex hybrid set composed of seven species which differed so little in electrophoretic mobility that resolution of the bands was not achieved. This finding in conjunction with the sedimentation equlibrium experiments on different types of regulatory subunit preparations indicates that the Zn-regulatory subunits exist in a rapidly reversible monomer-dimer equilibrium.

Additional evidence that the regulatory dimers existed as structural entities within ATCase molecules came from reconstitution experiments involving native catalytic subunits and regulatory subunits in which the two polypeptide chains in each dimer were covalently cross-linked to each other [15, 16]. Cross-linking was achieved by reacting the Zn-regulatory subunits with dimethyl pimelimidate [29]. Since this reaction, under the conditions employed, led to the formation of aggregates. COHLBERG fractionated the modified regulatory subunit preparation by chromatography on Sephadex G-100; in this way he obtained a derivative consisting predominantly of cross-linked dimers which were then used with native catalytic subunits for reconstitution of ATCase-like molecules [15]. The steps in this experiment are summarized in Fig. 8 which shows cellulose acetate electrophoresis patterns of the various controls, the unfractionated and Sephadex-purified cross-linked regulatory subunit, and the reconstituted ATCase-like materials. The final preparation was subjected to sucrose gradient centrifugation to isolate the 11 S material which, as seen in Fig. 8, had the same electrophoretic mobility as native ATCase. This result, like that from the hybridization experiments, provides conclusive evidence that the regulatory polypeptide chains in each Zn-regulatory dimer remain associated upon incorporation into ATCase and that the dimers of the regulatory chains are integral structural entities of the ATCase molecules [15, 16].

One other approach has led to the same conclusion about the relationship between the regulatory polypeptide chains within ATCase. Cross-linking native ATCase followed by SDS-polyacrylamide electrophoresis produced a major band corresponding to cross-linked regulatory dimers [30]. Hence the regulatory chains must have been located so closely topographically in intact ATCase molecules as to permit cross-linking by a relatively small bifunctional reagent.

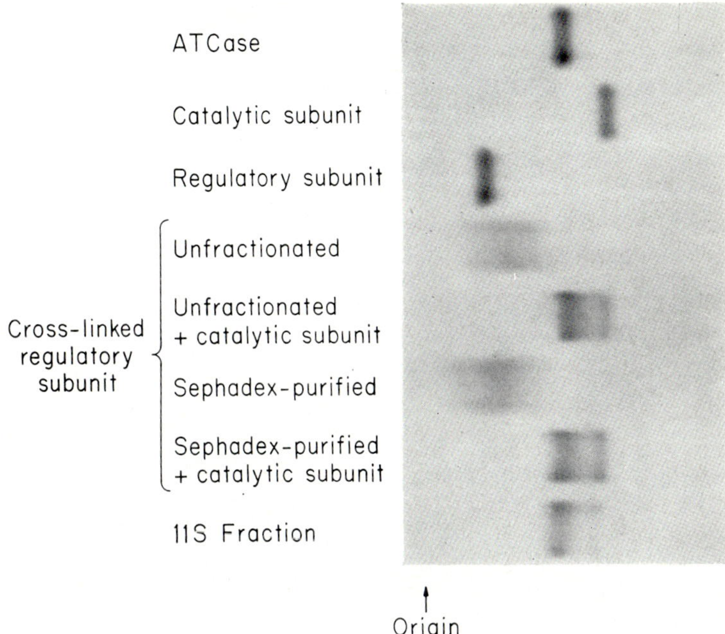

Fig. 8. Reconstitution of ATCase from cross-linked regulatory subunits and catalytic subunits. Electrophoresis on cellulose acetate membranes was conducted for 15 min at 250 V in 0.1 M Tris-HCl, pH 8.0, containing 10 mM MET and 0.2 mM Zn Cl_2. The upper three channels contained native ATCase, catalytic subunit, and Zn-regulatory subunit. The fourth channel shows a sample of the unfractionated cross-linked regulatory subunit. In the fifth channel is shown a pattern resulting from mixing the unfractionated cross-linked regulatory subunit with catalytic subunit. The presence of reconstituted ATCase is evident. The sixth pattern represents the purified cross-linked regulatory subunit obtained by Sephadex chromatography. Below it, in the seventh pattern, is the result of mixing the Sephadex-purified cross-linked regulatory subunit and catalytic subunit. At the bottom is shown the electrophoresis pattern for the 11 S fraction obtained by sucrose density gradient centrifugation of reconstituted ATCase containing cross-linked regulatory subunits (from Cohlberg [15])

A Model of ATCase

We now proceed to consider the arrangement of the six catalytic and six regulatory polypeptide chains within ATCase in the form of two trimers and three dimers, respectively. This arrangement

Fig. 9. Schematic model showing the bonding domains within the regulatory and catalytic subunits of ATCase. At the top are shown small tail-less monkeys which illustrate individual regulatory polypeptide chains. The individual chains in these dimers are related to each other through a twofold rotation axis by an isologous association made up of identical contributions from each polypeptide chain of the pair. This is illustrated by having the monkeys bonded through their left arms. In the lower diagram is shown the corresponding representation of the catalytic chains which are related through a threefold symmetry axis. The bonding domains here are illustrated schematically as a heterologous association by bonding between left and right arms (from Pigiet [25])

must be such that the ATCase complex possesses both threefold and twofold axes of symmetry as observed in crystallographic studies of ATCase and its derivatives [9, 11].

Stable regulatory dimers are readily formed by isologous association of asymmetric polypeptide chains; such dimeric structures would have a twofold axis of symmetry [1]. In such an arrangement, illustrated by the smaller tailless monkeys [31] in Fig. 9, the regions of bonding involve identical sets of residues (left hand to left hand in the model) on the two polypeptide chains. These contacts are designated as the r:r bonding domains. Closed stable trimers can be formed with all the intra-subunit bonding domains satisfied by heterologous association among the c:c bonding domains [1]. A structure of this type, illustrated schematically in Fig. 9 by the larger tailless monkeys with bonding between left and right hands, would have a threefold axis of symmetry.

Since the catalytic subunits show little tendency to aggregate and the addition of Zn-regulatory subunits leads rapidly to reconstitution of ATCase we can conclude that the regulatory subunits serve as cross-bridges between the two catalytic subunits. The bonding between the different types of polypeptide chains, designated r:c domains, can be visualized in our model as a link between the right hand of a small monkey and the right armpit of a large monkey. A partially assembled model of ATCase is shown in Fig. 10 with three such r:c bonding domains. Each regulatory dimer in the model would have a free right hand having an affinity for an armpit of a larger monkey; thus the completion of the assembly requires only the addition of a second trimer of large tailless monkeys. Symmetry considerations dictate an isologous association of catalytic subunits; hence the two trimers must face each other as in Fig. 11. Once again the association between a regulatory and a catalytic chain involves a right hand and a right armpit. In this assembled model we have six c:c heterologous bonding domains within the two catalytic trimers, three isologous r:r bonding domains within the three regulatory dimers, and six r:c bonding domains linking the 12 polypeptide chains.

Although the model in Fig. 11 is useful in highlighting the symmetry of ATCase and in delineating the various types of bonding domains, it is too superficial to provide answers to a series of questions about the real structure of the native enzyme. Are the two catalytic trimers in an eclipsed or staggered arrangement relative to each other? Are the regulatory subunits sandwiched between the two catalytic subunits or do they extend to the outside of the molecule? Is there direct physical contact between the two catalytic trimers and among the three regulatory dimers? Does each regulatory dimer link two catalytic chains which are directly under one another or does a regulatory dimer link catalytic chains which are 120° apart?

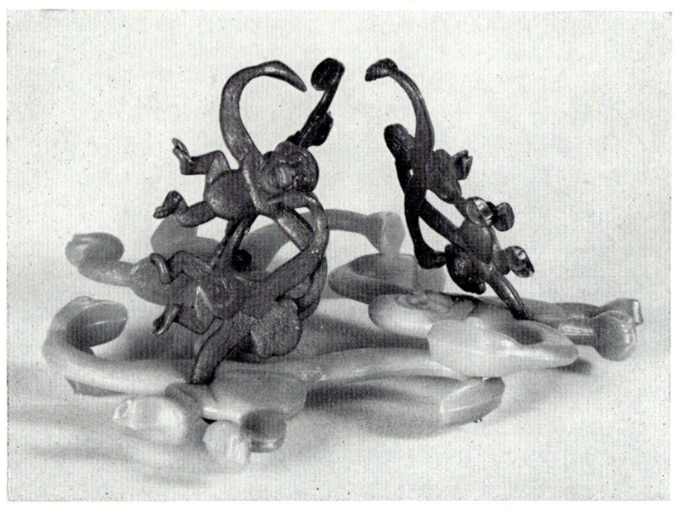

Fig. 10. Schematic representation of the domains of bonding between the catalytic and regulatory subunits in native ATCase. In this partially assembled schematic model of ATCase the bonding domains between regulatory and catalytic chains are illustrated by having the right arm of a smaller monkey linked to the right armpit of a larger monkey (from Pigiet [25])

Tentative answers to most of these questions come from recent electron micrographs of the catalytic subunit and ATCase [17]. Negatively stained preparations of the catalytic subunit showed a structure with a contour close to that of an equilateral triangle with edge lengths of 90 to 95 Å. Micrographs of ATCase showed a structure composed of an inner, solid equilateral triangle (appearing identical in size and form to the catalytic subunit) and a circumscribing triangle rotated by 60°. These micrographs of the intact enzyme dried in a very thin stain, pre-deposited upon a hydrophilic carbon surface, showed the threefold axis of symmetry normal to the plane of the specimen film. When, however, the enzyme was dried in thick stain it exhibited two prominent halves (with shapes

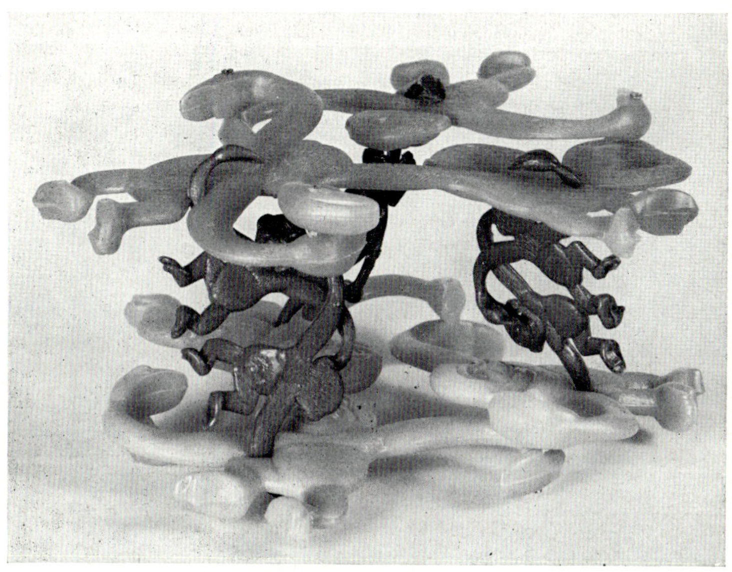

Fig. 11. Schematic representation of the arrangement of the regulatory and catalytic chains in native ATCase. In this fully assembled model of native ATCase are shown the spatial relationships among the two catalytic and three regulatory subunits. The threefold symmetry characteristic of the isolated catalytic subunits is preserved in this assembled molecule. Also, the twofold symmetry axis characteristic of the isolated regulatory subunit is maintained in the assembled structure across the r:r domains of bonding. As shown in this structure, native ATCase may be conceived structurally as a dimer of trimers. The two sets of larger tail-less monkey (representing the catalytic subunits) face each other as in an isologous relationship (from Pigiet [25])

somewhat like segments of a circle), indicating that the enzyme dried with its threefold axis parallel to the specimen film. There was a space, estimated to be in the range of 20 to 40 Å, between the two segments. In addition, some of the particles showed faint extensions of material beyond the ends of the segments. This material, which probably serves to connect the two segments, appears to correspond to the arms of the outer triangle and is thought to represent the regulatory subunits [17, 16].

On the basis of the electron micrographs and a variety of physical chemical data, we have constructed a model of ATCase [16] which

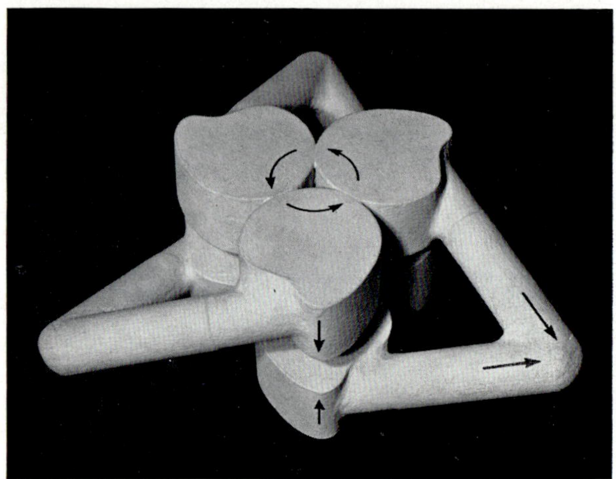

Fig. 12. Model for the arrangement of the polypeptide chains in ATCase. Each catalytic subunit is shown as a triangular array of three asymmetrically shaped catalytic polypeptide chains. Arrows on the upper face indicate the heterologous association of the c:c bonding domains; arrows on the side faces of the catalytic chains indicate the isologous relationship of the catalytic trimers. Each regulatory subunit is shown as a pair of cylindrical regulatory chains oriented to one another at an acute angle; arrows on the cylinders illustrate the isologous association of the r:r bonding domains (from COHLBERG et al. [16])

is shown in Fig. 12. Each catalytic subunit is represented as a slightly oblate structure having a triangular arrangement of the three asymmetrically shaped polypeptide chains. The heterologous association of the polypeptide chains within the subunits is indicated by the arrows on the upper face of one catalytic subunit. Arrows pointing toward each other on the side faces of the catalytic subunits illustrate the isologous relationship between the two catalytic trimers. Each regulatory subunit is shown as a pair of long cylinders each of which represents a regulatory polypeptide chain; the arrows pointing toward each other illustrate the isologous association at the r:r bonding domains. As shown in Fig. 12, one regulatory chain in each dimer bonds to one of the upper catalytic chains with the other regulatory chain in the same dimer being bonded to a lower catalytic chain displaced by 120°. In this way identical sets of amino acid residues in the two catalytic chains are involved in the r:c bonding domains; i.e., the residues located on the left-hand side of each chain in the upper trimer are identical to those on the right-hand side of each catalytic chain in the lower trimer. We cannot distinguish, of course, between this arrangement and an alternative structure in which the upper regulatory chain in a dimer bonds to the right-hand side of a catalytic chain in the upper trimer and the lower regulatory chain in the same dimer is bonded to the left-hand side of a catalytic chain in the lower trimer.

The model shown in Fig. 12 is particularly interesting in terms of recent x-ray diffraction studies on ATCase which have led to an electron density map at 5.5 Å resolution [11]. From these investigations it appears that the catalytic subunits exist as a stack of eclipsed triangles when viewed down the threefold axis. In contrast the regulatory chains are not in the eclipsed form; instead they appear to be rotated about the dimer axis so that they appear almost adjacent when viewed down the threefold axis [11]. As yet it is not clear how close are the two catalytic trimers in the intact enzyme, but the overall dimensions deduced from the x-ray diffraction studies [11] are in good agreement with those estimated from the elctron micrographs [17]. In constructing the model shown in Fig. 12, we have deliberately used two cylindrical particles in a V-shaped arrangement to represent the regulatory subunits. In this way we have only one r:c bonding domain between a single regulatory polypeptide chain and a single catalytic chain. The cork ball model [11] proposed on the basis of the x-ray diffraction studies, though having many of the same characteristics as that shown in Fig. 12, is not as definitive as the wooden model in showing only a single r:c domain between each regulatory and each catalytic chain. We, of course, do not know the exact size and shape of each regulatory cylinder in the V-shaped dimers and their rod-like shape and the resulting open space (between catalytic and regulatory chains) may be exaggerated.

It is of interest to speculate about the rotation model of GERHART [6] in terms of Fig. 12. If the upper trimer were rotated 60° in a clockwise direction (upon the addition of the appropriate ligands) the regulatory subunits would assume a more eclipsed arrangement while the catalytic subunits would become staggered. Efforts to test this idea by electron microscopy as yet have been inconclusive.

Thus far there is no direct evidence as to the location of the six zinc ions in ATCase and their role is only partially elucidated. As shown by NELBACH et al. [14] these zinc ions are essential for the stabilization of the quaternary structure of the intact enzyme and they could be implicated, directly or indirectly, in either the r:r or the r:c bonding domains (or both). The metal ions do not seem to be involved in the c:c bonding domains since catalytic subunits lacking zinc have been found to exist as stable, active trimers even at concentrations as low as 1 μg/ml. Four possibilities remain as to the location and role of the zinc ions in ATCase. First, they may be located in the r:r domains with their sole effect being to strengthen these bonding domains. Second, they may be located in the r:c bonding domains where they influence only the association between regulatory and catalytic chains. Third, they may be located at either one of these domains which they influence directly while affecting the other type indirectly. Fourth, they may be located at neither domain while still enhancing the association at either or both types of bonding domains indirectly through a conformational change. In this regard it is of interest to note that the amino acid sequence of the regulatory chain [8] contains a tetrapeptide, -cys-lys-tyr-cys-, and a hexapeptide, -cys-pro-asp-ser-asn-cys-, which would be likely candidates for strong binding sites for metal ions. Indeed, the regulatory chains do bind zinc and other metal ions strongly [13, 14]. Accompanying the binding of zinc ions to the apo-regulatory subunits is a marked alteration in the circular dichroism spectrum [15] as shown in Fig. 13. Thus there may be a conformational change in the regulatory subunits upon the addition of zinc ions with a consequent effect on the bonding domains between two regulatory chains and between a regulatory and a catalytic chain.

In considering the role of the zinc ions in stabilizing the structure of ATCase we must assess the direct evidence for their effect in strengthening the r:r bonding domains. For the apo-regulatory subunit COHLBERG [15] found that the monomer-dimer association constant (evaluated from the sedimentation equilibrium experiments) was only about 10^4 M^{-1}. In contrast, the association constant was between 10^7 and 10^{11} M^{-1} for the Zn-regulatory subunits

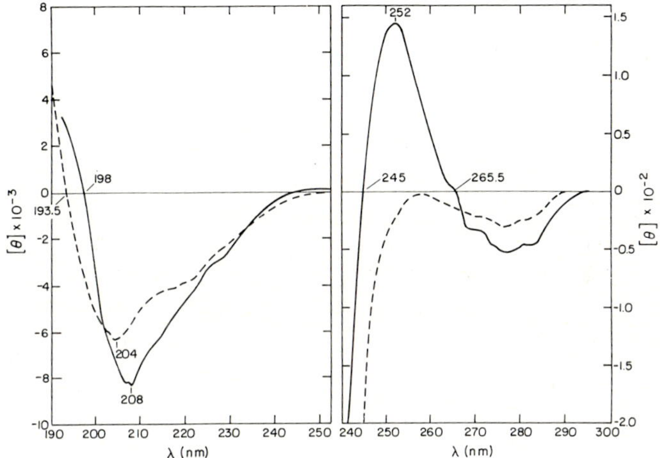

Fig. 13. Circular dichroism spectra of the Zn- and apo-regulatory subunits. Spectra of Zn- (——) and apo-regulatory (----) subunits were determined in 0.04 M phosphate at pH 7.0 containing 0.2 mM Zn Cl_2 and 1 mM EDTA, respectively, for the two types of subunits. MET was present at 10 mM for measurement of the near UV spectra and 1 mM for the corresponding far UV spectra. Data are expressed in degrees-cm²/decimole of residues based on a mean residue molecular weight of 112 (from COHLBERG et al. [16])

estimated from the sedimentation equilibrium studies [15, 16] and the rate of hybrid formation [28]. Thus the addition of zinc ions led to a change in the free energy of association of regulatory chains of −4 to −10 kcal/mole. Since there are three such r:r domains in each ATCase molecule it is clear that the considerable increase in the affinity at the r:r domains upon the addition of zinc ions could be responsible, in part, for the marked enhancement of the stability of the ATCase molecules. But, is this sufficient or must we consider also that zinc ions affect the stability of the r:c bonding domains? No quantitative estimates of the effect of zinc ions on these domains are available as yet. However, apo-regulatory subunits show little or no tendency to associate with catalytic subunits [14] and ATCase (containing zinc ions) shows no tendency to dissociate at concentrations as low as 3 μg/ml [32]. These observations provide some indication that zinc ions contribute to strengthening of the r:c bonding domains.

Changes in the Quaternary Structure of ATCase

All models that have been proposed to explain allosteric transitions in oligomeric proteins are based on the postulate that the protein molecules exist in different conformational states and that the relative populations of these states are influenced by the absence or presence of stereospecific ligands [1, 2]. Hence it is pertinent to examine the physical properties of ATCase (and its subunits) in terms of different probes which are sensitive to conformational changes. Many such probes are available, but here we will focus primarily on the sedimentation coefficient which is related to the weight, shape, and volume of the hydrodynamic unit. Since ATCase shows so little tendency to dissociate (unless treated with certain mercurials or denaturing agents) a change in its sedimentation coefficient upon the addition of a ligand provides a measure of alterations in the volume or shape of the protein molecules. Such changes in the quaternary structure can then be compared with changes in the secondary or tertiary structure as revealed by a spectral probe which is sensitive to local environment.

The various intra- and inter-subunit hybrids provide an excellent opportunity to evaluate some aspects of the functional behavior of the oligomers in terms of their structural features. Isolated catalytic subunits do not exhibit the homotropic effects characteristic of native ATCase, as shown in Fig. 14. Thus it is of interest to examine the behavior of a hybrid ATCase-like molecule composed of native regulatory subunits, one native catalytic subunit, and one inactive (succinylated) catalytic subunit. Would such a molecule exhibit cooperativity and would it undergo conformational changes like those of the native enzyme?

Before attempting to evaluate results from studies on the inter-subunit hybrids of ATCase we first examine the characteristics of the intra-subunit hybrids of the catalytic subunits. PIGIET obtained each member of the hybrid set, C_{nnn}, C_{nns}, C_{nss}, and C_{sss}, in pure form by DEAE-Sephadex chromatography [25]. Results of enzyme assays and difference sedimentation experiments [25] are presented in Table 2. The specific activities are in the ratio of 3:2:1:0 for the various species, C_{nnn}, C_{nns}, C_{nss}, and C_{sss}. Thus the enzyme activity is directly proportional to the number of native polypeptide

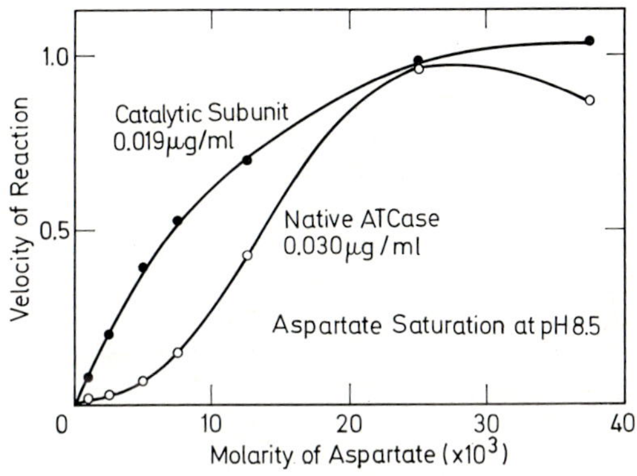

Fig. 14. Comparison of native ATCase and the isolated catalytic subunit with respect to their enzymic activity. On the ordinate (in arbitrary units) is given the reaction velocity at variable aspartate concentrations (on the abscissa). Carbamyl phosphate was present at 3.6×10^{-3} M. The reaction was measured at pH 8.5, 28 °C, with the concentration of ATCase and catalytic subunit adjusted to give equal concentrations of active sites (from CHANGEUX and GERHART [33])

chains in the different trimers. Upon the addition of carbamyl phosphate (as compared to phosphate) the sedimentation coefficient increased about 0.2% (corresponding to only 0.01 S). This extremely small change can be accounted for in terms of the increased buoyant weight of the protein as a result of binding of the ligand [19, 35]. Since the same value of $\Delta s/s$ was obtained for all the species, we can conclude that carbamyl phosphate binds to succinylated as well as native catalytic chains. However, the addition of succinate, a competitive inhibitor of aspartate, produces changes which are directly proportional to the content of native polypeptide chains in the oligomer [25]. These changes (about 1% in $\Delta s/s$ depending upon the pH of the solution) are significantly greater (about threefold) than the increase in sedimentation coefficient which is expected on the basis of the increment in molecular weight [35]. This increase in sedimentation coefficient upon the addition of both ligands can be attributed to a conformational change in the catalytic subunit

Table 2. *Independence of chains in catalytic subunits*[a]

Species	Enzyme activity[b] units/mg	$\Delta s/s$ in %[c]	
		CAP vs. P_i	Succ + CAP vs. Glut + P_i
C_{nnn}	52	+ 0.21	+ 1.48
C_{nns}	35	+ 0.20	+ 1.10
C_{nss}	17	+ 0.22	+ 0.67
C_{sss}	0	+ 0.22	+ 0.31

[a] All these measurements by PIGIET [25] were made on preparations in 0.2 M glycylglycine buffer (pH 8.2) containing MET (2 mM) and EDTA (0.2 mM).

[b] Enzyme activities correspond to millimoles of carbamyl aspartate formed per hour per mg of protein.

[c] Values of $\Delta s/s$ were obtained by the difference sedimentation technique [34] in which the protein in one solvent is compared directly with the same protein in another solvent. The principal buffer constituents were maintained constant and only the ligands (carbamyl phosphate, phosphate, succinate and glutarate at 5 to 10 mM) differed in the pairs of solutions. Some of the experiments involved buffers at pH 7 containing 0.5 M KCl so that charge effects from succinylated chains were reduced. The solutions at pH 8.2 also contained 0.5 M KCl.

which results in a more compact or isometric form [35]. Thus we can conclude from the kinetic behavior and the difference sedimentation experiments that the polypeptide chains in the catalytic subunits act independently. An oligomer composed of one native chain and two inactive chains contracts (or changes its shape) only one-third as much as the native catalytic subunit. In this regard the response of the oligomeric catalytic subunits represents simply the sum of its constituent parts [25].

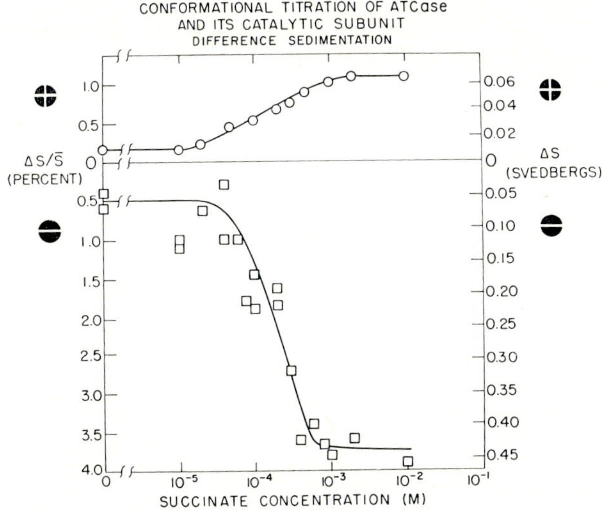

Fig. 15. Conformational titration of catalytic subunit and ATCase. The difference in sedimentation coefficient, $\Delta s/s$, in percent is plotted vs. succinate concentration for both the catalytic subunit and ATCase. The scale (on the right) for Δs in Svedbergs is different for the two proteins. The data on ATCase were obtained from simultaneous experiments with wedge and plain window cells and are illustrated by □. Each point represents the difference in sedimentation rate between the sample and reference solution. Both solutions contained ATCase at 4.2 mg/ml in 0.04 M phosphate buffer (pH 7.0). In the sample were 1.8×10^{-3} M dilithium carbamyl phosphate and a given concentration of succinate. In the reference was an additional 1.8×10^{-3} M phosphate and an equivalent amount of glutarate. Data for catalytic subunit were obtained by difference sedimentation with Rayleigh optics and are represented by ○. For each experiment both sectors of a double-sector ultracentrifuge cell contained catalytic subunits at a concentration of 8 mg/ml in 0.04 M potassium phosphate buffer (pH 7.0), 2 mM mercapto-ethanol, and 0.2 mM EDTA. One sector contained 2×10^{-3} M dilithium carbamyl phosphate and a given concentration of succinate. The other sector contained 2×10^{-3} M additional phosphate and an equivalent concentration of glutarate. All experiments were performed at 20 °C and the rotor speed was 60,000 rpm (from KIRSCHNER and SCHACHMAN [35])

In the light of the tightening of the quaternary structure of the catalytic subunits upon the addition of ligands it is especially interesting to recall the effect of these ligands on intact ATCase molecules [36]. Fig. 15 shows titration curves [35, 36] for both catalytic subunits and for ATCase in terms of the change in sedimentation coefficient as a function of succinate concentration (at a constant level of carbamyl phosphate). Whereas the ligands promote a 1.0% *increase* in the sedimentation coefficient of the catalytic subunit, they cause a 3.6% *decrease* for ATCase. Thus there is a swelling (or elongation) of the ATCase molecules upon the addition of the same ligands which promote a contraction of the isolated catalytic subunits. Just as the enzymic behavior of the intact ATCase molecules (Fig. 14) is not merely the sum of the activities of its parts, so the conformational change in ATCase does not correspond to the changes in the various isolated parts. Indeed the conformational change of the isolated catalytic subunits (contraction) differs not only in direction and magnitude from the change in ATCase (expansion) but also the saturation curves for binding of ligands differ appreciably. There are at least two possible interpretations of these difference sedimentation experiments. On the one hand, it is possible that the changes observed in the isolated catalytic subunits are different from those which the subunits undergo when they are constrained in native ATCase molecules. On the other hand, the catalytic subunits, free and in ATCase complexes, may suffer the same changes in tertiary and quaternary structures but in ATCase these changes are linked to alterations in quaternary structure which are larger and opposite in direction. Spectroscopic evidence [19, 25], some of which is presented below, leads us to favor the latter alternative; hence we conclude from the hydrodynamic results in Fig. 15 that the change in ATCase does not reflect simply the sum of the changes in the catalytic subunits but rather that an additional and different type of transition occurs in ATCase which involves changes in the packing or architecture of the whole complex.

Striking evidence of the effect of incorporating a catalytic subunit into an ATCase-like structure comes from PIGIET's studies of the intersubunit hybrid, C_N C_S [R]. Even though one catalytic subunit contained inactive polypeptide chains, the complex showed a sigmoidal response to aspartate concentration [25]. Treating the kinetic data in a form which provided a quantitative index of the cooperativity, PIGIET obtained a Hill coefficient, n_{Hill}, of 1.3; for native ATCase, n_{Hill} is 1.6, and of course the isolated, unmodified catalytic subunit had a Hill coefficient of 1.0. Thus the quaternary constraint of incorporating a non-cooperative catalytic subunit into an ATCase-like structure leads to significant cooperativity [25]. Difference sedimentation experiments also showed this quaternary constraint. As seen in Table 3, the values (+ 0.48%) for $\Delta s/s$ for C_S C_S [R] in the presence of one (carbamyl phosphate) or both specific ligands (carbamyl phosphate and succinate) are merely twice the value for the isolated catalytic subunit. If both catalytic subunits are native, the value of $\Delta s/s$ caused by the addition of carbamyl phosphate is not +0.48% but rather −0.45%. With the

Table 3. *Effect of quaternary constraint*[a]

Species	$\Delta s/s$ in %[b]	
	CAP vs. P_i	Succ + CAP vs. Glut + P_i
C_N	+ 0.19	+ 1.10
C_S	+ 0.23	+ 0.24
C_N C_N [R]	− 0.45	− 3.60
C_N C_S [R]	− 0.04	− 1.70
C_S C_S [R]	+ 0.48	+ 0.46

[a] These data taken from PIGIET [25] came from studies at pH 7.0 in buffers containing 0.5 M KCl.

[b] The concentrations of carbamyl phosphate and phosphate were 5 mM and of succinate and glutarate the concentration was 2 mM.

hybrid, $\Delta s/s$ is −0.04%, a value intermediate between those for the completely inactive and fully active complexes. Similar changes in $\Delta s/s$ occur upon the addition of the second ligand. Even though the hybrid species, C_N C_S [R], contains only one active catalytic subunit it swells appreciably ($\Delta s/s = -1.70\%$) upon the addition of both carbamyl phosphate and succinate. Comparable results were obtained with hybrids containing native and succinylated catalytic chains in each subunit [25]. The hybrid, C_{nss} C_{nss} [R], for example, showed slight cooperativity with $n_{Hill} = 1.12$ and the value of $\Delta s/s$ upon the addition of carbamyl phosphate and succinate was 0.11%.

These results demonstrate clearly the effect of quaternary constraint. Molecules possessing the architecture of ATCase exhibit homotropic effects even when two of the chains in each catalytic subunit are inactive. Recently NAGEL has shown that cooperativity can be abolished completely in ATCase-like molecules containing two native catalytic subunits if the regulatory chains are extensively succinylated [28]. Thus the regulatory chains play a crucial role in mediating the allosteric transition. Preliminary studies with other derivatives containing less extensively succinylated regulatory subunits (and completely native catalytic subunits) suggest that homotropic effects can be maintained while heterotropic effects are largely abolished [28].

Linkage between Local and Gross Conformational Changes in ATCase

As shown above, the addition of specific ligands to the isolated catalytic subunits promotes conformational changes with each chain responding independently. Since these same ligands also cause conformational changes in ATCase we would like to know how, and to what extent, the catalytic subunits within ATCase molecules change upon the addition of the ligands. Are the ligand-promoted conformational changes in the catalytic subunits within intact ATCase molecules similar to those which the isolated catalytic subunits undergo? How are the conformational changes in ATCase linked to the conformational changes in the catalytic subunits within the intact enzyme molecules? Tentative answers to these questions came from KIRSCHNER's studies of chemically modified derivatives of the catalytic subunits (and reconstituted ATCase-like molecules) which contain a sensitive spectral probe whose absorption properties change markedly upon the addition of ligands [19].

By treating the catalytic subunit with tetranitromethane in the presence of the substrates, KIRSCHNER obtained enzymically active derivatives containing about one nitrotyrosyl group per polypeptide chain [19]. This derivative had a maximum in its absorption spectrum at 430 nm. Succinate alone had no effect on the spectrum at pH 7 and there was only a slight change in the spectrum when carbamyl phosphate was added. In contrast, upon the addition of both carbamyl phosphate (4 mM) and succinate (2 mM) there was a 14% decrease in the absorbance at 430 nm and an increase in absorption at 360 nm with an isosbestic point at 390 nm. Spectrophotometric titration of the unliganded species gave a pK of 6.25 which was raised to 6.62 upon the addition of the ligands. Studies with a variety of different ligands (in pairs and singly) showed that the absorbance at 430 nm, the absorption in the ultraviolet region of the spectrum [37], and the sedimentation coefficient changed in

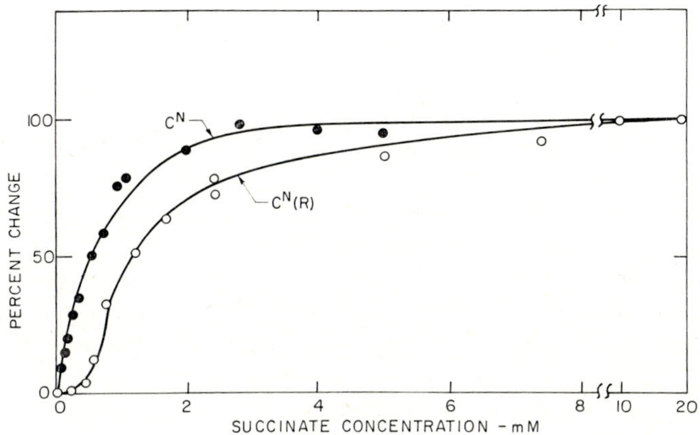

Fig. 16. Succinate titration of spectral change in nitrated catalytic subunit (C^N) and in nitrated ATCase (C^NR). All measurements were made by difference spectroscopy and the results are plotted on the ordinate as the percent of maximal change in the difference absorbance measurement vs. succinate concentration on the abscissa. The maximal change in absorbance at 430 nm was 15% for C^N and 16% for C^NR. The buffer in both cases was 0.04 M phosphate containing 2 mM MET, 0.2 mM EDTA, and 4×10^{-3} M carbamyl phosphate (from KIRSCHNER [19])

a parallel fashion. The spectral change of the nitrated catalytic subunit (C^N) occurs only with those ligand combinations which appear to produce conformational changes as measured by the other methods [35, 37]. It is possible, nonetheless, that the actual spectral change is produced by a direct effect of the bound dicarboxylic acid. The titration of the spectral change, shown in Fig. 16, showed saturation at a relatively low concentration of succinate and the curve was readily fit by a Michaelis-Menten type function with a dissociation constant of 3.7×10^{-4} M [19]. Similar values were obtained from difference sedimentation titration (Fig. 15), titration of the change in the optical rotatory dispersion [25], direct binding experiments by equilibrium dialysis, and enzyme kinetic studies which yielded the inhibition constant for succinate under the same conditions (buffer and temperature) [19]. Hence the conformational changes in the isolated catalytic subunits seem to be closely linked to binding. Moreover, these results taken in conjunction with the enzymic behavior (Fig. 14) and observations on the intra-subunit hybrids (Table 2) show that each polypeptide chain in the isolated catalytic subunits responds independently with respect to binding, enzyme activity, and local and gross changes in conformation.

With ATCase the binding of succinate [38] and conformational changes [36] are clearly separable, i.e., weakly linked, in terms of the concentration of succinate necessary to saturate the binding sites and cause a change in conformation. Hence we must inquire whether the conformational changes in the catalytic subunits within ATCase are weakly linked to binding. Alternatively, as indicated by the difference sedimentation studies, the conformational changes within the catalytic subunits may occur at the ligand concentrations necessary for saturation but these conformational changes are weakly linked to a different transition involving entire ATCase molecules. Since the nitrotyrosyl chromophore responds to local environment and difference sedimentation provides a measure of gross changes in the ATCase molecule, we can obtain information relevant to these alternatives by examination of ATCase molecules containing nitrated catalytic subunits.

Reconstituted ATCase-like molecules, $C^N R$, containing two nitrated catalytic subunits and native regulatory subunits were prepared by KIRSCHNER [19] and they showed properties similar to those of native ATCase [19]. Although V_{max}, CTP inhibition, Hill coefficient, and $\Delta s/s$ upon the addition of ligands were all slightly less for $C^N R$ than for native enzyme, the principal allosteric phenomena (cooperativity and feedback inhibition) were preserved. Hence studies were conducted to determine the effect of ligands on the spectral response of the nitrotyrosyl groups within ATCase-like molecules. As with C^N, the addition of both carbamyl phosphate and succinate to $C^N R$ caused a 13% decrease in the absorbance at

430 nm (some change in the absorbance occurred upon reconstitution due to the altered environment of the chromophore in C^NR as compared to C^N). Unlike C^N which responded in a hyperbolic fashion to varying concentrations of succinate (at a constant level of 4 mM carbamyl phosphate), C^NR showed distinctly sigmoidal behavior with a Hill coefficient of 1.68. As seen in Fig. 16, the half-titration level for C^NR was shifted to higher succinate concentration compared to that for C^N. Thus the incorporation of the nitrated catalytic subunits into ATCase led to an alteration in the nature and extent of the response of the chromophore to varying ligand concentration. This sigmoidal dependence of the absorbance of C^NR on succinate concentration is very similar to the ligand binding curve for ATCase [38]. Moreover, they both resemble the sigmoidal response of enzyme activity to substrate concentration (Fig. 14). In a similar vein, the hyperbolic titration curve for the absorbance of C^N as a function of succinate concentration is directly analogous to the ligand saturation curve for the catalytic subunit [38] and to the dependence of enzyme activity on substrate concentration (Fig. 14). Thus the contrasting behavior between C^N and C^NR is another manifestation of quaternary constraint.

Upon their incorporation into ATCase-like molecules the nitrated catalytic subunits exhibit a decreased response of their chromophores to low concentrations of succinate. Increasing the ligand concentration leads to parallel effects on enzymic activity, saturation of binding sites, and local conformational changes. These results suggest that the conformational changes within the catalytic subunits in intact ATCase molecules are sequential and that the binding of each succinate molecule makes a separate and equivalent contribution to the conformational change of each catalytic chain. However, as seen in Fig. 17, these local strongly linked conformational changes within the catalytic subunits in ATCase appear to be weakly linked to the gross conformational changes in ATCase revealed by the alteration in the sedimentation coefficient. At a succinate concentration sufficient to promote only 10% of the maximal change in absorbance at 430 nm the change in sedimentation coefficient was 50% of the total change. There is also a marked difference in the shape of the titration curves of $(\Delta A/A)_{430}$ and $\Delta s/s$ vs. succinate concentration. For the former the Hill coefficient, as indicated above, was 1.68 whereas the latter had a Hill coefficient of only 0.91. It thus appears from these studies with the nitrated derivative of ATCase that binding of ligands is accompanied by sequential, local changes in the conformation of the catalytic chains which are weakly linked with a concerted, gross change in the quaternary structure of the intact molecules. This gross change in conformation corresponding to a loosening of the structure may involve alterations in the orientation of the catalytic subunits

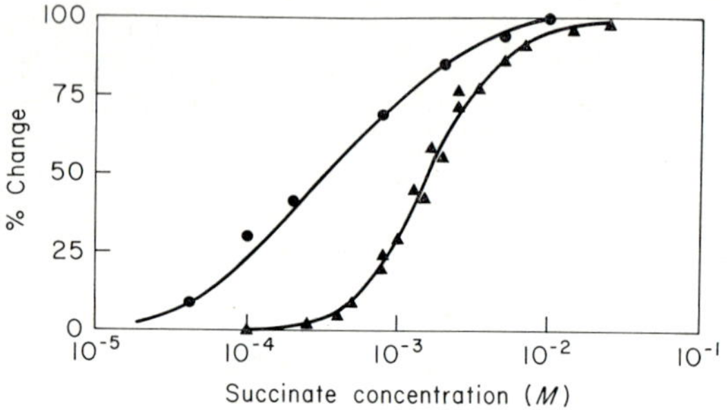

Fig. 17. Effect of succinate on the sedimentation coefficient and spectrum of nitrated ATCase. Difference sedimentation data are represented by ● and changes in absorbance at 430 nm are represented by ▲. In each experiment the concentration of carbamyl phosphate was 2×10^{-3} M in one sector plus a given concentration of succinate; the other sector contained 2×10^{-3} M phosphate and a corresponding concentration of glutarate. The data are plotted as percent maximal change on the ordinate vs. succinate concentration on the abscissa (from KIRSCHNER [19])

relative to each other and the regulatory subunits or to conformational changes in the regulatory subunits themselves. In view of the linkage between local changes in the conformational change of the catalytic subunit and the gross change in ATCase, we might expect that changes in the gross conformation of ATCase should affect the nitrotyrosyl groups on the catalytic subunits. In this regard it is of interest that the binding of CTP is accompanied by a change in the absorbance of the nitrotyrosyl groups on the catalytic subunits even though CTP binds to the regulatory subunits.

The data in Fig. 17 shows clearly that the gross conformational change in ATCase occurs at a lower concentration of succinate than is required to promote an equivalent percentage change in the local conformation of the catalytic subunits. However, we have no direct evidence demonstrating whether the binding of ligand to one catalytic subunit can promote local conformational changes in the other. Through appropriate reconstitution experiments it should be possible to obtain this information. If we had an ATCase-like molecule containing one inactive, nitrated catalytic subunit (to which ligands do not bind) and one native catalytic subunit, would the spectrum of the nitrotyrosyl groups in the inactive subunit change upon the binding of ligands to the active catalytic subunit? An answer to this question and other related questions dealing with the regulatory subunits would add materially to our understanding of the mediating role of the regulatory subunits and of the linkage between local and gross conformational changes in the intact enzyme molecules.

Summary

Evidence is reviewed showing that aspartate transcarbamylase from *Escherichia coli* is composed of six catalytic and six regulatory polypeptide chains. The catalytic chains were shown to exist in the intact enzyme as two trimers. This conclusion was based largely on intra- and inter-subunit hybridization experiments involving native and succinylated subunits. Hybridization at the subunit level produced a 4-membered hybrid set when a mixture of native and succinylated subunits was subjected to dissociation in 2 M guanidine hydrochloride and then dialyzed to permit reassociation of the polypeptide chains. The addition of regulatory subunits to a mixture of native and modified catalytic subunits yielded an inter-subunit hybrid set comprising three members, thereby demonstrating that each enzyme molecule contains two catalytic subunits. Analogous experiments with native and succinylated regulatory subunits mixed rapidly with native catalytic subunits gave a 4-membered hybrid set as expected for a complex containing three regulatory subunits. Moreover these hybridization experiments with different types of regulatory subunits along with sedimentation equilibrium studies showed that the isolated regulatory dimers existed in a monomer-dimer equilibrium. Additional evidence for the existence of the regulatory subunits as dimers within the intact enzyme molecules came from reconstitution experiments with catalytic subunits and covalently cross-linked regulatory subunits.

A model is proposed for the arrangement of the two catalytic trimers and the three regulatory dimers. This model, based on electron micrographs and other physical chemical evidence, possesses one threefold and three twofold symmetry axes in accord with evidence from crystallographic studies. The two catalytic subunits, each having a triangular arrangement of the three polypeptide chains associated by three heterologous bonding domains, are superimposed above one another in an eclipsed configuration without being in direct physical contact. The three regulatory dimers, each a V-shaped structure consisting of two cylindrically shaped monomers associated by isologous bonding domains, extend to the outside of the molecule and serve to interconnect the catalytic subunits. Catalytic chains in the upper trimer are linked by the regulatory dimers to catalytic chains in the lower trimer which are displaced 120° from those above.

Studies of the conformational changes of the individual, purified hybrids of the catalytic subunit along with their enzymic activities showed that the polypeptide chains acted independently. These conformational changes are strongly linked to the binding of ligands. An additional and different type of transition occurs in the intact enzyme upon the addition of these same ligands. This is seen even for the hybrid species containing one native and one inactive (succinylated) catalytic subunit. The hybrid molecule shows cooperativity in marked contrast with the isolated catalytic subunit which is non-cooperative in its enzymic behavior. Thus quaternary constraint is demonstrated in a molecule containing only one active catalytic subunit along with an inactive catalytic subunit and native regulatory subunits. By incorporating a chromophore onto the catalytic chains within intact active enzyme molecules it was possible to show that the conformational changes in these chains in the intact enzyme are similar to those in the isolated catalytic subunits. However these local changes in conformation were found to be weakly linked to a different transition which affects the gross architecture of the entire enzyme molecule. This gross conformational change, which was shown by hydrodynamic studies to be a loosening or swelling of the whole molecule, is concerted whereas the local change corresponding to a tightening or contraction of the catalytic subunits is sequential.

Acknowledgement. It is a pleasure to acknowledge the innumerable and invaluable contributions of J. A. COHLBERG, M. W. KIRSCHNER, G. M. NAGEL, V. P. PIGIET, Jr., and Y. YANG to the work described in this review. Many of the ideas and most of the experimental results derive from their research. All of the work stems directly from the pioneering studies of J. C. GERHART whose continued interest, suggestions, criticisms and collaboration have been indispensable.

References

1. MONOD, J., WYMAN, J., CHANGEUX, J.-P.: J. molec. Biol. **12**, 88—118 (1965).
2. KOSHLAND, D. E., Jr., NEMETHY, G., FILMER, D.: Biochemistry **5**, 365—385 (1966).
3. PERUTZ, M. F.: Nature (Lond.) **228**, 726—734 (1970).
4. YATES, R. A., PARDEE, A. B.: J. biol. Chem. **221**, 757—770 (1956).
5. GERHART, J. C., PARDEE, A. B.: J. biol. Chem. **237**, 891—896 (1962).
6. — Curr. Top. Cell. Regul. **2**, 275—325 (1970).
7. — SCHACHMAN, H. K.: Biochemistry **4**, 1054—1062 (1965).
8. WEBER, K.: Nature (Lond.) **218**, 1116—1119 (1968).
9. WILEY, D. C., LIPSCOMP, W. N.: Nature (Lond.) **218**, 1119—1121 (1968).
10. MEIGHEN, E. A., PIGIET, V., SCHACHMAN, H. K.: Proc. nat. Acad. Sci. (Wash.) **65**, 234—241 (1970).
11. WILEY, D. C., EVANS, D. R., WARREN, S. G., MCMURRAY, C. H., EDWARDS, B. F. P., FRANKS, W. A., LIPSCOMB, W. N.: Cold Spr. Harb. Symp. quant. Biol. **36**, 285—290 (1971).
12. ROSENBUSCH, J. P., WEBER, K.: J. biol. Chem. **246**, 1644—1657 (1971).
13. — — Proc. nat. Acad. Sci. (Wash.) **68**, 1019—1023 (1971).
14. NELBACH, M. E., PIGIET, V. P., Jr., GERHART, J. C., SCHACHMAN, H. K.: Biochemistry **11**, 315—327 (1972).
15. COHLBERG, J. A.: Berkeley: Ph. D. Thesis, University of California 1972.
16. — PIGIET, V. P., Jr., SCHACHMAN, H. K.: Biochemistry **11**,3396—3411 (1972).
17. RICHARDS, K. E., WILLIAMS, R. C.: Biochemistry **11**, 3393—3395 (1972).
18. GERHART, J. C., HOLOUBEK, H.: J. biol. Chem. **242**, 2886—2892 (1967).
19. KIRSCHNER, M. W.: Berkeley: Ph. D. Thesis, University of California 1971.
20. WEBER, K.: J. biol. Chem. **243**, 543—546 (1968).
21. HERVE, G. L., STARK, G. R.: Biochemistry **6**, 3743—3747 (1967).
22. CHANGEUX, J.-P., GERHART, J. C., SCHACHMAN, H. K.: Biochemistry **7**, 531—538 (1968).
23. STEITZ, T. A., WILEY, D. C., LIPSCOMB, W. N.: Proc. nat. Acad. Sci. (Wash.) **58**, 1859—1861 (1967).
24. CHANGEUX, J.-P., GERHART, J. C., SCHACHMAN, H. K.: In: Regulatory mechanisms in nucleic acid and protein biosynthesis (KONINGSBERGER, V. V., BOSCH, L., Eds.) Biochim. biophys. Acta Libr. **10**, 344—350 (1967).
25. PIGIET, V. P., Jr.: Berkeley: Ph. D. Thesis, University of California 1971.
26. GILBERT, G. A.: Disc. Faraday Soc. **20**, 68 (1955).

27. Nagel, G. M., Schachman, H. K., Gerhart, J. C.: Fed. Proc. **31**, 423 Abs. (1972).
28. — — (in preparation).
29. Davies, G. E., Stark, G. R.: Proc. nat. Acad. Sci. (Wash.) **66**, 651—656 (1970).
30. — — Private communication (1972).
31. Green, N. M.: Nature (Lond.) **219**, 413—414 (1968).
32. Schachman, H. K., Edelstein, S. J.: Biochemistry **5**, 2681—2705 (1966).
33. Changeux, J.-P., Gerhart, J. C.: In: The Regulation of enzyme activity and allosteric interactions, Fed. Eur. Biochem. Soc., 4th Meeting (Kvamme, E., Pihl, A., Eds.) Vol. **I**, 13—38. New York: Academic Press 1968.
34. Kirschner, M. W., Schachman, H. K.: Biochemistry **10**, 1900—1919 (1971).
35. — — Biochemistry **10**, 1919—1926 (1971).
36. Gerhart, J. C., Schachman, H. K.: Biochemistry **7**, 538—552 (1968).
37. Collins, K. D., Stark, G. R.: J. biol. Chem. **244**, 1869—1877 (1969).
38. Changeux, J.-P., Gerhart, J. C., Schachman, H. K.: Biochemistry **7**, 531—538 (1968).

Discussion

E. Helmreich (Würzburg): The data shown in the table present another example of quaternary restraint. Dr. K. Feldmann in our laboratory has studied muscle phosphorylase bound to CNBr activated Sepharose [Proc. nat. Acad. Sci. (Wash.) **69**, 2278 (1972)]. The first entry in the table is the control experiment. It shows the activity of sepharose bound phosphorylase dimer *a* and gives the residual activity (about 10%) after dissociation of dimer *a* with imidazol citrate. B, denotes Sepharose bound monomer *a*. S is the soluble subunit. PLP is pyridoxal phosphate. On addition of soluble monomers *a*—prepared by reaction of soluble phosphorylase dimer *a* with p-chloromercuribenzoate—to Sepharose bound monomers *a* and subsequent removal of p-chloromercuribenzoate with β-mercaptoethanol the original activity of the Sepharose bound dimer *a* is regained. The second series of experiments represents the properties of a Sepharose bound hybrid dimer *a*. The soluble subunit of this phosphorylase derivative was modified not in the protein but in the cofactor essential for activity. The phosphorylase *a* derivative contained a cofactor which was esterified at the 5′-phosphate group (pK_2) of pyridoxal-P. The pyridoxal-P monomethylester phosphorylase is completely inactive. If this intrinsically inactive phosphorylase monomer was added to the Sepharose bound monomer activity was induced in the latter and the hybrid formed had one half the activity of the natural dimer *a*. This demonstrates:

1. that monomeric phosphorylase *a* has little or no activity, but has a site which becomes active on interaction with the inactive subunit;

2. that the phosphorylase derivative which carries the inactive cofactor has retained structural complementarity, since it still can induce activity, but cannot become active itself;

3. that it may be advantageous for hybridization experiments to chemically modify the cofactor rather than the apoenzyme. The last entry in the slide is again a control experiment which indicates, that induction of activity is specific and that heterologous proteins known to form complexes with phosphorylase such as serum albumin and protamine sulfate are ineffective.

Interaction between Monomers a Containing Pyridoxal-P and Inactive Analogs

Preparations	Activity [μ-moles $P_i \cdot min^{-1}$]	
	−AMP	+1 mM AMP
Dimer *a* ($PLP^B \cdot PLP^S$)	0.44	0.52
Monomer *a* (PLP^B)	0.05	—
Dimer *a* ($PLP^B \longleftrightarrow PLP^S$)	0.46	0.52
Dimer *a* ($PLP^B \cdot PLP^S$)	0.63	0.79
Monomer *a* (PLP^B)	0.06	—
Hybrid dimer *a* ($PLP^B \leftarrow$ PLP-Ester^S)	0.33	0.42
Dimer *a* ($PLP^B \cdot PLP^S$)	0.60	0.72
Monomer *a* (PLP^B)	0.05	0.06
Monomer *a* {$\sum$ + serum albumin	0.05	0.06
Monomer *a* {$\sum$ + protamine sulfate	0.05	0.06

Lazdunski (Marseille): There have been reports of anticooperativity for CTP binding to native ATCase. Moreover, as far as I can remember, no more than 3 to 4 moles of succiniate can bind to the enzyme. Dr. Schachman, could you comment cn this negative cooperativity in terms of subunit-subunit interactions in native ATCase? Could you also explain the relevance of this property for the catalytic mechanism of the enzyme.

Schachman (Berkeley): As you could see from my earlier comments, we have sufficient difficulty in attempting to explain the positive cooperativity exhibited by ATCase in its binding of substrates and their analogs. Thus, I do not relish additional problems such as explaining negative cooperativity in terms of subunit interactions. In my view, it is premature to speculate on this matter until there is clarification of the experimental data. At present the results from various laboratories appear conflicting, but it should be recognized that different techniques were employed, the ATCase preparations differed, and the conditions of pH, buffers, and temperature varied considerably.

In their studies of the binding of CTP by ATCase, Winlund and Chamberlin found two classes of sites (three sites in each class) differing markedly in their affinities. Hammes and his colleagues, however, using the method of continuous variations, found six binding sites for BrCTP with no evidence for differences in affinity. In still another study, Buckman found three tight binding sites for CTP and six weaker sites. Early work from our laboratory revealed two classes of CTP binding sites in ATCase as well as evidence for differences in affinity for the sites in the isolated regulatory subunit.

The situation with regard to the number of succinate binding sites now seems clear. Although our early studies yielded two sites for the isolated catalytic subunit and about four on ATCase, more recent work with other preparations has given higher values. Collins and Stark have provided definitive answers with their end-point titration of the binding of the transition state analog, N-(phosphonacetyl)-L-aspartate. They have shown that the catalytic subunit possesses three strong binding sites and ATCase has six such sites. Also, Rosenbusch and Weber found three succinate binding sites in the isolated catalytic subunit. As yet, the only evidence for a sigmoidal saturation curve for ATCase comes from our early studies which gave low values for the number of sites. Although the number of sites, for some unexplained reason, was low, we still feel that the shape of the saturation curve is correct. The data of Kirschner for the spectral titration of nitrated ATCase provides some support for this view.

Quaternary Structure of Proteins

Subunits to Superstructures: Assembly of Glutamate Dehydrogenase

R. Josephs, H. Eisenberg and E. Reisler

Department of Polymer Research
The Weizmann Institute of Science, Rehovot, Israel

With 24 Figures

I. Introduction

The last 10 years have seen the development of a large body of knowledge dealing with problems of the assembly of biological macromolecules into specific ordered structures. The ideas which have emerged from these studies may be regarded as an extension of the well established principle that the sequence of amino acid residues in a polypeptide chain is the main determinant of the complex folding of the chain into the native protein conformation. The concept of "self-folding" of the polypeptide chain was a natural starting point for dealing with the question of how macromolecules assemble themselves into large ordered structures for it introduced the idea that seemingly minor alterations in the microstructure of molecules can induce profound changes in their macrostructure.

In the present context self-assembly refers to the ability of a molecular species to undergo spontaneous interaction resulting in the formation of larger more complex structures of specific geometry. The level of organization involved need not be restricted, and may range in complexity from genesis of the quaternary structure of an enzyme from its subunits to the formation of bacterial flagella. Nor is the number of components restricted to a single molecular species but may include the interaction of multiple species as in the assembly of tobacco mosaic virus and muscle filaments and even extends to very complex systems consisting of many dozens of different proteins and nucleic acids such as the ribosome. The essential feature characterizing all of these systems as "selfassemblying" is that under proper conditions the individual components, when mixed together, undergo a spontaneous interaction generating the original structure from which they were derived.

Obviously phenomena of self-assembly are not restricted to biological systems, but here they take on special significance in that they represent the mechanism of the primary steps in the morphogenesis of whole organisms. It is not yet clear to what level of organization morphogenesis is controlled by self-assembly alone, however, to date the principle seems to be valid for some very complex structures indeed and the ultimate limits remain to be established.

An aspect of self-assembly which has received much attention is how the effect of environmental conditions (i.e. pH, temperature, ionic composition) control the extent and mode of assembly. That these conditions do play such an important role leads one to speculate that they may be one of the principal means by which self-assembly may be controlled.

Investigation of the control of self-assembly provides an exciting insight into the dynamic nature of the interactions governing many biological processes. For instance, when a virus such as tobacco mosaic virus infects a cell the protein coat is cast off and the viral nucleic acid directs synthesis of new viral protein and nucleic acid. These in turn reassemble into whole virus particles identical to the original infecting particle. What initiates the disassembly of the virus upon infection? Do intracellular conditions change so as to promote reassembly of the newly synthesized virus proteins and nucleic acid or is there an equilibrium between assembled and unassembled components which can be affected by conditions within the cell? Alternatively if the cell gives a signal to the unassembled components to assemble what is the nature of the signal?

These and similar questions have been investigated and partially answered for a number of different systems. In the succeeding pages we propose to describe studies of one system, the enzyme glutamate dehydrogenase, which we have examined in some detail and which in our view presents a particularly interesting example of self-assembly in that some of the control mechanisms for assembly are understood and that the assembly process itself has been characterized at several different levels of molecular organization.

II. Organization of Glutamate Dehydrogenase

A. Subunit Structure

The properties of the most primitive level of organization, namely the unfolded random polypeptide chain may be readily examined in concentrated (about 5 to 7 molar) solutions of guanidine hydrochloride. Ultracentrifuge [1] and light scattering [2] studies of the unfolded chain in guanidine solutions have yielded molecular weights (about 53,500) of the subunit polypeptide chain in close agreement with that obtained [3] from the amino acid sequence for the subunit (about 56,000).

For reasons which we shall consider more fully below, determination of the number and arrangement of subunits in the active form of the molecule has presented considerable difficulties for almost two decades. As a result of several independent lines of investigation a model has been proposed [4—6] for the quaternary structure of the active form of the enzyme. The main features of the model depict the molecule as consisting of six subunits arranged with 32 point group symmetry. Combined solution studies and electron microscopy have established that the particle is 100 to 130 Å in length and about 80 Å in diameter. The subunits are arranged in the form of a triangular antiprism and electron micrographs of isolated molecules in different orientations display profiles which closely accord with projections computed from the model (Fig. 1 and 2).

B. Linear Polymers

The next higher level of organization involves association of the individual molecules to form polymeric species of high molecular weight. That association between molecules occurs has been known from the very earliest studies on glutamate dehydrogenase; the key observation being [7] that the sedimentation coefficient increases with increasing protein concentration. It was only recently however that the nature of the association products was established. Three lines of evidence support the view that the molecules undergo end-to-end association resulting in the formation of linear polymers: 1. A plot of molecular weight against the radius of gyration (determined from light scattering) yields a straight line (Fig. 3). The range of linearity [2, 5] encompasses the whole range of measured molecular weights and extends to a molecular weight above 3 million

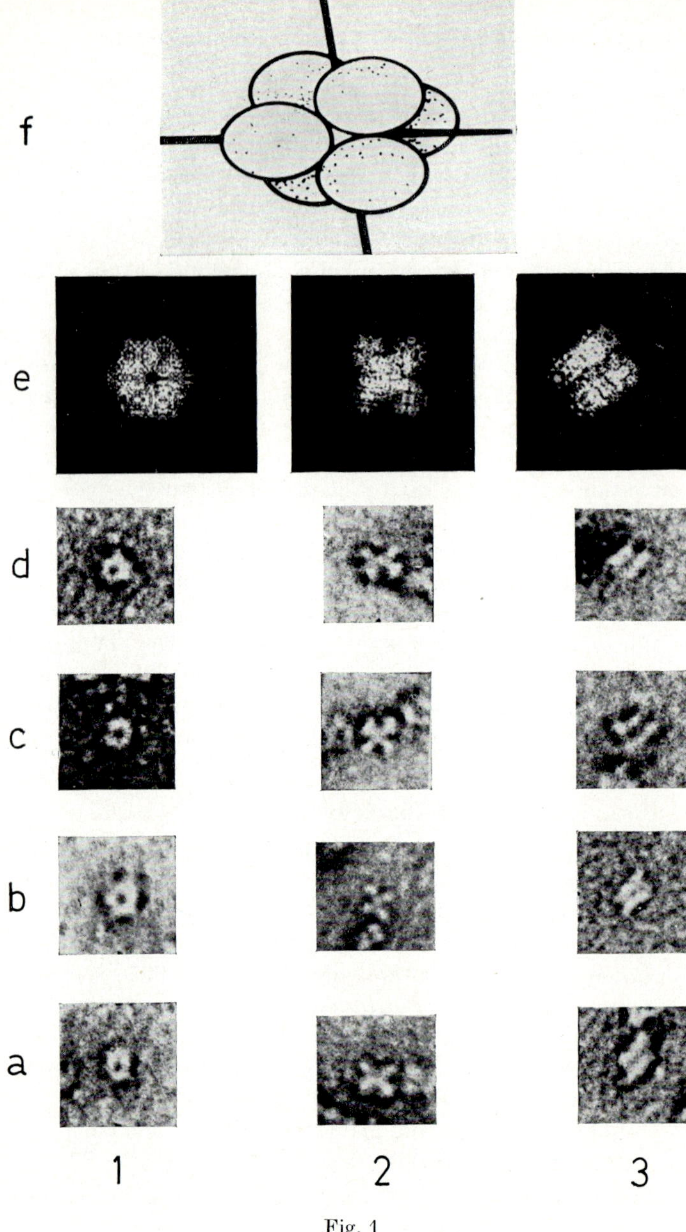

Fig. 1

representing a polymer containing about ten enzyme molecules. 2. The mass per unit length and radius of gyration of the cross-section of the association products of glutamate dehydrogenase may be determined from measurements of low angle X-ray scattering. Measurements carried out over a thirty-fold concentration range show

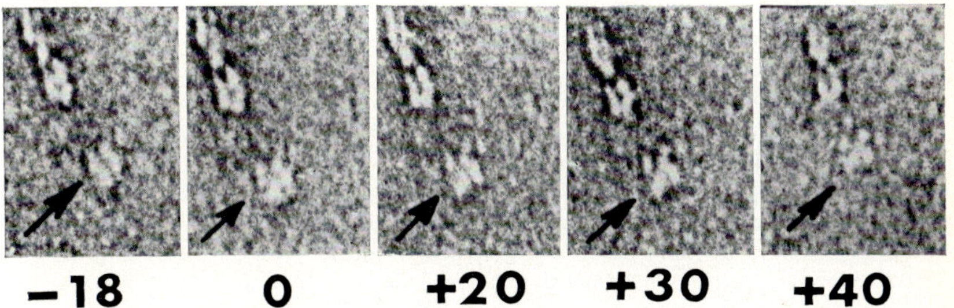

Fig. 2. Electron micrographs of a molecule of glutamate dehydrogenase tilted about the axis indicated by the arrow. The tilt angle in degrees is indicated beneath each frame. As the particle is tilted from − 30° to + 40° there is a clear change from two rows of density (row 3, Fig. 1) to a cross pattern (row 2, Fig. 1) (from Ref. [6])

that both the linear mass and cross-sectional radius of gyration are constant, independent of the molecular weight of the enzyme which varies from about 300,000 to above several million [8] (Fig. 4); the clear implication is that a highly extended linear structure is involved. 3. Electron micrographs of polymers of glutamate dehydrogenase reveal that they are indeed linear structures and that

---

Fig. 1. Columns a) to d), electron micrographs of isolated molecules of glutamate dehydrogenase in various orientations. Column e), computed projections of the model. These projections are to be compared with micrographs to the left on the same row. Column f), model for glutamate dehydrogenase. The subunits are elliptical (axial ratio 1:1.5). The vertical axis is the 3-fold axis of each molecule, and the horizontal axis represents one of the three identical 2-fold axes. The positions of the other 2-fold axes are defined by operation of the 3-fold axis, (from Ref. [6])

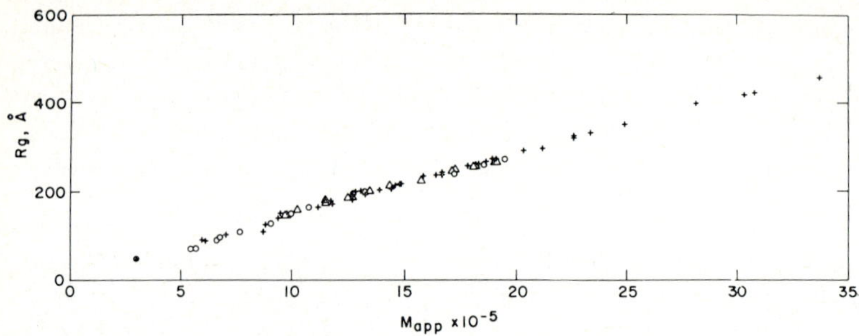

Fig. 3. Radius of gyration, R_g, from light scattering of glutamate dehydrogenase solutions *vs.* apparent molecular weight, M_{app}; (○) sodium phosphate buffer 0.2 M, pH 7, 10^{-4} M EDTA, 25 °C, enzyme concentration range 0.5 to 11 mg/ml; (△) same buffer, various temperatures (10 to 30 °C), same range of enzyme concentrations; (+) same buffer, saturated with respect to toluene, enzyme concentration range 0.03 to 0.5 mg/ml; (●) calculated value for the oligomer (from Ref. [5])

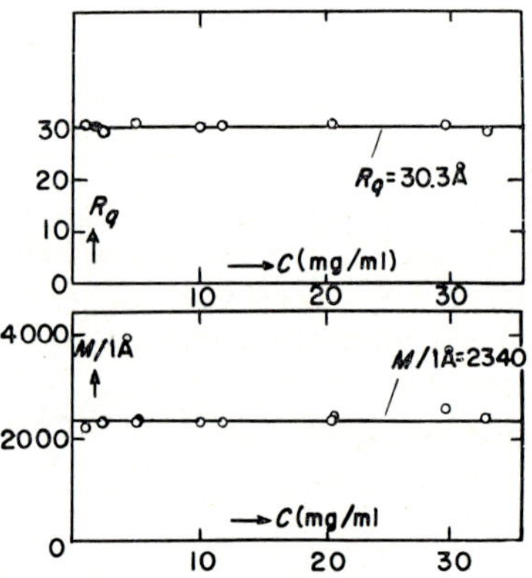

Fig. 4. Radius of gyration (R_g) of the cross-section and mass per 1 Å length of the particles (M/1 Å) as a function of protein concentration (from Ref. [8])

Fig. 5. Polymer chains of glutamate dehydrogenase. The length of these polymers is somewhat variable, and chains consisting of up to ten or more molecules can be observed. The spacing between molecules is 100 to 120 Å and the diameter of the polymer is 80 to 90 Å (from Ref. [6])

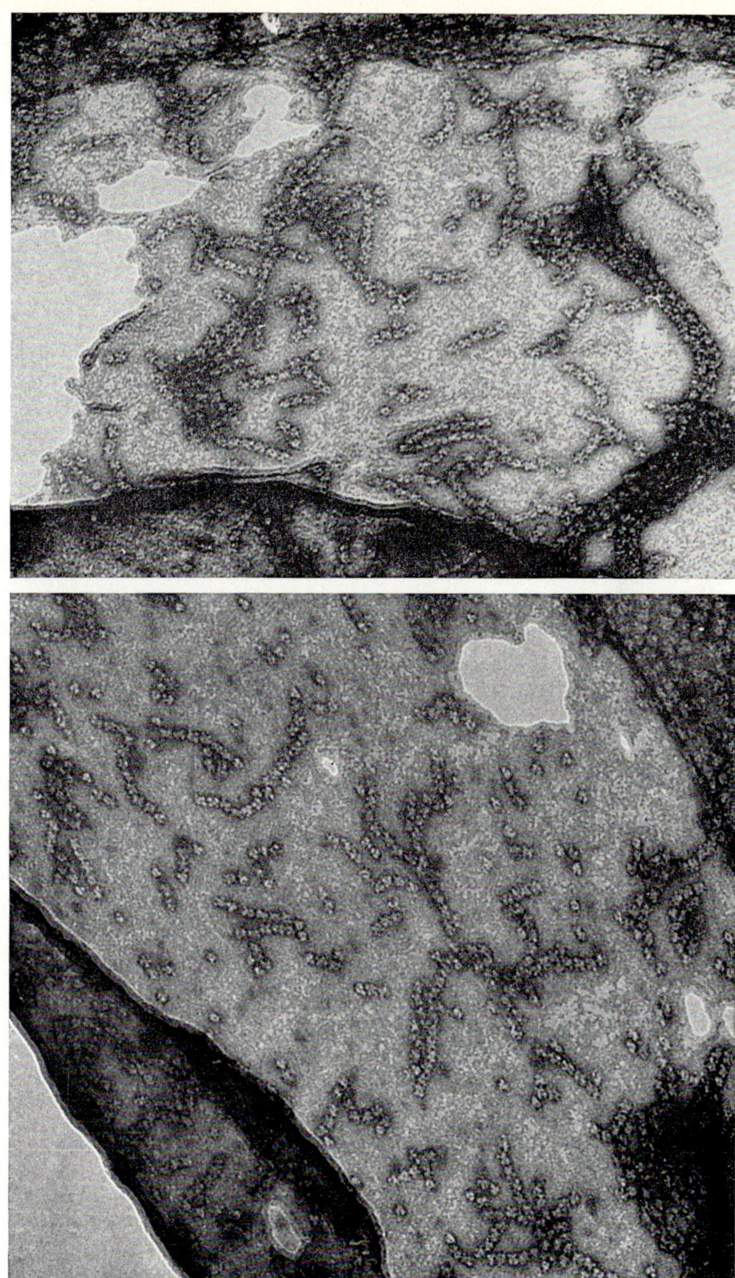

Fig. 5

individual molecules may be delineated along their length [6] (Fig. 5). Thus a number of different lines of evidence converge in establishing the linear nature of the polymers.

C. Higher Order Structures

The linear polymers which were described in the preceding paragraph show ordered growth in one dimension only, i.e. along their length. Under appropriate conditions linear polymers may form more complex associates which exhibit two dimensional and three dimensional growth as well and we shall consider each of these in turn.

The structures containing two dimensional order consist of rows of linear polymers associated in side-by-side array to form sheets [9]. The prominence of the linear polymers is unmistakable; electron micrographs clearly revealing the individual polymer chains within the sheet (Fig. 6). The separation between chains and spacing of molecules along the length of the chains exactly correspond to that observed for isolated linear polymers. Moreover, the polymer chains extend along the entire length of the sheet and interruptions in their continuity are not observed. Within the sheet the polymer chains are not straight, but rather present a wavy appearance which repeats after four molecules suggesting that some degree of flexibility may exist between neighboring molecules.

In addition to the sheets, linear polymers of glutamate dehydrogenase generate entirely different types of structures. These are tubes consisting of four linear polymer chains associated in helical array. Several species of tubes differing in the relative disposition of the polymer chains have been identified and examined in detail [10][1]. In the simplest tube the position of each of the four polymer chains is related to that of its neighbors by a four-fold rotation axis which is coincident with the axis of the tube (Fig. 7). In a second form of tube the polymer chains are staggered by one quarter of a molecule corresponding to an axial displacement of each chain of 25 Å and an azimuthal rotation of 10 degrees relative to its neighbor (Fig. 7 b). The relation between the positions of the

[1] The disposition of the enzyme molecules on the different tube lattice was derived by analysis of the optical diffraction patterns shown in Figs. 7 a and b. The reader is referred to Ref. [10] for the details of the analysis.

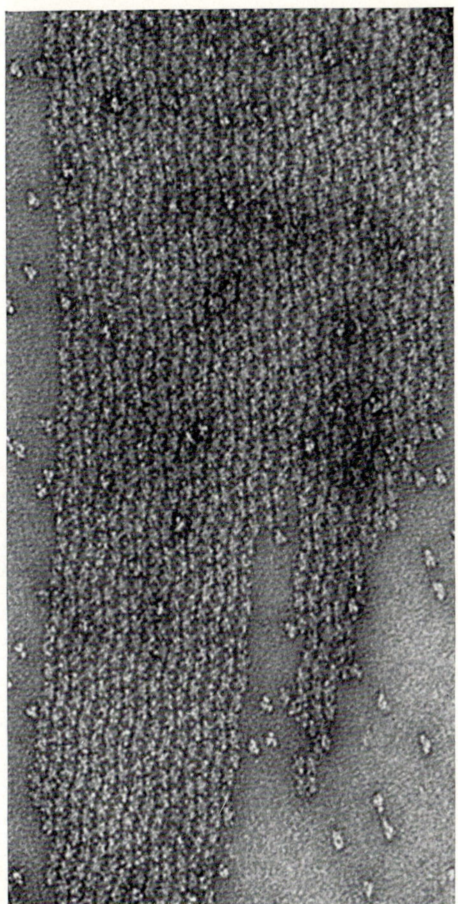

Fig. 6. Two dimensional sheet composed of linear polymer chains of glutamate dehydrogenase. The distance between chains is 80 to 85 Å and the spacing between molecules along the length of the chains 100 to 115 Å (from Ref. [9])

molecules in the tubes shown in Fig. 7 can be readily seen in Fig. 8, in which lattices of each type of tube are constructed on the surface of a cylinder.

Other more complex variations in the relative disposition of the polymer chains within the tube were observed as well, but the two

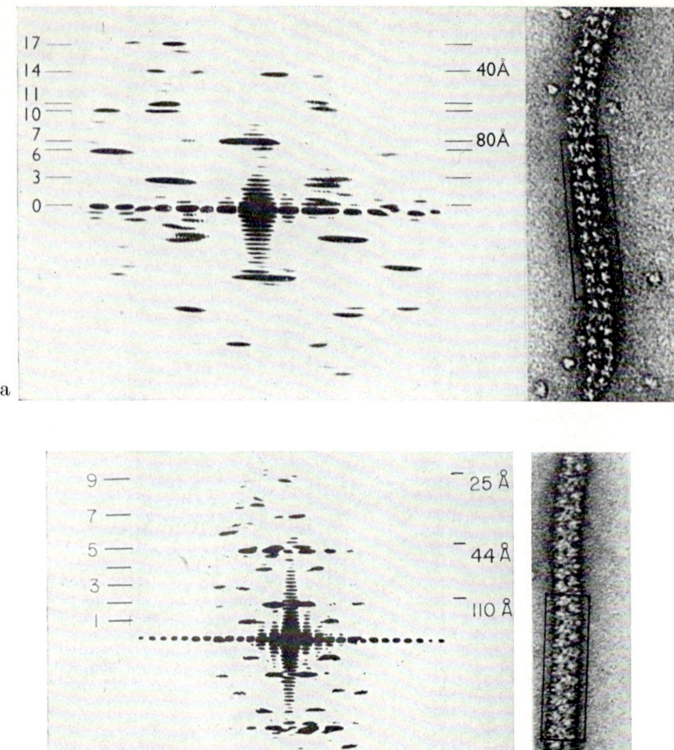

Fig. 7. Helical tubes of glutamate dehydrogenase and the optical diffraction patterns derived from the segment of the tube enclosed by the rectangular box. Layer line numbers are indicated on the left of the diffraction and spacings to the right. Tube diameter is 200 Å. Each tube consists of four polymer chains helically twisted about a common axis (from Ref. [10]). a) In this tube the polymer chains are symmetrically related by a 4-fold rotation axis coincident with the axis of the tube. b) In this tube the polymer chains are staggered by one quarter of a molecule

examples presented here suffice to establish that association of the chains within the tube is not invariant, but a rather considerable degree of flexibility exists in forming these structures.

Although different types of tubes are formed by varying modes of association of the four polymer chains about their respective helical axis it is of interest to note that in all cases the helical path

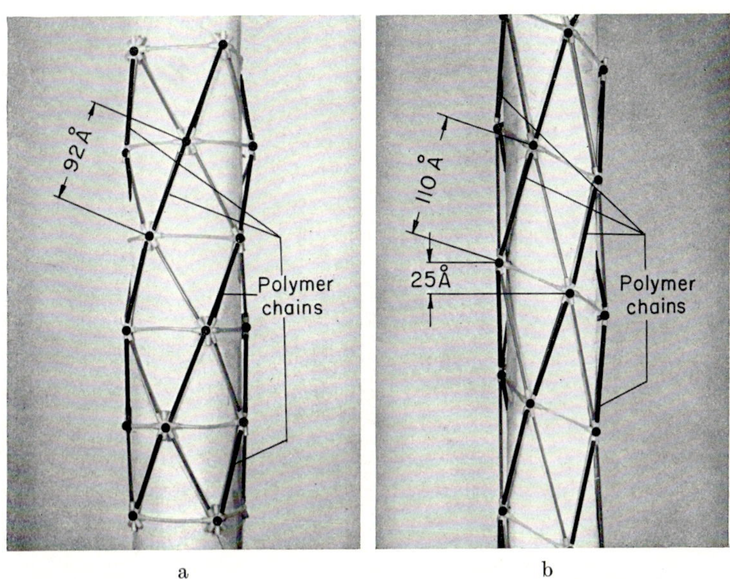

Fig. 8. Photograph of the tube lattices shown in Fig. 7. The positions of the connectors (●) indicate the location of the enzyme molecules in the tube lattice and the dark lines represent the polymer chains. In as much as only one side of the tube lattice is visible not all four polymer chains are simultaneously shown. a) Lattice showing location of molecules in the tube shown in Fig. 7 a. b) Lattice showing location of molecules in the tube shown in Fig. 7 b. Note that in frame a) the molecules in each chain are at the same axial heigth whereas in frame b) the axial distance between molecules in adjacent chains is 25 Å. This occurs because the chains are staggered by one-quarter of a molecule (adapted from Ref. [10])

described by each of the four chains is nearly the same. The helical path of each chain contains nine molecules per turn of the helix. The radius of the helix is 100 Å and the polymer chain makes an angle of about 28 degrees to the helix axis. As in the case of the two dimensional sheets adjacent molecules along the polymer chain do not lie in straight lines but rather make an angle of 19 degrees with each other.

D. Role of the Linear Polymer in Forming Higher Order Structures

One of the obvious conclusions deriving from studies of these higher order structures is that considerable flexibility exists in the packing arrangements of the polymer chains. Whole chains may be shifted relative to one another as in the tubes or flattened out as in the sheets. The long range geometry of the polymer chains is consequently susceptible to being drastically altered by small deformations of the bonding between adjacent enzyme molecules along the chain, the energy for which is obtained by favorable interactions between chains. The dominant feature present in all of these structures is that the polymer chain represents the unit of assembly and once incorporated into the superstructure it retains its continuity; the absence of single enzyme molecules from intermediate positions along the length of a chain is rarely observed.

III. Control of Assembly

At higher levels of organization the nature of the superstructure formed by association of the polymer chains is affected by small changes in environment. A prerequisite for the formation of higher order structures is association of the individual enzyme molecules to form the polymer chains. Association to form linear polymers thus plays a key role in the control of the formation of the more complex structures described above. We now turn to examine the nature of the association reaction in greater detail.

The length of the polymers (and indeed whether they may form at all) is under two types of control. a) In buffered salt solutions at neutral pH the size of the polymeric species increases with increasing concentration of enzyme. b) However, the state of aggregation of the enzyme may be dramatically altered by the addition of small amounts of NADH along with an additional regulatory agent. Of the large variety of compounds which are effective we mention here GTP which, in conjunction with NADH leads to near although not complete dissociation of the polymers. The magnitude of both effects is illustrated in Fig. 9 which presents a plot of the molecular weight of glutamate dehydrogenase as a function of enzyme con-

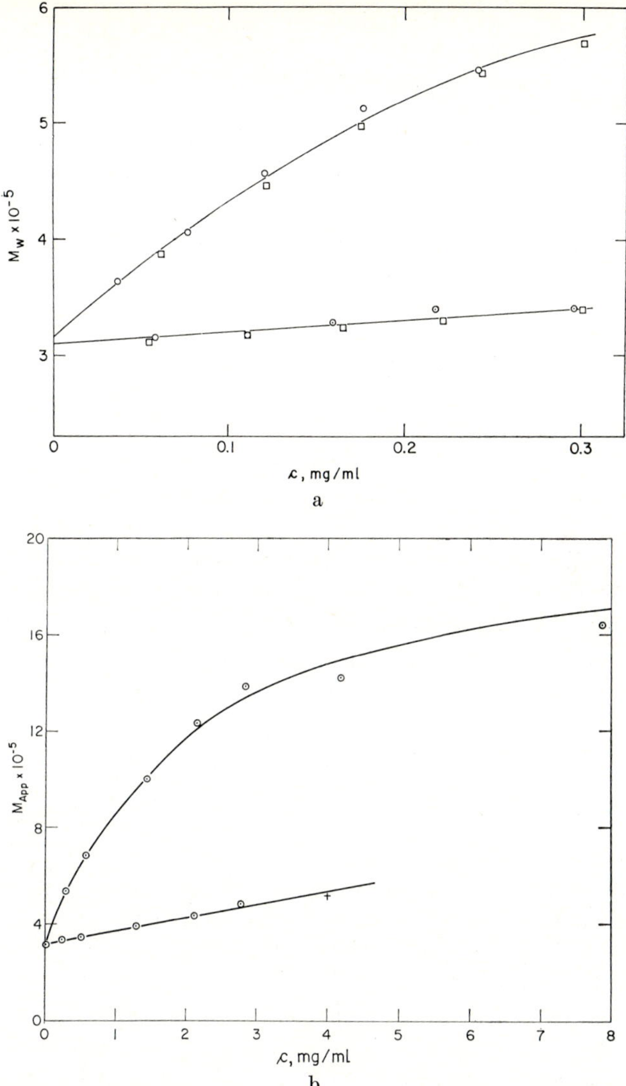

Fig. 9 a and b. a) Apparent weight-average molecular weight of glutamate dehydrogenase as a function of enzyme concentration. Upper curve, dialyzed against 0.2 M-phosphate buffer, 10^{-3} M EDTA, pH 7; lower curve, same as above plus 10^{-3} M GTP, 10^{-3} M NADH. Light scattering experiments at 25 °C, at 546 mμ. Crosses and circles represent independent experiments. b) Apparent weight-average molecular weight of GDH. Same as a) but over higher concentration range (from Ref. [2])

centration both in the presence and absence of NADH and GTP. It should be noted that lower concentrations of GTP and NADH cause a proportionately smaller drop in molecular weight of the polymers. Thus although we shall discuss them separately, we shall wish to keep in mind both mechanisms for controlling the formation of the linear polymers appear to operate simultaneously.

A. Concentration Dependent Polymerization[2]

The linear nature of the polymerization reaction, and the existence of a molecular two-fold axis of symmetry normal to the direction of chain growth leads us to believe that the free energy of adding a single enzyme molecule to a polymeric species is independent of the chain length and the same irrespective as to which end of the chain the monomer is added. In a formal sense this type of polymerization mechanism corresponds to an open-ended linear condensation polymerization of bifunctional monomers (i.e. the enzyme molecule), and the equations describing this type of process have been well known for many years [11].

If the reaction is described by the equation

$$P_i + P_1 \rightleftarrows P_{i+1} \tag{1}$$

then the association constant K_i is given by

$$K_i = \frac{c_{i+1}}{c_i\, c_1} \tag{2}$$

where P_i is a polymer of degree of polymerization i and c_i is the concentration (in mg/ml) of P_i. The reaction written above describes the addition of a single enzyme molecule to a polymer of i monomers. On the basis of assumptions made above all the K_i are equal ($\equiv$ K). It can then be shown that for $Kc_1 < 1$, which is always true,

$$(M_w/M_1)^2 - 1 = 4\,Kc \tag{3}$$

where M_w is the weight average molecular weight and $c = \sum_1^i c_i$. The value of K may be evaluated from a plot of $(M_w/M_1)^2 - 1$ against c. Such a plot is shown in Fig. 10. The experimental points were derived from overlapping ultracentrifuge and light scattering expe-

[2] The studies discussed in sections III A and III C were undertaken in 0.2 M Na phosphate buffer, pH 7, 10^{-4} M EDTA.

riments and the straight lines calculated for the indicated values of K [12]. At concentrations below 0.4 mg/ml, the data correspond to a value of K between 2.1 and 2.0 mg/ml. At higher protein concentrations the deviations from the calculated curves may be ascribed to the contribution of a positive virial coefficient.

The analysis of the model given here for the polymerization process allows calculation of the distribution of particle sizes. From

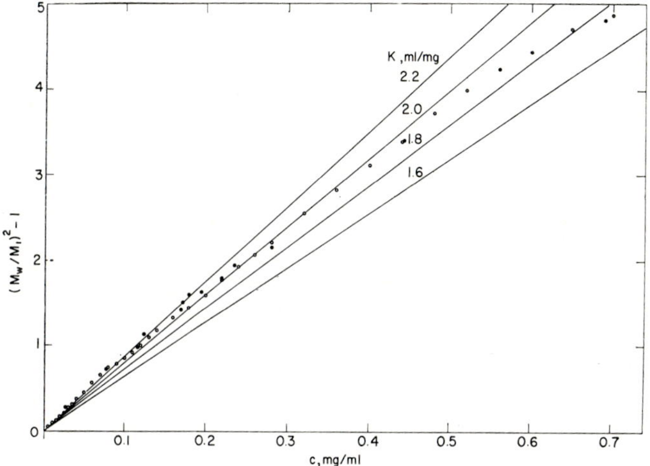

Fig. 10. Plots of $[(M_w/M_1)^2 - 1]$ *vs.* c; M_w values averaged at each concentration from data from overlapping runs (from Ref. [12])

the theoretical distribution one may then calculate the number average (M_n) and z average (M_z) molecular weight as a function of concentration. The interrelation between these molecular weight moments and M_w for this polymerization model is given by

$$2M_n = M_w + M_1$$
$$2M_z = 3M_w - M_1^2/M_w \qquad (4)$$

Fig. 11 presents plots of the three molecular weight moments against enzyme concentration along with ideal curves calculated for the indicated values of K. The excellent concordance between the calculated and experimental values of K for the three molecular

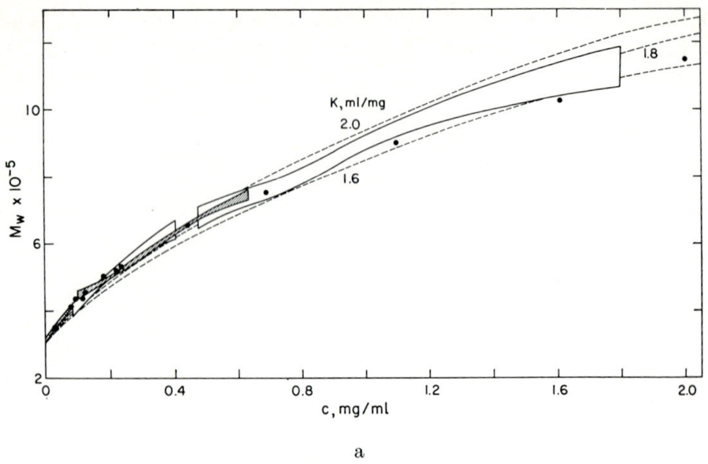

a

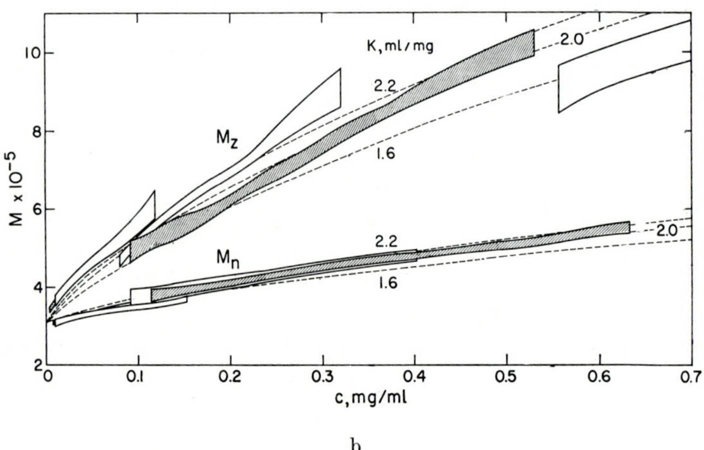

b

Fig. 11. a) Weight-average molecular weight (M_w) *vs.* concentration in phosphate buffer; boxes, computed values from equilibrium sedimentation runs with estimated error; filled circles, light scattering results, also at 20°; curves, calculated according to reversible infinite linear association with single value of association constant K (from Ref. [12]). b) z-average (M_z) and number-average (M_n) molecular weights from equilibrium sedimentation; curves and symbols as in a) (from Ref. [12])

weight averages provides strong confirmation for the assumed mechanism of polymerization in that it correctly predicts both the concentration dependence of the molecular weight and the distribution of sizes.

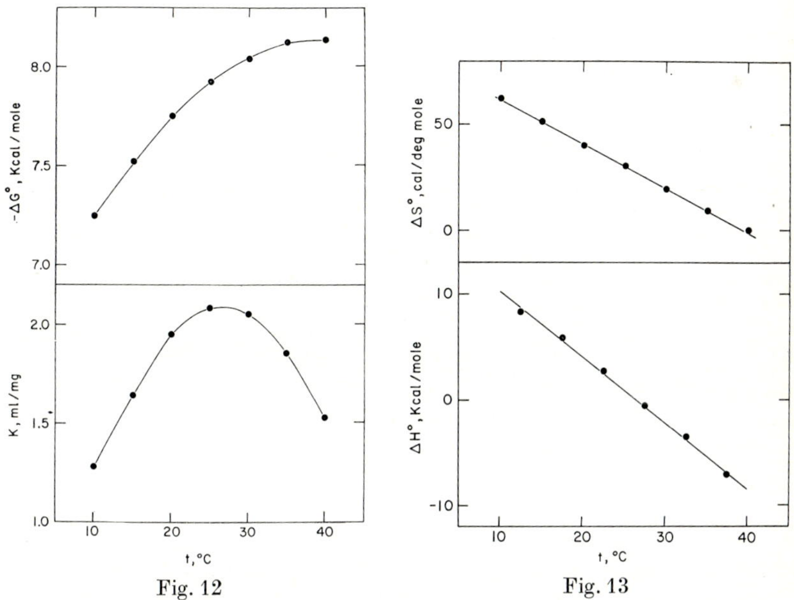

Fig. 12

Fig. 13

Fig. 12. Equilibrium constant, K, and standard free energy, $\Delta G°$, of glutamate dehydrogenase association reaction, calculated from light scattering data (from Ref. [13])

Fig. 13. Standard enthalpy, $\Delta H°$, and entropy, $\Delta S°$, of glutamate dehydrogenase association reaction, calculated from light scattering data (from Ref. [13])

The association of glutamate dehydrogenase is temperature dependent and the equilibrium constants exhibit a shallow maximum around 28 °C (Fig. 12); the standard free energy $\Delta G°$ of the reaction is negative and increases towards a limiting value with increasing temperature [13]. Inspection of Fig. 13 reveals that the association is entropy driven at low temperatures, presumably due to interactions with solvent, while at higher temperatures the association is driven by the favorable enthalpy change.

B. Control by Allosteric Effectors

We now come to an aspect of our subject which by the nature of its complexity is less well understood than the association reaction forming the linear polymers[3]. An unusual feature of glutamate dehydrogenase assembly is that the process appears to be controlled

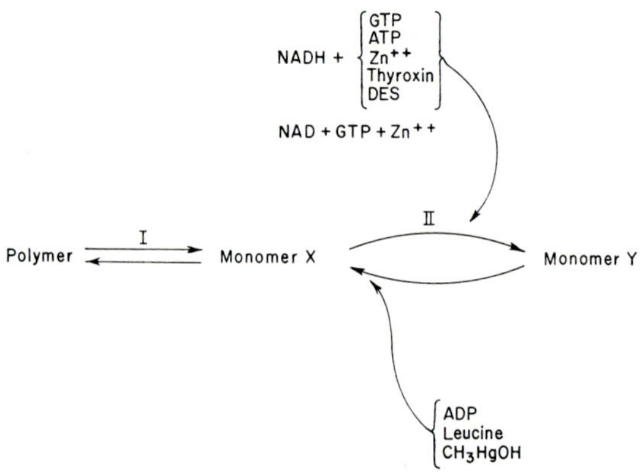

Fig. 14. Scheme depicting control of association of glutamate dehydrogenase (from Ref. [14])

by the active state of the enzyme. Two active states have been suggested; one which catalyzes the oxidation of glutamic acid to α-ketoglutaric acid and the other which catalyzes oxidation of alanine to pyruvic acid. A number of different agents (see below) stimulate one or the other of these activities, but it is believed that only that form of the enzyme exhibiting glutamic acid dehydrogenase activity which undergoes association. The interrelationship between these forms is schematically depicted in Fig. 14.

[3] Much of the discussion in this section has been reviewed by Stadtman [14], Frieden [15, 16] and Tomkins and collaborators [17, 18] and the reader is referred to these works for a more detailed exposition of these authors' views.

Reaction I is the concentration dependent association discussed above. Monomer x has high glutamic acid dehydrogenase activity and may polymerize; monomer y acts on alanine and may not polymerize. The formation of monomer y is controlled by the addition of allosteric effectors some of which are indicated in Fig. 14. Each form of the monomer is in equilibrium with the other and the relative distribution of each species is controlled by the amounts of effector present. An example of this phenomenon has already been demonstrated in Fig. 9 which shows the dramatic reduction in the molecular weight by GTP and NADH. Other compounds acting in conjunction with NADH have been reported by various laboratories to have equally dramatic effects. Among these are included the steroid hormone diethylstilbestrol which can cause dissociation at a hormone concentration 5×10^{-5} M. Other hormones, less active but still effective in dissociating the linear polymers are estradiol, progesterone and testosterone. In most cases dissociation requires the presence of NADH and is accompanied by an inhibition of glutamate dehydrogenase activity and activation of alanine dehydrogenase activity. Another hormone, unrelated to those discussed above but displaying an effect similar to the steroid hormones is thyroxine and some of its derivitives. Here again the maximum effect is observed in the presence of NADH. A remarkable feature of the inhibition caused by the above mentioned hormones on GTP is that it can be reversed by addition of ADP.

While most of the effects caused by GTP and the hormones just mentioned occur at physiological concentrations, their biological significance in controlling the activity of GDH *in vivo* is still a subject of controversy.

C. Agents Enhancing Polymerization

Before leaving the subject of control of polymerization we should briefly mention an additional class of compounds which in contrast to those mentioned in the preceding paragraph, enhance polymerization but are without effect on the biological specificity of the enzyme. These compounds include the aromatic hydrocarbons toluene and benzene. For example, association of glutamate dehydrogenase is greatly enhanced upon saturation of enzyme solutions with toluene [19]. In the presence of toluene association constants have been calculated [20] from the slopes of a plot [according to Eq . (3)] of $(M_W/M_1)^2 - 1$ against enzyme concentration (Fig. 15). The resulting plots obtained at several different temperatures are linear and the association constants are 10 to 25 times greater than in solutions not containing toluene.

From the temperature dependence of the association constants we derive that ΔG° in solutions saturated with toluene is almost

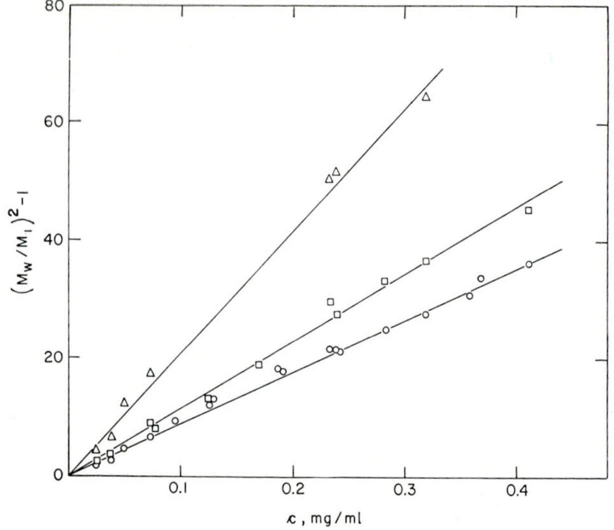

Fig. 15. Plots of $[(M_w/M_1)^2—1]$ *vs.* c in solutions saturated with respect to toluene; ○, 25 °C; □, 20 °C; △, 10 °C (from Ref. [20])

constant with temperature (and equal to —9.3 Kcal/mole); consequently the reaction in solutions saturated with toluene is driven by a favorable enthalpy term corresponding to the behavior in the high temperature branch in the enzyme system in the absence of toluene [20].

A plot of the radius of gyration against molecular weight is continuous with that obtained in the absence of toluene (Fig. 3) and the relationship between M_n, M_w and M_z determined by ultracentrifugation follows that predicted by Eq. (4), leading us to conclude that the associates formed in the presence of toluene are of the same linear nature as those formed in its absence.

Solubility studies of toluene in buffered glutamate dehydrogenase solutions, saturated with respect to toluene, may be interpreted to indicate that 8 to 13 moles of toluene are preferentially bound per 53,500 g of enzyme; below saturation conditions considerably less toluene is preferentially bound to the enzyme. The enzymatic activity of glutamate dehydrogenase is not affected by toluene binding and it can therefore be concluded that activity is independent of degree of association. In the presence of the effectors GTP and NADH (see below) which strongly inhibit both enzymatic activity and enzyme association, preferential binding of toluene was not observed [20].

IV. Fixed Polymers

Very early in the study of the physical properties of the linear polymers of glutamate dehydrogenase it was recognized that the concentration dependent polymerization equilibrium did not facilitate investigation of those properties of the system which are dependent upon the length of the polymers since for every different concentration one obtained a different distribution of lengths. In order to obviate this complexity we sought to covalently crosslink the polymers with glutaraldehyde and then study the products. Our choice of this reagent was predicted on the well known success obtained with it in the preservation of the fine structure of biological materials for electron microscopy. An additional rationale for choosing this reagent was that its reaction with proteins has been partially characterized and several studies of the properties of the crosslinked products have been published [21—23].

A. Fixation with Glutaraldehyde

Glutaraldehyde reacts with proteins to form intra- and intermolecular covalent crosslinks which are stable in the presence of dissociating agents such as guanidine hydrochloride or urea. If carried out under mild conditions crosslinking of proteins in solution as well as in crystals may often be accomplished without loss of biological activity [24, 25].

The crosslinking of glutamate dehydrogenase polymers reaction was carried out at low concentrations of both protein and glutaraldehyde in order to minimize formation of intermolecular crosslinks. Protein concentrations were about 1 mg/ml and the mole ratio of crosslinking agent to protein was 75:1. The reaction proceeded for 10 days at 5 °C. Under these conditions the major fraction of the crosslinks formed are along the length of single polymer chains (see below).

Because low protein concentrations were employed it was found advantageous to carry out the crosslinking reaction in solutions saturated with toluene which as was noted above (section III C) has the effect of greatly enhancing the polymerization, but does not cause any changes of either the biological activity or the linear nature of the polymers.

B. Characterization of Crosslinked Enzymes

When solutions of glutaraldehyde-fixed polymers of glutamate dehydrogenase are examined by ultracentrifugation or light scattering profound changes in their physical-chemical properties are observed, the most striking of which is that the concentration dependent association-dissociation reaction observed in solutions of native enzyme is no longer evident. The molecular weight of solutions of fixed enzyme remains constant even after the dilution to the lowest concentrations at which the molecular weights can be measured. Furthermore, neither removal of toluene nor addition of 10^{-3} M NADH + 10^{-3} M GTP results in any change of molecular weight, whereas either of these operations results in very drastic reductions in molecular weight of unfixed solutions of polymers.

The length distributions of particles in solutions of native glutamate dehydrogenase is very broad; at high degrees of association it approaches the so-called "most probable distribution" which characterizes open-ended linear polymerization equilibria. For such a process the number average molecular weight (M_n) is related to the weight average molecular weight (M_w) by $2M_n = M_w$. It is not unreasonable to assume that after fixation polymer solutions will retain at least this degree of polydispersity, and perhaps even display a greater degree of heterogeneity due to side reactions such as some interchain crosslinks being formed as well.

The high degree of polydispersity as well as the possibility that interchain crosslinks had formed branched as well as linear species present formidable difficulties for the interpretation of physical chemical data and for these studies it was thought advisable to fractionate the polymer solutions into fractions of narrow weight distribution.

Efficient fractionation was achieved by applying a solution of fixed polymers (concentrated by ammonium sulfate precipitation) to a Sepharose 6 B column. A typical chromatogram along with the molecular weight of each fraction is shown in Fig. 16. The chromatogram exhibits several noteworthy features. There is a peak appear-

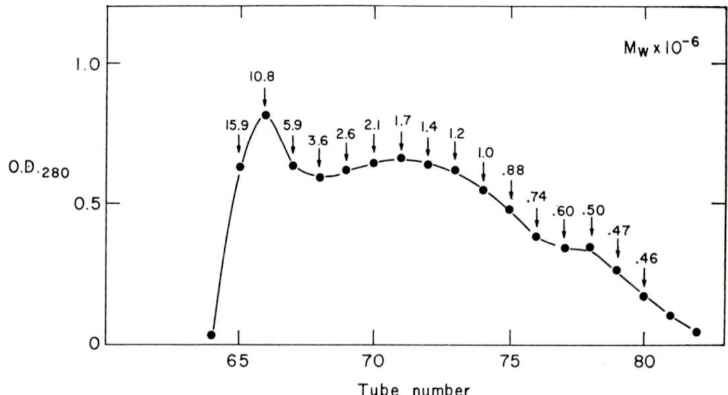

Fig. 16. Chromatogram of fixed glutamate dehydrogenase fractionated on sephrose 6 B. The molecular weight ($\times 10^{-6}$) of each fraction is given

ing in the void volume containing fractions of molecular weight between 5 and 15 million. This material was examined by electron microscopy and found to consist of polymers which had undergone interchain crosslinks (Fig. 17). The number of enzyme molecules in particles which have undergone interchain crosslinking may vary from 10 to 20 as in Fig. 17 a to even several hundred as shown in Fig. 17 b.

If the fractions eluting in the void volume are discarded and the remaining material pooled, concentrated and rerun, it is possible to obtain preparations consisting of only linear polymers.

Fig. 18 shows electron micrographs comparing refractionated fixed polymers with native polymers of glutamate dehydrogenase.

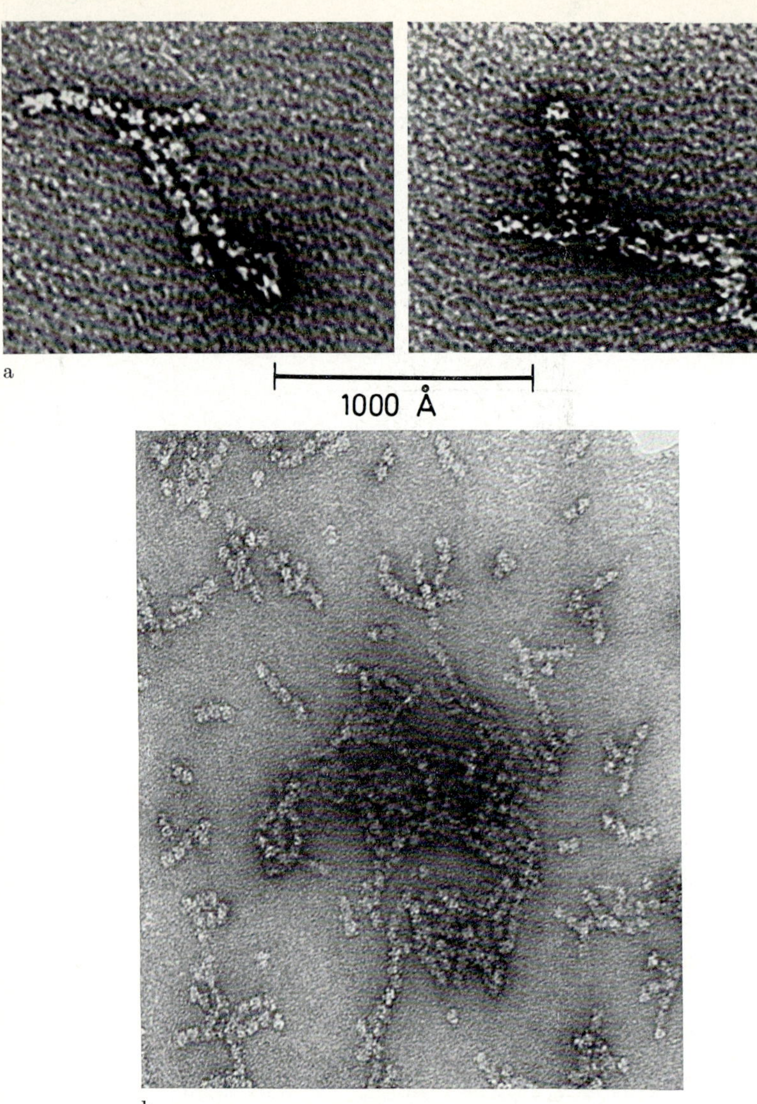

Fig. 17 a and b. a) Electron micrograph of fixed glutamate dehydrogenase showing the presence of branched polymers consisting of several dozen molecules. b) An electron micrograph of a branched polymer in which a very large number of molecules are crosslinked into a single aggregate. Aggregates of this size are commonly observed in the high molecular fractions in the "void volume" of the chromatogram, for example fraction 65 of Fig. 16. Branched polymers are believed to be caused by interchain crosslinks formed during fixation

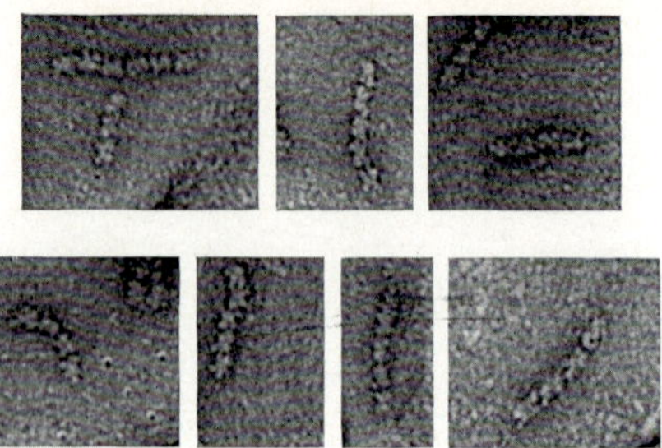

Fig. 19. High magnification electron micrograph of selected particles of fixed glutamate dehydrogenase

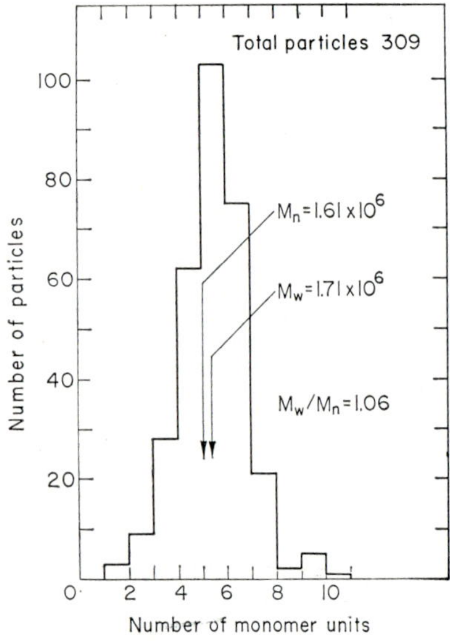

Fig. 20. Typical histogram obtained for rechromatographed fixed enzyme. The histogram shows the number of monomer units per particle of cross-linked enzyme. M_n and M_w were calculated assuming the molecular weight of each monomer unit was 320,000

The appearance of the particles in the two micrographs is virtually the same and it is of interest to note that, although both fields contain a very high density of particles, yet no aggregates of the type shown in Fig. 17 are evident. Micrographs taken at low magnification simultaneously display a large number of particles so as to give an impression of the general appearance of fixed and unfixed (or native) polymers. Higher magnification micrographs Fig. 19 confirm that the appearance of the fixed polymers is indistinguishable from that of untreated enzyme (c.f. Fig. 5).

Weight average molecular weights of different fractions of fixed polymers vary from as low as 450,000 corresponding to a large proportion of monomeric units to as high as 2.0 million, the weight of a six unit particle. Typical histograms of the number of monomer

Table 1. *Comparison of molecular weights obtained by light scattering and electron microscopy*

Fraction	Light Scattering	Electron Microscopy
C 1—78	0.48	0.45
E 2—34	0.88	0.87
E 2—28	1.44	1.26
E 5— 9	1.63	1.52
E 5— 8	1.96	1.71

units per particle against particle frequency (Fig. 20) show the distribution characteristically obtained. The ratio of the number average weight to the weight average molecular weight calculated from such histograms is between 1.05 and 1.15 representing a very considerable improvement of the homogeneity of these fractions over the original solution. Furthermore, the weight average molecular weights determined by light scattering are in good accord with those obtained from histograms of the distribution particles per chain (Table 1), providing additional verification of the success of the fractionation procedure.

C. Enzymatic Properties of the Fixed Polymers[4]

One of the most intriguing features of our investigation of the properties of the fixed polymers relates to a comparison of the

[4] The studies discussed in this section where undertaken in 0.05 M Na phosphate buffer, pH 7.6 at 25°C.

of the polymer has no discernable effect on the enzymatic activity. In fact, under no conditions examined to date have we been able to detect any difference in enzymatic properties of polymers of different length.

As indicated above the activity and substrate specificity of glutamate dehydrogenase may be influenced by a wide variety of agents.

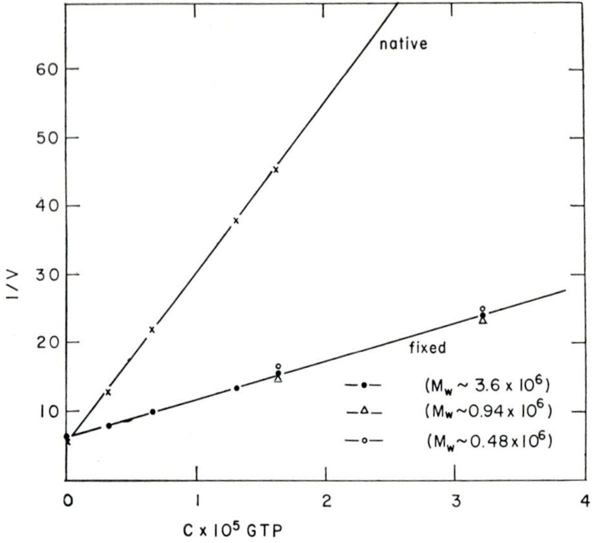

Fig. 23. Plots of the reciprocal of the initial velocity as a function of GTP concentration. The plots compare the activity of three fractions of fixed enzyme (molecular weights are given in the figure) with that of the native enzyme (enzyme concentration 0.0016 mg/ml; α-ketoglutarate 8.3 mM; NH_4Cl 20 mM; NADH 0.1 mM)

Here we wish to restrict ourselves to considering the effects of two of these.

It has long been known that the oxidation of NADH by ammonia and α-ketoglutaric acid to glutamic acid and NAD is inhibited by GTP. This inhibition is illustrated in Fig. 23 which shows a plot of the reciprocal of the reaction velocity against the concentration of GTP present in the assay mixture. Both native and fixed enzyme are inhibited, but to greater different extents. Over the GTP concentration range examined there is an approximately nine-fold inhibition of the native enzyme whereas inhibition of the fixed enzyme is only three-fold.

The second agent which we consider is the coenzyme NADH. Increasing the concentration of NADH from very low levels results

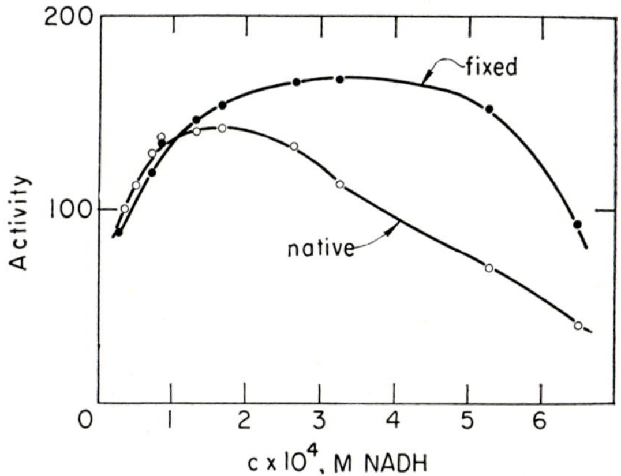

Fig. 24. Plot comparing inhibition of activity of native and fixed enzyme by high concentrations of NADH (enzyme, α-ketoglutarate and NH_4Cl concentrations, same as in Fig. 23)

in an increase in enzymatic activity — as is commonly observed when substrate or coenzyme concentration is increased. However, at moderate concentration (2×10^{-4} M), the coenzyme inhibits enzymatic activity and at high concentrations of NADH enzymatic activity falls to quite low values (Fig. 24). The fixed enzyme is affected in a similar manner, except that inhibition sets in at higher concentrations of NADH and is not as severe.

D. Discussion

On the preceding pages we have described various aspects of the assembly of glutamate dehydrogenase. The key intermediate is clearly the linear polymer chains which can associate with one

Discussion

H. Sund (Konstanz): I would like to add some remarks with respect to our own results on glutamate dehydrogenase which we obtained in recent years.

1. From X-Ray small-angle data we calculated the diameter and the length of the glutamate dehydrogenase oligomer to 84 Å and 126 Å, respectively [Sund, H., Pilz, I., Herbst, M.: Europ J. Biochem. **7**, 517 (1969); Pilz, I., Sund, H.: Europ J. Biochem. **20**, 561 (1971)]. This result is in agreement with the data obtained with the electron microscope.

2. The investigation of the association-dissociation equilibrium of the beef liver enzyme by light scattering measurements over a wide range of protein concentration (up to 50 mg/ml) has shown that this equilibrium can be described as an open equilibrium ($M_i + M_h \rightleftharpoons M_{i+h}$, i and h $\geqq 1$, M = Oligomer) with identical equilibrium constants for all steps and without limit. At 20 °C and in M/15 phosphate buffer, pH 7.6, the equilibrium constant was found to be 9.0×10^5 M^{-1} (corresponding to a free energy change of 7.8 kcal/mole for each consecutive step in the association reaction) and the second virial coefficient to be 9 nmole $\times$ l $\times$ g^{-2}. The electrostatic contribution to the association process follows from the dependence of the sedimentation behaviour on the pH at low and high ionic strength [Krause, J., Markau, K., Minssen, M., Sund, H.: In: Pyridine Nucleotide-Dependent Dehydrogenases (Ed. by H. Sund), p. 279. Berlin-Heidelberg-New York: Springer 1970; Markau, K., Schneider, J., Sund, H.: Europ J. Biochem. **24**, 393 (1971)].

3. From modification experiments with 5-diazo-1H-tetrazole and glyoxal it follows that lysine residues are involved in the binding of the substrate α-ketoglutarate and of the nicotinamide part of the coenzyme. These lysine residues are different from the lysine residue in position 97 which reacts with pyridoxal 5′-phosphate. This modification does not influence the hydrodynamic properties and does not alter the ability to dissociate upon treatment with GTP and NADH. Irradiation of glutamate dehydrogenase in the presence of pyridoxal 5′-phosphate probably causes an addition reaction between a Schiff base intermediate with pyridoxal 5′-phosphate and an imidazole group. Simultaneously histidine residues are oxidized. This modification reaction influences primarily the association sites [Deppert, W., Hucho, F., Sund, H.: Europ J. Biochem. (in press)].

4. The rat liver glutamate dehydrogenase does not associate in phosphate buffer at pH 7.6. Upon treatment with benzene or at low pH and low ionic strength, association occurs as with the beef liver enzyme. In addition to this the rat liver enzyme hybridizes with the enzyme from beef liver. From our results taken together with those obtained with enzymes from animals which are lower on the phylogenetic scale than rats, it can be concluded that the ability to associate was developed during an early state of evolution and that the rat enzyme later lost this ability [Ifflaender, U., Sund, H.: FEBS Letters **20**, 287 (1972)].

Subunit Interactions in Haemoglobin

A. D. McLachlan, M. F. Perutz and P. D. Pulsinelli

MRC Laboratory of Molecular Biology, Hills Road, Cambridge, Great Britain

With 7 Figures

Introduction

During the last 2 years many of the cooperative effects in haemoglobin have been explained in terms of the molecular structure [1]. Here we wish to survey the established facts about the coupling between the subunits and to describe new experiments which describe how structural changes in the protein alter the electronic state of the haem iron atom.

Before describing the molecule we consider some general features of subunit association which are relevant to allosteric proteins and may be important in haemoglobin.

Subunit Association and Cooperative Effects

A tetramer of four identical subunits will normally have 222 symmetry [2] and there will be three kinds of contact region, across the three two-fold axes (Fig. 1 a). Why is this tetrahedral arrangement so common, and what advantages does it offer?

Monod [3] distinguishes two types of pairing between subunits, isologous and heterologous (Figs. 1 b and c). Isologous pairing gives pairs of complementary contacts across a two-fold symmetry axis. It is the mode most likely to evolve, as one mutation can lead to two contact regions in a single step. Many proteins form dimers, perhaps for this reason. The path of evolution to a tetramer could then be as shown in Fig. 1. First the dimer forms, giving one bonding region, and one two-fold axis. Later pairs of dimers couple, initially giving a 222 tetramer with two strong and two weak bonding regions. At this stage evolution can take a new path towards a structure with lower symmetry. Two ways are possible. One is to

Ligands Control the Oxygen Affinity

Haemoglobin carries oxygen to the tissues [5, 6] and brings back CO_2 and Bohr protons [7, 8] to the lungs. The oxygen affinity increases with the number of oxygen molecules combined, and decreases in the presence of CO_2 or hydrogen ions. In the human red cell the molecule is also regulated by 2,3-diphosphoglycerate (DPG) which binds to the deoxy form [9] and lowers the affinity.

The physiological functions are cooperative and thermodynamically linked [10], so that 2,3-diphosphoglycerate, for example, enhances the Bohr effect [11, 12], and competes with CO_2 binding [13]. All these effects are controlled by subunit interactions. They are absent in dimers [14, 15] or free chains [16].

Structure of the Molecule

The molecule is a tetramer of two α and two β chains which bind oxygen to four haem iron atoms [5, 17]. Each chain consists mainly of α-helical regions. The structures of the chains are remarkably similar in all species, ranging from the simple monomers of myoglobin [18] or worm and insect haemoglobin [19, 20], to weakly cooperative tetrameric lamprey haemoglobin [21], birds, fish and mammals. The haem group is linked covalently on one side to a histidine on helix F (F8). On the other, the oxygen links to the iron and is hydrogen bonded to a histidine on helix E (E7).

The tetramer has two-fold symmetry and the subunits are arranged tetrahedrally. The gross features of the oxy form resemble a structure with 222 symmetry. The regions of contact between the α and β chains are of two types. One, called $\alpha_1 - \beta_2$ is close to the haems. It is here that the chains slide past one another a distance of 3 to 5 Å when haemoglobin is deoxygenated, converting the molecule into its low affinity form. This region is nearly the same in all species. Mutations here often affect the sliding and weaken the cooperative effect. The other $\alpha_1 - \beta_1$ contact region is much more extensive and rather variable in sequence. Oxyhaemoglobin tetramers are quite loosely held together and easily split across the $\alpha_1 - \beta_2$ interface to form $\alpha_1\beta_1$ dimers. The $\alpha_1 - \beta_1$ contact splits less readily and gives rise to free α and β chains.

The oxy structure is held together mainly by van der Waals contacts. There are no hydrogen bonds linking $\alpha_1 - \alpha_2$ or $\beta_1 - \beta_2$, and four or perhaps five joining $\alpha_1 - \beta_1$. The amino and carboxyl ends of the chains are free, and their positions in the crystal structure are not well defined. The penultimate tyrosines (HC2) are also ill-defined. They spend part of the time in a surface pocket next to helix F with their hydroxyl groups hydrogen-bonded to Val FG5 CO, and for part of the time they are free.

Changes on Deoxygenation

In the deoxy form the oxygen binding site is empty and the iron atoms are five-coordinated. The overall tertiary structure of each chain changes little, with average atomic shifts of only 1 to 2 Å, but each iron atom moves out of the haem plane [1, 24] and in the β chain Val E11 obstructs the oxygen combining site. There are large movements at the amino and carboxyl ends of the chains. The subunits slide past one another at the $\alpha_1 - \beta_2$ and $\alpha_2 - \beta_1$ contacts, giving a new quaternary structure which is less symmetrical. There is now a large space between the β_1 and β_2 chains close to the two-fold axis.

The deoxy tetramer is held firmly together by salt bridges, and the C-terminal groups, including all four penultimate tyrosines, are now fixed in position. One set of $\alpha_1 - \alpha_2$ bridges links the α-amino group of Val 1 of each α chain to the terminal carboxyl of Arg 141, while the guanidinium of Arg 141 also links to Asp 126 of the other chain. A second set of $\alpha_1 - \beta_2$ and $\alpha_2 - \beta_1$ bridges links the terminal carboxyl group of His 146β to the side chain of Lys 40α. At the same time the imidazole group of His 146β swings round towards helix F and links to the carboxyl group of Asp 94β, forming a bridge within the surface of each β chain.

The Bohr effect comes mainly from the increased proton affinity of the α-amino groups of the α chains and the imidazoles of His 146β [25, 26].

The α-amino groups of the β chain remain unbridged. They can however bind CO_2 in the deoxy form.

The movements of the β subunits also allow one molecule of 2,3-diphosphoglycerate to bind specifically to the deoxy tetramer [9, 26], in a cavity lined with positively charged groups. Arnone [27] has recently calculated difference electron density maps for deoxyhaemoglobin with and without DPG, and demonstrated that

The deoxy structure is also constrained in a different sense, also anticipated by MONOD, WYMAN and CHANGEUX. The difference in oxygen affinity between the forms could, in principle, be produced in two ways, either by increasing the affinity of the oxy structure above that of the free α or β chains, or by lowering the affinity of the deoxy structure. In fact the oxy tetramer, the free chains [16], and the monomeric myoglobin and insect haemoglobins, all have high affinity [6]. So it is the deoxy form which is unique, having an unusually low affinity. It is therefore correct to call deoxyhaemoglobin a tense structure.

Changes at the Haem Iron Atom Trigger the Transition

How does combination with oxygen alter the quaternary structure and break the salt bridges? HOARD and WILLIAMS [31, 32, 33] showed that the answer lies in the electronic states of the iron atom, and in the structure of the porphyrin group.

The haem iron can exist in four states (Table 1), ferrous or ferric, each with high or low spin. The low-spin ferrous iron in oxyhaemoglobin has an Fe-N bond length of about 2.01 Å which fits into an unstrained planar porphyrin. The unique feature of 5-coordinated high-spin ferrous iron in deoxyhaemoglobin is that it has the longest Fe-N bond length to the four porphyrin nitrogens and the longest axial Fe-N bond to histidine F8. The iron cannot fit into the centre of the ring, so the structure distorts, and the iron atom moves about 0.7 Å out of the haem plane. Ferric methaemoglobin has a smaller displacement [18] of about 0.3 Å, while in LAMPREY CN and insect CO haemoglobins [20, 21] the iron lies in the haem plane.

Thus one important function of the porphyrin ring is as a firm base to hold the iron atom, which transforms a small change of ionic radius into a much larger movement perpendicular to the haem plane. Another is that the conjugated π-electron system maintains a delicate energy balance between the electronic states of the iron, so that high and low spin, in both ferrous and ferric, are very close [33].

When the salt bridges in the deoxy structure ease the tyrosines into their pockets the F helix is distorted and histidine F8 shifts away from the haem plane, lifting the iron atom. Conversely, the in-plane oxy haem iron pulls the histidine, squeezes the tyrosine out of its pocket, and weakens the C-terminal region, so that the salt bridges break. Once a sufficient number of oxygens are bound the quaternary structure clicks into the relaxed form [1].

Table 1. *Iron-porphyrin nitrogen distances and out of plane distortions (Ångstrom units) in haemoglobins and metal porphyrins. The distances are from* Hoard [31]. *The distortions are taken from protein X-ray data* [1, 18, 20, 21]. *Distances from metal to the axial ligand also change. Typical values* [31] *are: ferric high spin 1.842* (OCH_3), *low 1.957; ferrous high 1.90, low 2.01* (*estimated*)

	$Fe^{+++}(d^5)$	$Fe^{++}(d^6)$
High Spin	S=5/2 2·073 Cl 2·087 Fe−O−Fe F, H_2O — Horse, Myo 0·3	S=2 ≈2·14 DEOXY Horse Insect 0·5−0·7
Low Spin	S=1/2 1·989 CN, OH, N_3 — Lamprey FLAT	S=0 ≈2·01 OXY CO — Lamprey, Insect FLAT
Porph	2·01 Metal 2·057 H_2	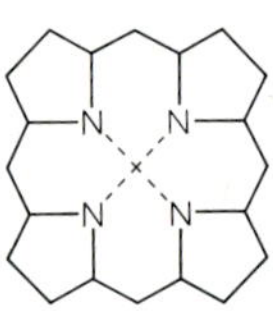 N N N N

The movement of the tyrosines appears to be an essential step in the transition. Moffat and Perutz [1, 34] studied a haemoglobin which was locked into the oxy quaternary structure by blocking Cys F9β with bismaleimidimethyl ether. They found that oxygenation of the intact α chains expelled their tyrosines. In several other modified or mutant haemoglobins interference with the tyrosine pocket is the common feature which destroys cooperative effects [28].

The Deoxy Haem is Under Tension

How do the salt bridges maintain a low oxygen affinity? One possibility is that the deoxy haem group prefers a non-planar structure in both the free chain and the constrained tetramer, but that when combination with oxygen pulls the iron atom into the haem plane, extra energy is required to expel the tyrosine and break salt bridges. This could be one source of the low affinity.

Another possibility is that the protein itself alters the electronic structure of the deoxy ferrous iron atom. In principle the iron can

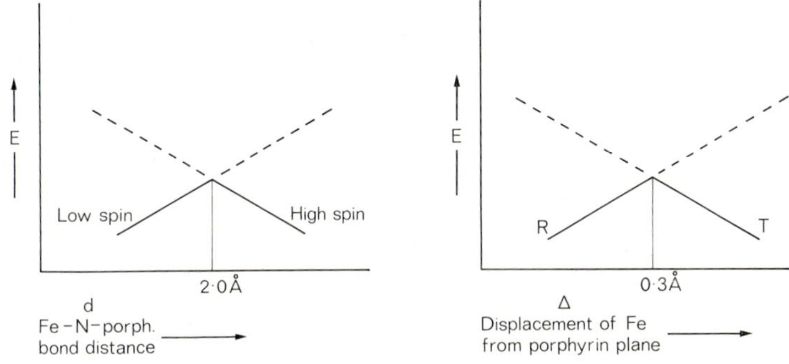

Fig. 2. Push-pull equilibrium of the spin states in haem iron. Energies of the high- and low-spin states are close, and cross over at an iron-porphyrin nitrogen bond length of about 2.0 Å. Tensions in the protein lift the iron out of the haem plane, stretching the bonds, and tilt the equilibrium from low to high spin

exist either in the in-plane low-spin state, with the d electrons in the d_{xy}, d_{yz} or d_{zx} orbitals, which should form strong bonds with an axial ligand; or in the non-planar high-spin state with d electrons in the $d_{x^2-y^2}$ or d_{z^2} orbitals. The high-spin state does not favour combination with oxygen. The protein could therefore lower the oxygen affinity of the haem by pulling the iron atom out of the haem plane and forcing it into the state of high spin (Fig. 2).

Perutz [35] now proposes that this second interpretation is the correct one. When separate unliganded α and β chains assemble into a deoxy tetramer there are characteristic changes in the Soret band of the ferrous haem optical spectra [6]. These are probably caused by changes in the iron-ligand distances. The protein constrains the iron, keeping the bond to His F8 under tension, and shifts the electronic spin equilibrium to the high-spin side. The reaction can be written

$$2\,\alpha^{r}(\text{low}) + 2\,\beta^{r}(\text{low}) \rightarrow \alpha_2{}^{t}\beta_2{}^{t}(\text{high})\ [\text{T}].$$

Here r and t indicate that the free chains have a relaxed tertiary structure with flat haems and free tyrosines, while the assembled chains have non-planar haems and are tense. Modified deoxy tetramers which only form the R state show spectra like the free chains because the tension is absent. Myoglobin and insect haemoglobin have high affinity in the deoxy state, and show only small displacements of the iron atom out of plane. This may be because the protein does not exert the same tension on the haem.

It is interesting that Smith and Williams long ago suggested that the protein may exert a tension on the *ferric* iron atom in methaemoglobin, which again has more high spin in the tetramer than the free chains.

The Balance between the R and T Forms

What is the sequence of steps which leads from the deoxy to the oxy structures as haemoglobin reacts with oxygen? Does one type of chain react before the other, and at what stage does the transition from R to T take place?

We have seen that in each α chain an added oxygen moves the iron atom and expels tyrosine HC2. This must pull arginine 141α and break its salt bridge to valine 1α. In each β chain the oxygen also expels a tyrosine and displaces valine E11 (67β). This movement probably breaks the salt bridge between His 146β and Asp 94β.

Such changes would make the release of Bohr protons run almost parallel with the uptake of oxygen, whichever chain reacted first, in agreement with experiment [36].

They would also weaken the T quaternary structure until at some stage it clicked over to the R form.

We have seen that the transition is essential. It is probable that the structures R and T, each with two-fold symmetry, are the only possible ones. Lopsided structures in which the $\alpha_1 - \beta_2$ contact is R and $\alpha_2 - \beta_1$ is T seem unlikely. We do not know whether there is any fixed sequence of events as oxygen binds. The balance of forces is so delicate that in solution there is probably a continuous series of dynamic equilibria between different tertiary and quaternary forms, which allows the transition to occur at any stage in the reaction. The equilibria are controlled by the concentrations of oxygen, hydrogen ions, CO_2 and DPG. There is evidence that in phosphate buffer the transition takes place after the third oxygen has bound [37], and that with n-butyl isocyanide as ligand the β chain combines first [38, 39].

Chemical modifications and mutations [22] which weaken the T structure raise the oxygen affinity, and those which weaken the R structure lower it, by biasing the allosteric equilibrium. We now give some examples which illustrate the importance of the salt bridges and the $\alpha_1 - \beta_2$ contacts.

Modifications and Mutations Shift the Balance

Kilmartin and Hewitt [29] removed the C-terminal groups of the α or β chains (Des-Arg 141α or Des-His 146β haemoglobin). They also displaced the β-chain histidine by coupling N-ethyl maleimide to the reactive SH group of Cys 93β (NES-haemoglobin). The truncated forms all have high affinity, but Des-Arg, Des-His and NES each still show cooperative oxygen binding at normal pH, with reduced values of n. However, in the doubly-changed (Des-Arg-Des-His) and (NES-Des-Arg) molecules haem-haem interaction disappears. Perutz and Ten Eyck [28] examined the crystal forms of all these derivatives. They found that all the singly modified forms possessed distinct oxy and deoxy quaternary structures, but the doubly modified deoxy derivatives crystallised in the R structure.

Deoxy Des-Arg haemoglobin shows how delicate the equilibrium is. In humans it crystallises as T, but in horse as R. The difference occurs because of variations in the lattice energies and solubilities of the two species.

The mutant haemoglobin Hiroshima [40], where His 146β becomes Asp, has half the normal Bohr effect and a high affinity because the salt bridge to Asp 94β cannot form. Another high affinity mutant is Bethesda [41] where Tyr 145β changes to His, and the connection from the haem iron to the C-terminus is broken.

The salt bridges clearly provide much of the motive force for the transition, while the $\alpha_1 - \beta_2$ contacts seem to be passive, acting as a two-way switch. The chains are dovetailed so that either of two projecting Thr groups of the α-chain C helix can engage in a groove in the β-chain FG corner. Each structure is locked by a hydrogen bond:

G4(β) Asn NH_2 → G1(α) Asp CO_2^- Oxy

C7(α) Tyr OH → G1(β) Asp CO_2^- Deoxy

All mutations in the $\alpha_1 - \beta_2$ contact region reduce haem-haem interaction, but three are of special interest because they bias the equilibrium by removing hydrogen bonds.

Haemoglobin Kansas [G4(β) Asp → Thr] has low affinity [42] because it weakens the oxy contacts. Kempsey and Yakima [G1(β) Asp → Asn, His] have high affinity and very weak haem-haem interaction, even though they form normal deoxy crystals [22].

2,3-Diphosphoglycerate Stabilises the Tense Structure

Benesch and Benesch have shown that under physiological conditions 2,3-diphosphoglycerate binds only to the T form, at the rate of one molecule per tetramer [9]. Bunn and Briehl [43] have also shown that DPG has no effect on the oxygen affinities of haemoglobins like F_I, F_{II} and A_{IC} which have altered β chain α-amino groups. Sheep and frog β chains also lack the first few residues of the amino end and do not interact with DPG. Thus these groups are essential for binding the phosphate. DPG also interacts with the positive charges of Lys 82β and His 143β. CO_2 normally links to the ends of the β chains, forming a negatively charged carbamate [7] ion -$NHCO_2^-$ which inhibits [13] the binding of DPG.

Diphosphoglycerate is an allosteric inhibitor molecule which delays the T → R transition until late in the oxygen binding sequence. According to Tyuma [44] it reduces the affinity of binding to the first three oxygens, but not the last, and it increases the free energy of interaction. Hill's constant may increase or decrease. If the equilibrium is naturally biased towards the T structure then DPG may tip the balance so that the R form only appears very late in the reaction, and n will be small [45].

The Transition Can Occur if Either the α or β Chain is Blocked

Mixed state haemoglobins can be prepared in which one type of chain, say α, is maintained in the ferric state by blocking the haem group with cyanide, while the other ferrous pair, β, is available for

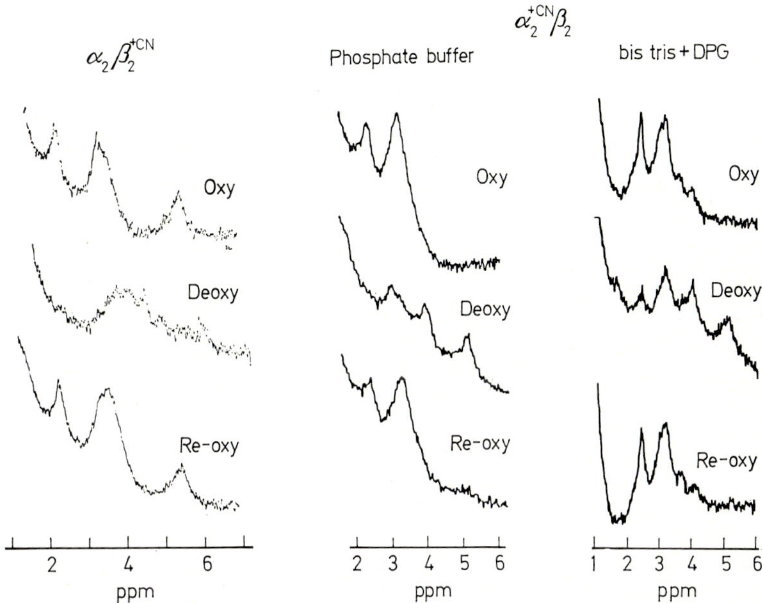

Fig. 3. NMR spectra of ferrous subunits in mixed state haemoglobins, from Ogawa and Shulman [46, 47]. In each case the deoxy spectrum shows increased paramagnetic resonance shifts. Left: normal α chains react with oxygen in the presence of inositol hexaphosphate. Centre: β chains react in phosphate buffer. Right: β chains react in the presence of 2,3-diphosphoglycerate

reaction with oxygen. Ogawa and Shulman [46, 47] have used the NMR spectra of the blocked ferric chains (Fig. 3) to demonstrate that deoxygenation of the ferrous chains is sufficient to cause a reversible R-T transition. The ($\alpha_2^{CN+++}\beta_2^{++}$) hybrid shows large spectral changes on deoxygenation in the presence of diphosphoglycerate or phosphate buffer. The other ($\alpha_2^{++}\beta_2^{CN+++}$) hybrid makes no transition with DPG, but it does in the presence of inositol hexaphosphate (IHP). These experiments show that removal of two oxygens from either type of chain shifts the equilibrium in favour of the constrained form, but the structure does not click over without the help of the phosphates.

The Transition Alters the Spin State of Ferric Iron

Pulsinelli and Perutz have crystallised a mutant human haemoglobin, M Milwaukee [48], which also has two chains blocked

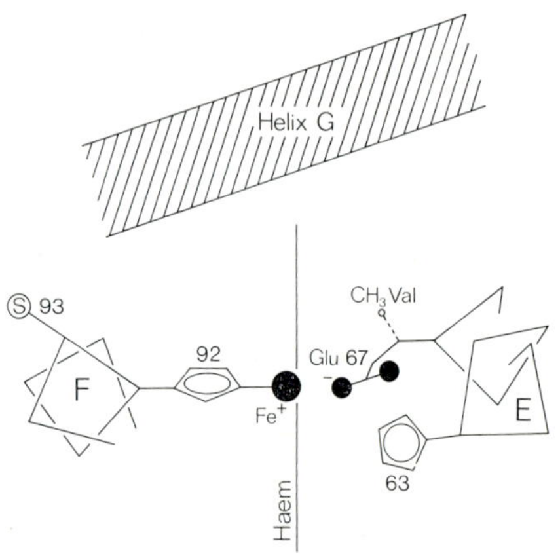

Fig. 4. Haemoglobin M Milwaukee. Valine 67 β of helix E is replaced by glutamic acid, which links to the β haem iron. There are no other changes in the protein structure. Iron is always ferric and cannot bind oxygen. α chains are normal

and shows weak haem-haem interaction. The iron atoms of the β chain are permanently combined with the carboxyl group of a glutamic acid (Fig. 4) which replaces Val 67β, and they are fixed in the ferric state. The α chains react normally, with a low affinity, and $n = 1.6$. The first point of interest about the mutant is that the oxy crystals are isomorphous with the normal R structure, and the deoxy are isomorphous with T, so that haem-haem interaction

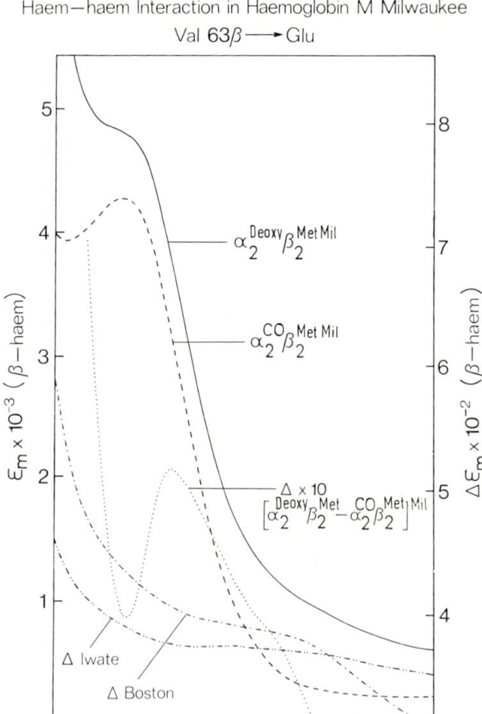

Fig. 5. Optical spectra of haemoglobin M Milwaukee in the ferric haem region near 620 nm. (——) α chains deoxy, β high spin Fe^{+++}. (----) α chains bound to CO, β low spin Fe^{+++}. (......) Difference spectrum magnified ten times. Notice the dip at 620 nm. All spectra taken in 0.1 M Na phosphate buffer at pH 6.5

takes place in the crystal. A difference electron density map of deoxy compared with normal human deoxy shows that the chains have the same tertiary structure, except for a small shift of the iron atom, and the changes at Glu 67β.

Perutz now studied the optical spectra of the ferric β haems in solution, to see whether the presence of a ligand at the α haem produced any change. When CO is bound there a spectral band at 620 nm becomes stronger and shifts to shorter wavelength (Fig. 5).

This band corresponds to one of the spectral bands of methaemoglobin A, and Perutz postulates that such a shift accompanies a change from high to low spin. Therefore the quaternary structure in solution influences the spin of the β haem:

$$\mathrm{R}\ [\alpha_2^{++O_2}\,\beta_2^{+++}(\mathrm{low})] \rightarrow \mathrm{T}\ [\alpha_2^{++}\,\beta_2^{+++}(\mathrm{high})].$$

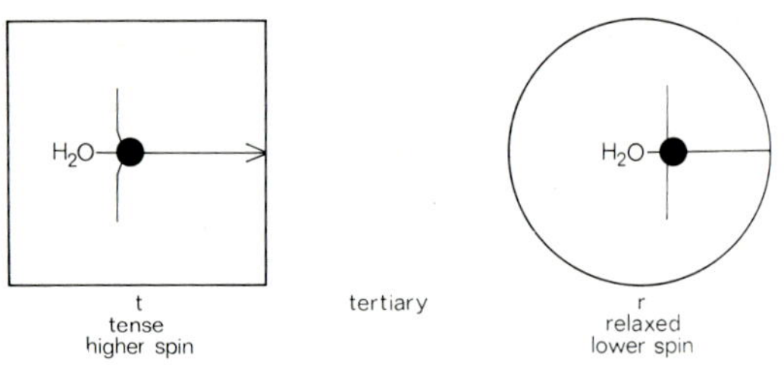

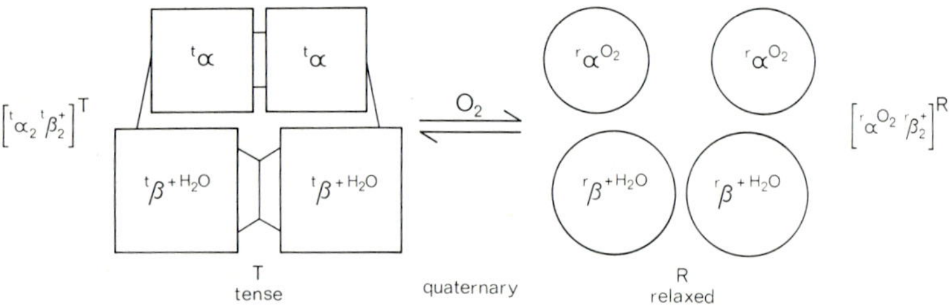

Fig. 6. Spin equilibrium of ferric haem iron. Above: tense tertiary structure of the chain favours high spin, relaxed favours low spin. Below: hybrid with β chains oxidised changes from T to R quaternary structure when the α haems combine with oxygen. The β haems also change from tense to relaxed conformations

The stereochemical reason for this transition lies in the tension produced by the protein at the deoxy haem (Fig. 6). There is a reciprocal push-pull relationship between the spin and the Fe-N bond lengths. As we have seen, high-spin ferric iron tends to project out of the haem plane [31], while low-spin haem is flat. Conversely, when the protein is relaxed it compresses the bonds and reduces

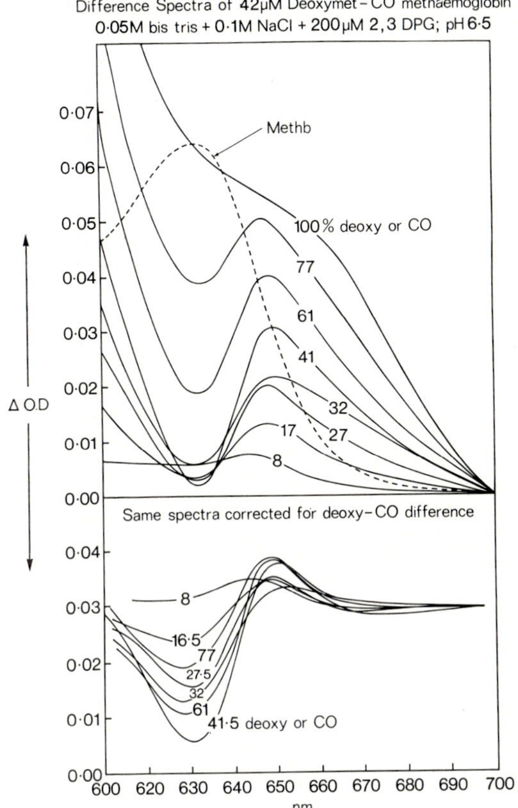

Fig. 7. Difference spectra (deoxy minus CO) for partially oxidised haemoglobin, showing changes in the ferric haem band at 630 nm. These reflect the high to low spin transition. Note that the differences are largest when about half of the haems are oxidised. Above: difference spectra. Below: spectra corrected for the changes in the ferrous haems as they react with CO. All at pH 6.5 with 2,3-diphosphoglycerate

the spin; when it is tense it stretches them and tilts the equlibrium towards high-spin. In this way a ligand binding at the α haem, acting through the quaternary structure of the protein, causes observable electronic changes in the β haem. The spectra give striking proof that haem-haem interaction extends all the way from one iron atom to another [49]. Lindstrom, Ho and Pisciotta [50] have observed parallel changes in the NMR spectrum of M Milwaukee.

Haemoglobin Milwaukee also illustrates how Hill's constant is reduced if the R-T equilibrium is too strongly biased. Inositol hexaphosphate pulls the tetramer over to the T structure and inhibits the spectral changes produced by binding CO. The spectrum of the T structure itself does not alter.

The Spin State of Ferric Haem Changes in Normal Haemoglobin

Does the same spin transition occur in the ferric haems of haemoglobin A? If so the optical spectra of ferric subunits in *partially* oxidised haemoglobin should show similar changes when the ferrous subunits combine with ligands. The corresponding spectral band of methaemoglobin A lies at 630 nm. Perutz has found that the expected differences appear on comparing the deoxy and CO forms of partly oxidised solutions in the presence of DPG or IHP (Fig. 7). The effect is largest when half of the haems are ferric and half are capable of combining with CO. Here again the spin change is of the type R(low) → T(high).

References

1. Perutz, M. F.: Nature (Lond.) **228**, 726 (1970).
2. Klug, A.: Symp. Int. Soc. Cell Biol. **6**, 1, New York: Academic Press 1968.
3. Monod, J., Wyman, J., Changeux, J.-P.: J. molec. Biol. **12**, 88 (1965).
4. Muirhead, H., Cox, J. M., Mazzarella, L., Perutz, M. F.: J. molec. Biol. **28**, 117 (1967).
5. Perutz, M. F., Muirhead, H., Cox, J. M., Goaman, L. C. G.: Nature (Lond.) **219**, 131 (1968).
6. Antonini, E., Brunori, M.: Hemoglobin and myoglobin in their reactions with ligands. Amsterdam: North-Holland 1971.
7. Roughton, F. J. W.: Biochem. J. **117**, 801 (1970).
8. Kilmartin, J. V., Rossi-Bernardi, L.: Nature (Lond.) **222**, 1243 (1969).
9. Benesch, R. E., Benesch, R., Renthal, R., Gratzer, W. B.: Nature (Lond.) New Biol. **234**, 174 (1971).
10. Wyman, J.: J. Amer. chem. Soc. **89**, 2202 (1967).
11. De Bruin, S. H., Janssen, L. H. M., Van Os, G. A. J.: Biochem. biophys. Res. Commun. **45**, 544 (1971).
12. Bailey, J. E., Beetlestone, J. G., Irvine, D. H.: J. chem. Soc. A 756 (1970).
13. Tomita, S., Riggs, A.: J. biol. Chem. **246**, 547 (1971).
14. Kellett, G. L.: Nature (Lond.) New Biol. **234**, 189 (1971).
15. Hewitt, J. A., Kilmartin, J. V., Ten Eyck, L. F., Perutz, M. F.: Proc. nat. Acad. Sci. (Wash.) **69**, 203 (1972).

16. Antonini, E., Bucci, E., Fronticelli, C., Wyman, J., Rossi-Fanelli, A.: J. molec. Biol. **12**, 375 (1965).
17. Muirhead, H., Greer, J.: Nature (Lond.) **228**, 516 (1970).
18. Watson, H. C.: Progr. Stereochem. **4**, 229 (1969).
19. Padlan, E. A., Love, W. E.: Nature (Lond.) **220**, 376 (1968).
20. Huber, R., Epp, O., Steigemann, W., Formanek, H.: Europ. J. Biochem. **19**, 42 (1971).
21. Hendrickson, W. A., Love, W. E.: Nature (Lond.) New Biol. **232**, 197 (1971).
22. Morimoto, H., Lehmann, H., Perutz, M. F.: Nature (Lond.) **232**, 412 (1971).
23. Rosemeyer, M. A., Huehns, E. R.: J. molec. Biol. **25**, 253 (1967).
24. Bolton, W., Perutz, M. F.: Nature (Lond.) **228**, 551 (1970).
25. Kilmartin, J. V., Wootton, J. F.: Nature (Lond.) **228**, 766 (1970).
26. Perutz, M. F.: Nature (Lond.) **228**, 735 (1970).
27. Arnone, A.: Nature (Lond.) **237**, 146, (1972).
28. Perutz, M. F., Ten Eyck, L. F.: Cold Spr. Harb. Symp. quant. Biol. **36**, 295 (1972).
29. Kilmartin, J. V., Hewitt, J. A.: Cold Spr. Harb. Symp. quant. Biol. **36**, 311 (1972).
30. Kellett, G. L.: J. molec. Biol. **59**, 401 (1971).
31. Hoard, J. L.: In: Structural chemistry and molecular biology, p. 573 (Rich, A., Davidson, N., Eds.) San Francisco: W. H. Freeman & Co. 1968.
32. Williams, R. J. P.: Fed. Proc. **20**, 5 (1961).
33. Smith, D. W., Williams, R. J. P.: In: Structure and bonding **7**, 1 (1970).
34. Moffatt, J. K.: J. molec. Biol. **58**, 79 (1971).
35. Perutz, M. F.: Nature (Lond.) **237**, 495 (1972).
36. Antonini, E., Schuster, T. M., Brunori, M., Wyman, J.: J. biol. Chem. **240**, PC 2262 (1965).
37. Gibson, Q. H.: J. biol. Chem. **245**, 3285 (1970).
38. Lindstrom, T. R., Olson, J. S., Mock, N. H., Gibson, Q. H., Ho, C.: Biochem. biophys. Res. Commun. **45**, 22 (1971).
39. Talbot, B., Brunori, M., Antonini, E., Wyman, J.: J. molec. Biol. **58**, 261 (1970).
40. Perutz, M. F., D. Pulsinelli, P., Ten Eyck, L., Kilmartin, J. V., Shibata, S., Iuchi, I., Miyaji, T., Hamilton, H. B.: Nature (Lond.) New Biol. **232**, 147 (1971).
41. Hayashi, A., Stamatoyannopoulos, G., Yoshida, A., Adamson, J.: Nature (Lond.) New Biol. **230**, 264 (1971).
42. Greer, J.: J. molec. Biol. **59**, 99 (1971).
43. Bunn, H. F., Briehl, R. W.: J. clin. Invest. **49**, 1088 (1970).
44. Tyuma, I., Shimizu, K., Imai, K.: Biochem. biophys. Res. Commun. **43**, 423 (1971).
45. Rubin, M. M., Changeux, J.-P.: J. molec. Biol. **21**, 265 (1966).
46. Ogawa, S., Shulman, R. G.: Biochem. biophys. Res. Commun. **42**, 9 (1971).

47. CASSOLY, R., GIBSON, Q. H., OGAWA, S., SHULMAN, R. G.: Biochem. biophys. Res. Commun. **44**, 1015 (1971).
48. PERUTZ, M. F., PULSINELLI, P. D., RANNEY, H.: Nature (Lond.) **237**, 259 (1972).
49. HAYASHI, A., SUZUKI, T., SHIMIZU, A., MORIMOTO, H., WATARI, H.: Biochim. biophys. Acta (Amst.) **147**, 407 (1967).
50. LINDSTROM, T. R., HO, C., PISCIOTTA, A. V.: Nature (Lond.) **237**, 263 (1972).

Discussion

G. BUSE (Aachen): Did not KILMARTIN claim that CO_2 can bind to the dissociated amino group (-NH_2) of Val NA 1 of both the α- and β-chains forming a carbamate? On the other hand in the deoxy- or T-structure of Hb, following Perutz' model, the -NH_3^+ of Val NA 1 (α) forms a salt bridge with the C-terminal carboxyl group of the other α-chain, and -NH_3^+ of Val NA 1 (β) binds with 2,3-DPG. Don't these results contradict each other?

A. D. MCLACHLAN (Cambridge): KILMARTIN and ROSSI blocked the α-amino groups of the α- and β-chains with cyanate. They then showed that the influence of CO_2 upon the Bohr effect is reduced. KILMARTIN now believes that CO_2 binds as carbamate ion to the uncharged -NH_2 groups of the β-chains. It is uncertain whether CO_2 binds to the α-chains.

In the T – structure the binding of CO_2 to the β-chains gives them a negatively charged N-terminal group, and prevents the binding of 2,3 DPG, as TOMITA and RIGGS have shown. So there is competition, but no logical difficulty.

G. BUSE (Aachen): What is than the physiological state of deoxy-Hb? Should we conclude that the fraction of CO_2 transported back to the lungs as Hb-carbamate is very small and that binding occurs only with the β-chains?

A. D. MCLACHLAN (Cambridge): Deoxy-haemoglobin can carry either CO_2 (as carbamate) or DPG, depending on the conditions. ROUGHTON has shown that CO_2 is carried by the red cell in two ways, as bound carbamate and as dissolved CO_2. The amount bound is a significant fraction of the total.

Approaches for Determining Protein Complexes

Interaction between Chymotrypsin (Trypsin) and the Pancreatic Trypsin Inhibitor (Kallikrein Inactivator)

R. Huber

Max-Planck-Institut für Eiweiß- und Lederforschung und Physikalisch-Chemisches Institut der Technischen Universität München, Germany

With 8 Figures

Introduction

The high velocity of enzymatic reactions is ascribed to the proximity of the catalytic groups and the substrate within the rigid enzyme substrate complex. Translational and rotational degrees of freedom are frozen out within the complex and the reactive centres favourably oriented [1—4]. Furtheron the geometric arrangement of the reactants involved in the catalytic reaction might be stabilized in a steric conformation closely resembling that of the transition state of the reaction [5, 6]. These factors may increase the velocity of enzymatic reactions by many orders of magnitude over their non-enzymatic analogs.

Essentially only static structures are observed by crystallographic methods. These static structures are not sufficient to describe an enzymatic reaction but they provide a structural framework for the interpretation of kinetic and functional studies.

The structure of stable enzyme inhibitor complexes often provides relevant information on the catalytic mechanism, particularly if the kinetics of the enzymatic reaction and of the inhibitor enzyme reaction have been determined in detail.

The natural proteinase inhibitors meet these requirements. They are in many respects closely similar to substrates and the study of their structure and the structure of their complexes with proteinases might throw light on the catalytic action of the proteinases.

Some Aspects of Structure and Function of the Serine Proteinases

A considerable number of proteolytic enzymes are serine proteinases, several of which have been studied extensively in their function and structure. The proteinases chymotrypsin, elastase and

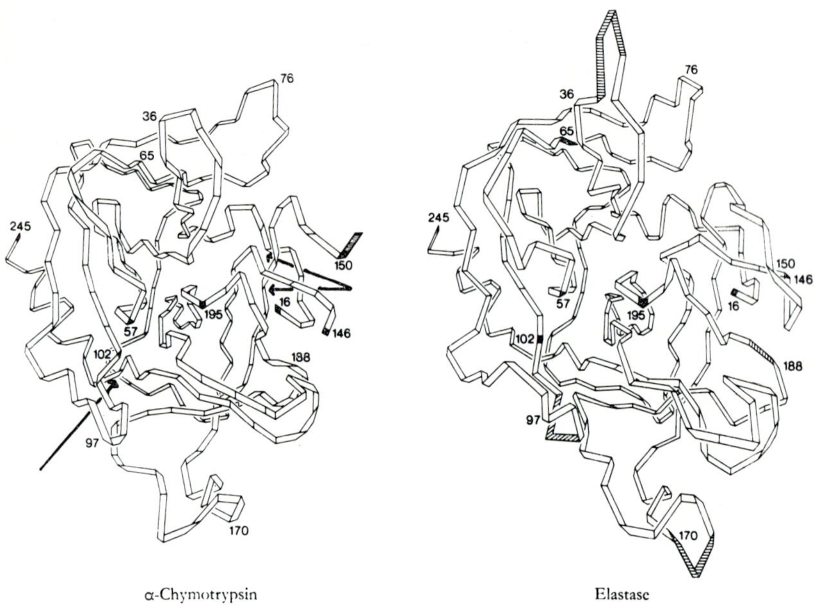

Fig. 1. Diagrams of the α-carbon conformations of α-chymotrypsin and elastase. Common chymotrypsinogen A numbering for both molecules. Sequence insertions in elastase are shaded. The positions of the additional disulfide bonds in trypsin (22—157) and (128—232) are indicated by arrows. Modified from D. Shotton [8]

trypsin have peptidase and esterase activity but differ in their specificity. Chymotrypsin specificity is directed mainly to the peptide bond on the C-terminal side of hydrophobic, preferentially aromatic amino acid residues. Elastase has a broad specificity for small residues with a preference for alanine bonds, and trypsin appears to split peptide bonds following exclusively basic residues.

The elucidation of their spatial structures showed that these proteinases are closely related [7—9]. This may be demonstrated by a comparison of the structures of chymotrypsin and elastase (Fig. 1). This schematic diagram shows the main chain conformations of α-chymotrypsin and elastase. The shaded areas in elastase are sequence insertions, which obviously do not disturb the overall chain folding [10] but form additional loops, as is clearly seen at the insertion of three amino acids 36 A to C in elastase (α-chymotrypsin enumeration). Trypsin is also very similar to α-chymotrypsin and elastase in its three-dimensional structure. A demonstrative aspect of this similarity is the additional disulfide bridges Cys (22—157) and Cys (128—232) in trypsin which may easily be built also in the model of α-chymotrypsin [9, 11]. These are indicated by arrows in the chymotrypsin drawing.

The similarity in the spatial arrangement of the residues involved in catalysis is also striking. Among these are the residues histidine (57), aspartic acid (102) and serine (195) which form a "charge relay" system [12]. The oxygen of serine (195) carries a partial negative charge induced by the buried aspartic acid (102) via the polarizable imidazole ring of histidine (57). In such a system the serine γ-oxygen atom can act as a nucleophile forming an acyl complex with the peptide substrate via a tetrahedral intermediate. The imidazole group could donate a proton to the leaving -NH group. De-acylation then occurs by a reversal of the acylation reaction. A water molecule replacing the nitrogen of the leaving group acts as a nucleophile, attacking the carbonyl carbon of the acyl enzyme (Fig. 2) [13, 14].

Considering the close similarity of the spatial relation of the crucial residues, an identical catalytic mechanism is almost certain for the three enzymes. What then causes the specificity difference observed?

The binding site of the amino acid side chain on the C-terminal side is a depression in the protein surface which is lined with different amino acids in the three proteins. This is schematically demonstrated in Fig. 3. In α-chymotrypsin it is a slit lined with apolar amino acids and well adapted for the accommodation of aromatic residues. In trypsin aspartic acid (189) is at the bottom of the hole. A very favourable charge-charge interaction with lysine or arginine side chains is possible. Insertion of neutral amino acid

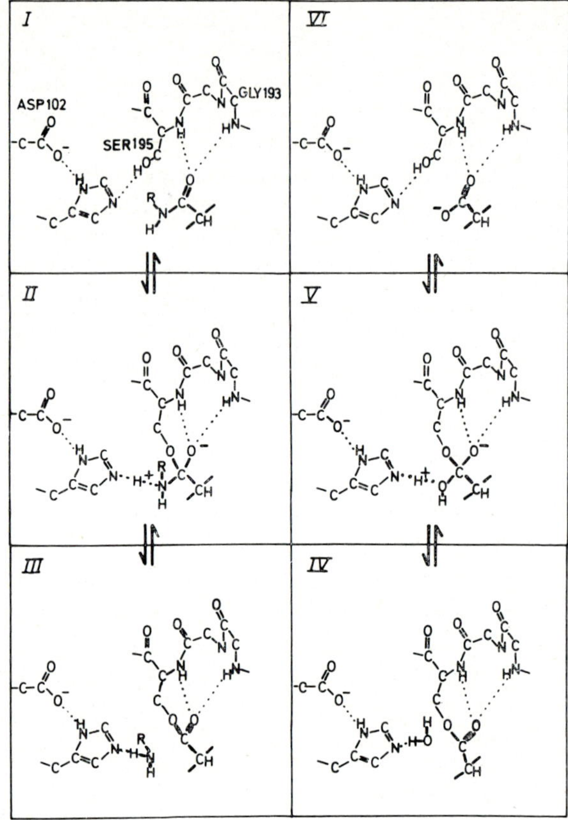

Fig. 2. Proposed sequence of steps in the hydrolysis of peptide substrates by serine proteinase (from R. Henderson, C. S. Wright, G. P. Hess, D. M. Blow [14])

side chains into the trypsin hole would only be possible at the expense of shielding the negative charge of aspartate (189), an energetically unfavourable situation. In elastase the hole is more or less closed by the substitution of the two glycines (216) and (226) by valine and threonine respectively. The preference for alanine substrates is therefore easily explained.

The catalytic mechanism described above and the probable position of the substrate have been deduced mainly from studies of complexes of chymotrypsin with amino acids, short peptides and analogs which bind on the amino terminal side of the sensitive bond. No information was available for the position of the peptide chain on the C-terminal side in the MICHAELIS complex.

Within these various complexes the enzyme structure was found to be largely invariant; only the minimal movements of a few atoms could be observed necessary for the formation of covalent bonds, i.e. the acyl bond in acylenzyme complexes. Chymotrypsin

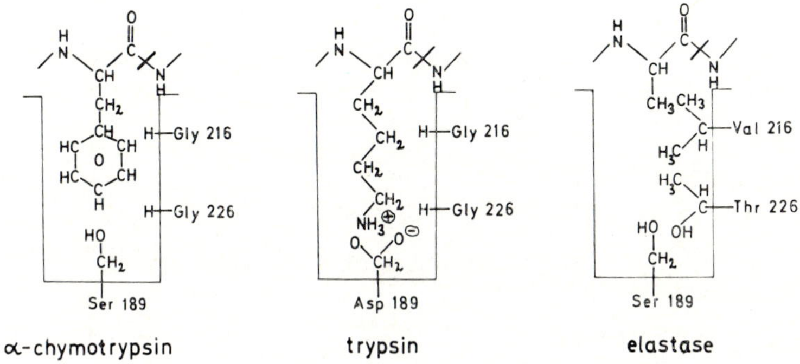

Fig. 3. Schematic representation of the binding pocket for the side chain on the C-terminal side of the sensitive peptide bond in α-chymotrypsin, trypsin and elastase (chymotrypsin enumeration)

behaves as a rigid enzyme. This encouraged us to study by model building the possible mode of interaction of chymotrypsin and of the serine proteinases generally with a natural inhibitor, the pancreatic trypsin inhibitor, whose structure has been determined [15, 16].

The inhibitor, too, is obviously a rigid structure, as deduced from its three-dimensional folding and from physico-chemical data which indicate an abnormal resistance to heat, alkali, acid and other denaturing agents [17].

Natural Proteinase Inhibitors

The term natural proteinase inhibitors refers to proteins which associate reversibly with proteinases. In the complexes the catalytic functions of the proteinases are competitively inhibited. A large number of proteinase inhibitors have been characterized from different sources in animals and plants [18]. Their physiological role is, in general terms, to limit the action of the proteinases in time and space. This function is particularly evident in the case of the pancreatic secretory inhibitors which are excreted together with the zymogens and inhibit their early activation.

The molecular weight of the inhibitors varies over a wide range, but there is a certain preference for molecular weights of about 6000 [19]. Most inhibitors form 1:1 complexes with proteinases, but 2:1 complexes and even 4:1 complexes have also been characterized. The latter are complexes with ovoinhibitors, capable of associating with two molecules of trypsin and chymotrypsin [20].

Molecular Properties of the Pancreatic Trypsin Inhibitor (Kallikrein Inactivator) (PTI)

A substance which inhibits kallikrein was first isolated by KRAUT, FREY and BAUER [21]. It also inhibits trypsin, chymotrypsin and plasmin to various degrees. It is inactive towards elastase and subtilisin. Its identity with the bovine pancreatic trypsin inhibitor of KUNITZ [22] was later established by sequence determination of the two materials [23, 24]. It is a molecule containing 58 amino acids. The sequence is shown in Fig. 4. It is characterized by a predominance of basic residues responsible for the high isoelectric point of the molecule and by a high cystine content.

Although a high proportion of disulfides appears to be generally characteristic of proteinase inhibitors, some extreme cases should be mentioned: a trypsin inhibitor from peanuts (seeds of Arachis hypogaea) [25] containing 10 half cysteins/48 amino acids, or the inhibitor from lima beans [26] containing 14 half cysteins/84 amino acids.

The high content of disulfides might account for the generally high stability towards denaturants. For example, PTI remains completely in its native conformation at 77 °C and pH 2 or in 6 M guanidium chloride. Between pH 2—1 and pH 10—11 some subtle conformational changes occur, probably involving only a few side chains [17]. This conformational stability might also be intimately connected with the observed stability of inhibitors towards proteolytic enzymes. It is not attacked by any protease except thermolysin above 60 °C [27]. As will be seen later, geometric fitting appears to be a stringent requirement in productive substrate enzyme interactions.

The atomic structure of the PTI molecule, as determined by crystal structure analysis, shows a compact, pear-shaped molecule with a relatively high content of secondary structure, in particular β-structure [28, 29].

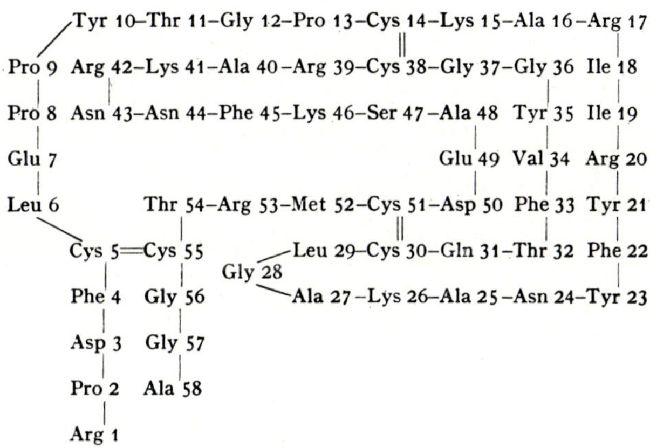

Fig. 4. Amino acid sequence of the pancreatic trypsin inhibitor (kallikrein inactivator) (PTI)

Fig. 5 shows the conformation of the main chain atoms of the molecule, with hydrogen bonds between main chain atoms indicated by dotted lines. This structure has two interesting features: the two-stranded antiparallel β-structure from amino acid alanine (16) to glycine (36) which is twisted to form a double helix with a pitch of 37 Å and 14 residues per turn, and a short segment of polyproline II structure including the chain leucine (6) to tyrosine (10).

The most interesting segment of the structure is, of course, the binding site to the proteinase. A trypsin inhibitor may be expected to bind specifically by lysine or arginine side chains to trypsin. Indeed, it has been shown that modifications of lysine (15) and

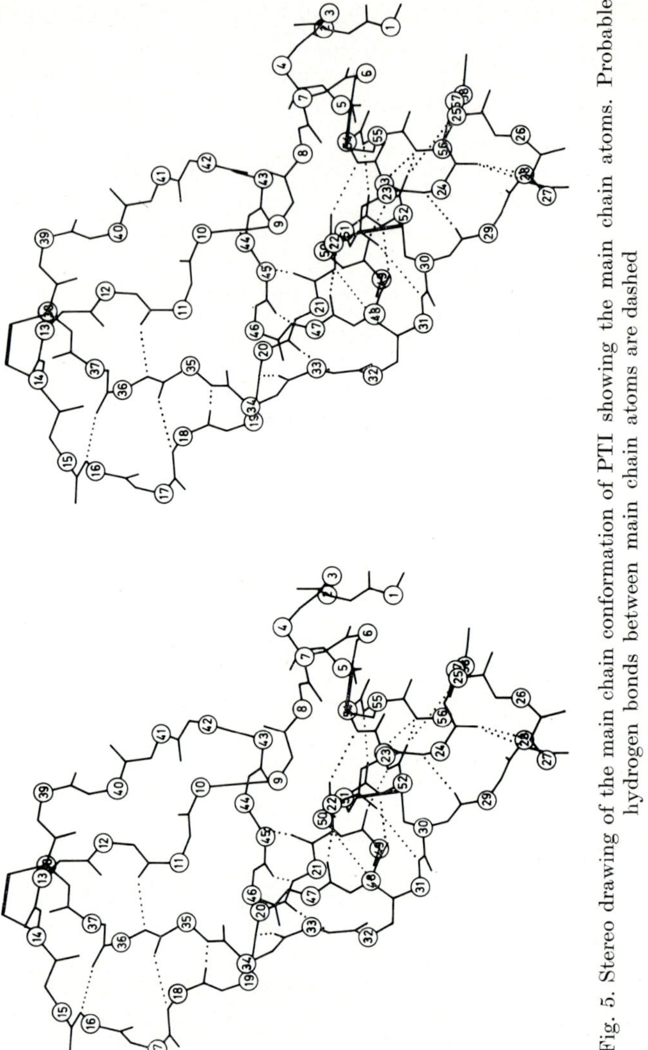

Fig. 5. Stereo drawing of the main chain conformation of PTI showing the main chain atoms. Probable hydrogen bonds between main chain atoms are dashed

lysine (15) only affect the inhibitory properties [30, 31]. Conversely, it could be shown that within the complex lysine (15) is shielded. Arginine residues do not seem to be involved in binding to trypsin,

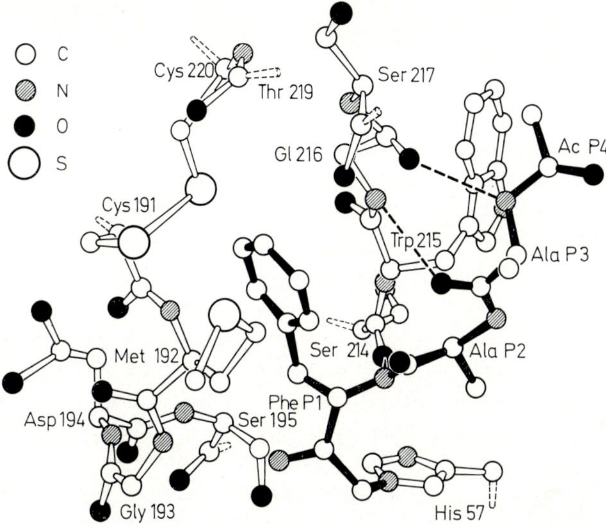

Fig. 6a. Acetyl-ala-ala-phe-chloro-ketone bound to γ-chymotrypsin [33]

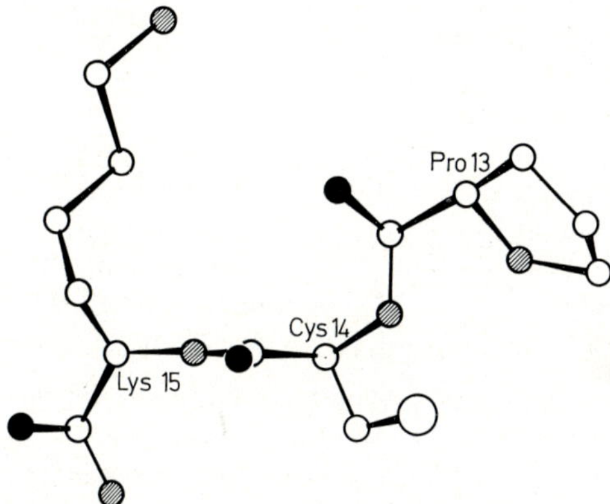

Fig. 6b. Conformation of the segment Pro (13) to Lys (15) of PTI as seen in approximately the same orientation as the chloro-ketone

as shown by chemical modification studies [32]. Lysine (15) is at the top of the molecule in a very exposed segment of the structure.

A remarkable observation is that the conformation of the polypeptide chain of the inhibitor probably involved in binding proline

(13) to lysine (15) is virtually identical with the conformation of the peptide chain of the acetyl-ala-ala-phe-chloroketone inhibitor when bound to γ-chymotrypsin [33] (Fig. 6a and b). This small peptide is covalently bound to histidine [57] and irreversibly inhibits the catalytic properties of the enzyme. This suggests that the inhibitor segment proline (13) — alanine (16) is frozen in the conformation of a substrate chain when bound to the enzyme. This recalls Pauling's remarks on the nature of effective inhibitors which should resemble the activated complex (5).

The inhibitor must have a high dipole moment, as deduced from the spatial distribution of its charged amino acid residues. Particularly near the binding region, only basic residues occur. Acidic residues are distant from lysine (15), [29]. It has been observed that K_M and k_{cat} of peptide hydrolysis may vary to a large extent. Positively charged residues surrounding the hydrolyzed peptide bond increase k_{cat} by several orders of magnitude over normal values [34]. Also in this respect PTI is an ideal inhibitor.

It is tempting to suggest that the conformation of the peptide chain around the sensitive bond is very similar in all natural inhibitors. Steric complementarity to the enzyme binding site appears to be a strict requirement for a good inhibitor. Inhibitors should either have the correct conformation at the contact area, or their polypeptide chain should be flexible to adapt to the enzyme. The latter appears improbable in view of the high stability and extensive cross-linkage of most natural inhibitors.

Mode of Action of PTI and Other Inhibitors

Susceptible Peptide Bond

The detection of a susceptible peptide bond appears to play an important role in the understanding of inhibitor action [35]. Many inhibitors, if not all, contain a particularly sensitive peptide bond which is hydrolyzed during incubation with catalytic amounts of proteinases [19]. Hydrolysis continues until an equilibrium mixture of inhibitor molecules with intact and cleaved peptide bonds (virgin and modified forms) is reached. Experiments with the soy-bean trypsin inhibitor demonstrated that it is indeed a true equilibrium, which may also be obtained from the modified form of the inhibitor through enzymatic resynthesis of the susceptible peptide bond [19, 25, 36—39].

It is interesting to note that the "double-headed" inhibitor from lima beans contains two sensitive peptide bonds in different regions of the amino acid sequence. A *leucyl-seryl* peptide bond is involved in the anti-chymotrypsin site, while a *lysyl-seryl* peptide bond is within the anti-trypsin site. Most remarkably, the sequences around these sensitive peptide bonds are strongly homologous, indicating sterical equivalence [26].

We may define an equilibrium constant K of hydrolysis $K_{hyd.} = \frac{C_M}{C_V}$ for the sensitive peptide bond of an inhibitor where C_M is the concentration of the modified species and C_V is the virgin form. $K_{hyd.}$ varies strongly for different inhibitors having values between ~ 0 to 3. Normal peptide bond hydrolysis is expected to proceed to 100% cleaved products corresponding to a very high constant of hydrolysis. The thermodynamic parameters of peptide hydrolysis in dilute solutions suggest that the driving force is the entropy gain upon bond cleavage.

One might expect that the low values of $K_{hyd.}$ for inhibitors are due to limited mobility of the new chain ends in the modified species. Indeed, it has been shown that the low $K_{hyd.}$ value of soybean inhibitor depends strongly on the native conformation of the molecule [19].

$K_{hyd.}$ of the PTI must be very small. No modified species with the lysine (15) — alanine (16) bond cleaved could be observed except in the case where the disulfide cystine (14—38) adjacent to the sensitive peptide bond has been reduced (see Fig. 5) [40, 41]. It is remarkable that chymotrypsin also splits the reduced form of the inhibitor at the lysine (15) — alanine (16) bond [42], an unexpected result in view of the specificity of chymotrypsin.

A value of virtually zero for $K_{hyd.}$ of the native form of PTI indicates that the peptide bond lysine (15) — alanine (16) is strongly stabilized. Fig. 7 shows the structure of PTI near the sensitive peptide bonds lysine (15) — alanine (16). The polypeptide chain from glycine (12) to isoleucine (18) and glycine (36) to arginine (39) are particularly close together. The potential new chain ends at lysine (15) and alanine (16) are held to the rest of the molecule by the disulfide cystine (14)—(38) on the one side and a hydrogen bond NH [Ala (16)] ... OC [Gly (36)] on the other side; this is the first hydrogen bond of the twisted pleated-sheet segment previously described (see also Fig. 5). Hydrophobic interactions of the residues alanine (16), isoleucine (18) and (19) should contribute to the stabilization energy. We may therefore conclude that peptide bond cleavage should not appreciably increase mobility of the polypeptide segments.

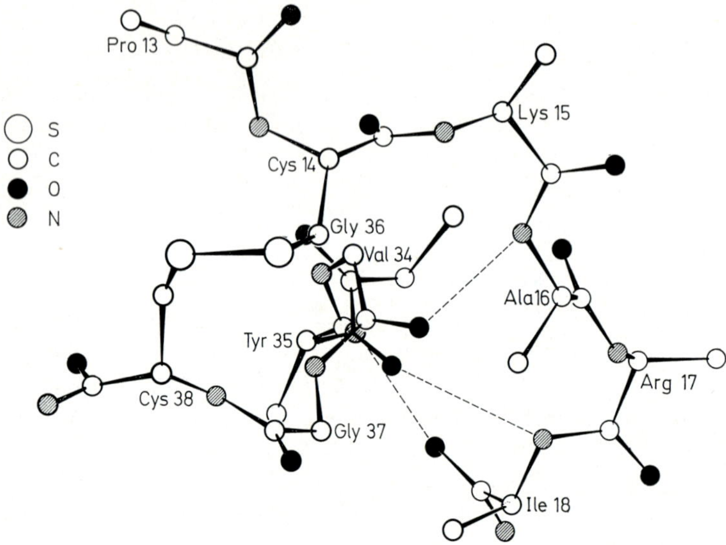

Fig. 7. Segment of PTI near Lys (15) — Ala (16). All main chain atoms including the sulfur atoms of Cys (14—18) are shown

Kinetics of Interaction of PTI and Other Inhibitors with Proteinases

The kinetics of interaction of soy-bean trypsin inhibitor with trypsin has been extensively characterized by LASKOWSKI, Jr., and his group [36].

The few data available for PTI suggest a close similarity in the mechanism of interaction of soy-bean trypsin inhibitor and PTI with proteinases. The pH dependences of the association rates of soy-bean trypsin inhibitor and PTI with trypsin are similar [36, 43], as are the pH dependences of the association constants and the dissociation rate constants [19, 36].

The second-order association rate constant of PTI with trypsin is approximately $3 \cdot 10^5$ l $M^{-1}sec^{-1}$ [43] at neutral pH, comparable to acylation rates (exactly k_{cat}/K_m) of very good substrates [34]. Its pH dependence between pH 5 and pH 8 also appears to follow the same curve as k_{cat}/K_m for substrates [44]. This is in accordance with the idea that PTI is a good substrate for the association process.

The kinetics of dissociation of inhibitor protease complexes and its pH dependence, however, are completely different from the kinetics of de-acylation of good substrates. The rate constant for de-acylation of good substrates is of the order of 10^2 sec^{-1} at neutral pH [45], while the dissociation rate constant for soy-bean trypsin inhibitor is $10^{-4}sec^{-1}$ and unmeasurably small for PTI. The estimated half-life of the complex is of the order of several weeks [36]. This is, of course, the reason why inhibitors are inhibitors.

The tentative model of the complex of PTI with proteinases to be described later suggests that the exclusion of water from the contact area prevents normal de-acylation.

Tentative Model of the Complex of PTI with Trypsin and Chymotrypsin

In summary, it appears permissible to assume that PTI is bound to trypsin or chymotrypsin with the side chain of lysine (15) within the specificity pocket and the sensitive peptide bond lysine (15) — alanine (16) at the catalytic site. The stable complex is probably an acyl enzyme. Indeed, when a model of the relevant part of the inhibitor was compared with the active site of α-chymotrypsin, it was evident that only one mode of binding was possible. The peptide bond lysine (15) — alanine (16) is situated at the catalytic site of the enzyme and the side chain of lysine (15) occupies the specificity pocket [16].

Stroud, Kay and Dickerson (1971) [9] reached a similar conclusion in their model studies with PTI and trypsin. Fig. 8 shows a view of the model of the inhibitor — α-chymotrypsin complex.

The enzyme surface at the catalytic site is completely covered by the inhibitor. Water is excluded. Approximately 400 contacts

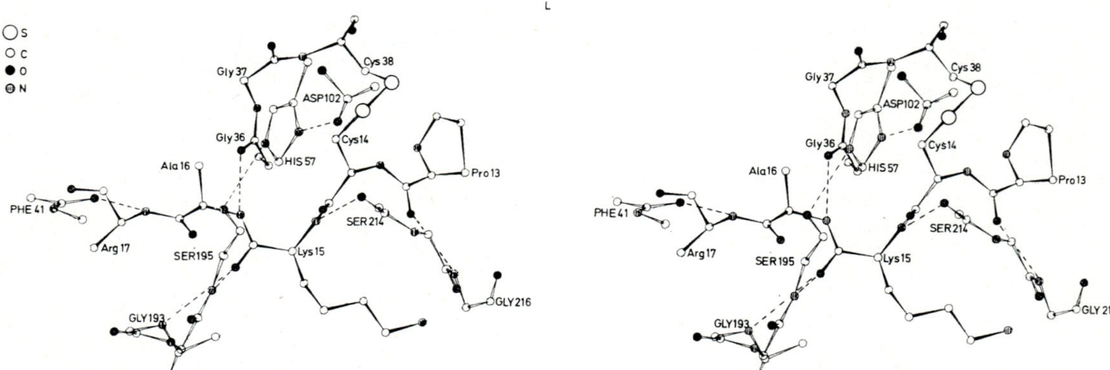

Fig. 8. Stereo drawing of the contact area of PTI and α-chymotrypsin in the tentative model of the complex. Open lines: α-chymotrypsin. Filled lines: PTI. Dashed lines: hydrogen bonds. Only the main chain atoms and a few side chains are shown

are made between enzyme and inhibitor, as shown in the table. These contacts are predominantly between apolar residues of the aliphatic parts of long polar side chains (see Table).

The mode of association with trypsin is very similar, due to the structural similarities between the two enzymes.

The association energy of PTI and trypsin is approximately 14 kcal [42]. It appears to be provided predominantly by hydrophobic and VAN DER WAALS interactions.

Table. *Van der Waals contacts between PTI and α-chymotrypsin*
Residues Involved

PTI	Chymotrypsin (Trypsin)	No. of contacts
Pro—13	Trp—215	2
Cys—14	His—57, Ile—99 (Leu), Trp—215	14
Lys—15	His—57, Ser—190, Cys—191, Met—192 (Gln) Ser—195, Trp—215, Gly—216, Gly—226	35
Ala—16	Cys—42, His—57, Met—192 (Gln), Gly—193, Ser—195	15
Arg—17	His—40, Phe—41, Trp—141, Gly—142, Leu—143 (Asn), Met—192 (Gln), Gly—193, Asp—194, Ser—195	66
Ile—18	Phe—41, Cys—58	3
Ile—19	Phe—39 (Tyr)	3
Gly—36	His—57	1
Gly—37	His—57	6
Cys—38	Ile—99 (Leu)	3
Arg—39	Ser—96, Leu—97 (Asn), Thr—98, Ile—99 (Leu)	69

It is not clear whether the suggested acyl bond between inhibitor and enzyme contributes to the association energy, as amide and ester linkages are of comparable energies [46]. It may well be, however, that the small structural perturbations accompanying amide-bond cleavage and acyl-bond formation provide a better fit between two molecules by increasing their interaction energy. Another possibility is that the peptide bond lysine (15) — alanine (16) is strained, whereas the acyl bond is not.

The NH_3^+ group of lysine (15) in the specificity pocket of trypsin may form a salt bridge with aspartate (189). The specificity pocket in chymotrypsin, however, is lined with apolar residues and the accommodation of lysine (15) is only possible at the cost of lowering its pK or burying its positive charges. The association energy with chymotrypsin is in fact lower by about 4 kcal [42].

The short segment of β-structure formed between enzyme and inhibitor within the complex is obvious. The hydrogen bonds are between CO(Pro 13I) ... NH(Gly 216) and NH(Lys 15I) ... CO(Ser 214) like those found in the acetyl-ala-ala-phe-chloromethylketone bound to γ-chymotrypsin, as already demonstrated in Fig. 6 (33)[1].

The contribution of these and other hydrogen bonds to the interaction energy is difficult to estimate but might be small, as all groups involved are in contact with water in the free molecules. But they are probably important for the correct steric arrangement of inhibitor and enzyme.

In the PTI-chymotrypsin complex the lysine (15) carbonyl group is oriented by the hydrogen bonds with the NH of serine (195) and glycine (193). Such an orientation might be particularly favorable for nucleophilic attack of the γ-oxygen atom of serine (195) at the carbonyl carbon atom.

As already described, PTI and other inhibitors are inhibitors by reason of an abnormally slow dissociation of the inhibitor-enzyme complex. The model suggests that the leaving group is held to the rest of the inhibitor molecule by the β-sheet interactions and hydrophobic bonds described before. Water is excluded, which is necessary for de-acylation. This is also in accordance with the observation on several inhibitors that dissociation to the inhibitor with intact peptide bond is kinetically favoured over de-acylation to the modified form [36].

References

1. Bruice, T. C., Brown, A., Harris, D. O.: Proc. nat. Acad. Sci. (Wash.) **68**, 658 (1971).
2. Page, M. I., Jencks, W. P.: Proc. nat. Acad. Sci. (Wash.) **68**, 1678 (1971).
3. Bruice, T. C.: Cold Spr. Harb. Symp. quant. Biol. **36**, 21 (1971).
4. Koshland, D. E., Jr., Carroway, K. W., Dafforn, G. A., Gass, J. D., Storm, D. R.: Cold Spr. Harb. Symp. quant. Biol. **36**, 13 (1971).
5. Pauling, L.: Amer. Sci. **36**, 51 (1948).
6. Lienhard, G. E., Secemski, I. I., Koehler, K. A., Lindquist, R. N.: Cold Spr. Harb. Symp. quant. Biol. **36**, 45 (1971).

[1] I means inhibitor

7. Birktoft, J. J., Blow, D. M., Henderson, R., Steitz, T. A.: Phil. Trans. **B 257**, 67 (1970).
8. Shotton, D.: In: Proceedings of the international research conference on proteinase inhibitors, pp. 47—55 (Fritz, H., Tschesche, H., Eds.). Berlin: Walter de Gruyter 1971.
9. Stroud, R. M., Kay, L. M., Dickerson, R. E.: Cold Spr. Harb. Symp. quant. Biol. **36**, 125 (1971).
10. Hartley, B. S., Shotton, D. M.: In: The enzymes, Vol. III, p. 323 (Boyer, P. D., Ed.). London: Academic Press 1971.
11. Sigler, P. B., Blow, D. M., Matthews, B. W., Henderson, R.: J. molec. Biol. **35**, 143 (1968).
12. Blow, D. M., Birktoft, J. J., Hartley, B. S.: Nature (Lond.) **221**, 337 (1969).
13. Henderson, R.: J. molec. Biol. **54**, 341 (1970).
14. — Wright, C. S., Hess, G. P., Blow, D. M.: Cold Spr. Harb. Symp. quant. Biol. **36**, 63 (1971).
15. Huber, R., Kukla, D., Rühlmann, A., Steigemann, W.: Cold Spr. Harb. Symp. quant. Biol. **36**, 141 (1971).
16. Blow, D. M., Wright, C. S., Kukla, D., Rühlmann, A., Steigemann, W., Huber, R.: J. molec. Biol. **69**, 137 (1972).
17. Vincent, J.-P., Chicheportiche, R., Lazdunski, M.: Europ. J. Biochem. **23**, 401 (1971).
18. Vogel, R., Trautschold, I., Werle, E.: Natural proteinase inhibitors. New York: Academic Press 1968.
19. Laskowski, M., Jr., Sealock, R. W.: In: The enzymes, Vol. III, pp. 375—473 (Boyer, P. D., Ed.). London: Academic Press 1971.
20. Woo-Hoe Liu, Means, G. E., Feeney, R. E.: Biochim. biophys. Acta (Amst.) **229**, 176 (1971).
21. Kraut, H., Frey, E. K., Bauer, E.: Hoppe-Seylers Z. physiol. Chem. **175**, 57 (1928).
22. Kunitz, M., Northrop, J. H.: J. gen. Physiol. **19**, 991 (1936).
23. Anderer, F. A., Hörnle, S.: J. biol. Chem. **241**, 1568 (1966).
24. Kassell, B., Laskowski, M., Sr.: Biochem. biophys. Res. Commun. **20**, 463 (1965).
25. Hochstrasser, K., Illchmann, K., Werle, E., Hössl, R., Schwarz, S.: Hoppe-Seylers Z. physiol. Chem. **351**, 1503 (1970).
26. Tan, C. G. L., Stevens, F. C.: Europ. J. Biochem. **18**, 515 (1971).
27. Wang, Tsun-Wen, Kassell, B.: Biochem. biophys. Res. Commun. **40**, 1039 (1970).
28. Huber, R., Kukla, D., Rühlmann, A., Epp, O., Formanek, H.: Naturwissenschaften **57**, 389 (1970).
29. — — — Steigemann, W.: In: Proceedings of the international research conference on proteinase inhibitors, pp. 56—65 (Fritz, H., Tschesche, H., Eds.). Berlin: Walter de Gruyter 1971.
30. Chauvet, J., Acher, R.: J. biol. Chem. **242**, 4274 (1967).
31. Fritz, H., Schult, H., Meister, R., Werle, E.: Hoppe-Seylers Z. physiol. Chem. **350**, 1531 (1969).
32. Keil, B.: FEBS Letters **14**, 181 (1971).

33. Segal, D. M., Powers, J. C., Cohen, G. H., Davies, D. R., Wilcox, P. E.: Biochemistry **10**, 3728 (1971).
34. Abita, J. P., Delaage, M., Lazdunski, M., Savrda, J.: Europ. J. Biochem. **8**, 314 (1969).
35. Finkenstadt, W. R., Laskowski, M., Jr.: J. biol. Chem. **240**, PC 962 (1965).
36. Laskowski, M., Jr.: In: Proceedings of the international research conference on proteinase inhibitors, pp. 117—134 (Fritz, H., Tschesche, H., Eds.). Berlin: Walter de Gruyter 1971.
37. Uy, R. L., Feeney, R. E.: Fed. Proc. **30**, 454 (1971).
38. Tschesche, H., Klein, H.: Hoppe-Seylers Z. physiol. Chem. **349**, 1645 (1968).
39. Fritz, H., Fink, E., Meister, R., Klein, G.: Hoppe-Seylers Z. physiol. Chem. **351**, 1344 (1970).
40. Wilson, K. A., Laskowski, M., Sen.: J. biol. Chem. **246**, 3555 (1971).
41. Kress, L. F., Laskowski, M., Sen.: J. biol. Chem. **242**, 4925 (1967).
42. Rigbi, M.: In: Proceedings of the international research conference on proteinase inhibitors, pp. 74—88 (Fritz, H., Tschesche, H., Eds.). Berlin: Walter de Gruyter 1971.
43. Pütter, J.: Hoppe-Seylers Z. physiol. Chem. **348**, 1197 (1967).
44. Hess, G. P.: In: The enzymes, Vol. III, pp. 213—248 (Boyer, P. D., Ed.). New York: Academic Press 1971.
45. Zerner, B., Bond, R., Bender, M.: J. Amer. chem. Soc. **86**, 3674 (1964).
46. Fersht, A. R.: J. Amer. chem. Soc. **93**, 3504 (1971).

Discussion

G. Pfleiderer (Bochum): We have isolated different proteases from invertebrates with tryptic and chymotryptic specificity. All but one were inhibited by the pancreatic inhibitor. Peptidases with other splitting specificities were not influenced. The only exception was a chymotrypsin isolated from hornet larves with a molecular weight of 12 to 13.000, as estimated by gelfiltration. The strong inhibition would predict a very similar inhibition mechanism and similar interactions between these different peptidases in spite of the fact that they are far apart from the evolutionary point of view.

R. Huber (München): I am quite sure about this. I would say that the interaction which I showed must be present; the two molecules must fit. One example: the pancreatic trypsin inhibitor does not inhibit subtilisin. It is well known that in subtilisin the arrangement of the catalytic residues is quite similar to chymotrypsin. Nevertheless it does not bind. You need the stereo-complementarity of the two molecules.

An Example of "Quasi-Irreversible" Protein-Protein Interaction: The Trypsin-Pancreatic Trypsin Inhibitor Complex

M. LAZDUNSKI and J. P. VINCENT

Centre de Biologie Moléculaire du Centre National de la Recherche Scientifique, 31, Ch. Joseph Aiguier, Marseille, France

With 1 Figure

Each of the partners in the trypsin-PTI[1] complex is well characterized. Both their covalent and their three-dimensional structures in the crystalline state are available (WALSH and NEURATH, 1964; STROUD et al., 1971; KASSELL and LASKOWSKI, Sen., 1965; CHAUVET et al., 1964; ANDERER and HÖRNLE, 1966; HUBER et al., 1970) and a considerable amount of work has been devoted to the analysis of the components of their active sites. Therefore, the trypsin-PTI complex appears to be an excellent model of heterologous protein-protein associations.

The association of trypsin (Ti) with PTI was first studied by GREEN and WORK (1953). It may be represented simply by the following equation:

$$\mathrm{Ti} + \mathrm{PTI} \underset{k_{-1}}{\overset{k_1}{\rightleftharpoons}} \mathrm{Ti} - \mathrm{PTI}$$

Values for the second-order rate constant of association k_1, for the first-order rate constant of dissociation k_{-1} and for the dissociation constant K of the enzyme-inhibitor complex are given in the table.

The rate constant of association between the native partners is $1.1 \times 10^6\ \mathrm{M^{-1}\ sec^{-1}}$. It is similar to those found with other systems of macromolecular interactions such as the association of iodoinsulin with its membranous receptor (CUATRECASAS et al., 1971), $3.5 \times 10^6\ \mathrm{M^{-1}\ sec^{-1}}$, or the association of leucyl-tRNA to leucine-

[1] PTI: pancreatic trypsin inhibitor.

tRNA synthetase, $0.9-1 \times 10^6$ M^{-1} sec^{-1} (ROUGET and CHAPEVILLE, 1971). One of the most striking conclusions of the data is the "quasi-irreversible" association of PTI with the trypsin receptor. The dissociation constant 6×10^{-14} M is, as far as we are aware, the lowest constant ever determined for such an effector-receptor interaction. It is well below dissociation constants for usual enzyme-

Table. *Kinetics and thermodynamic characteristics of the interaction of trypsin, pseudotrypsin, RCOM*trypsin, trypsinogen and chymotrypsin with PTI and its derivatives*

Receptor	Inhibitor	k_1 (M^{-1} sec^{-1})	k_{-1} (sec^{-1})	K (M)
Trypsin	PTI	$1.1 \cdot 10^6$	$6.6 \cdot 10^{-8}$	$6.0 \cdot 10^{-14}$
Trypsin	R*PTI	$3.2 \cdot 10^5$	$5.7 \cdot 10^{-4}$	$1.8 \cdot 10^{-9}$
Trypsin	RCAM*PTI	$1.3 \cdot 10^5$	$2.2 \cdot 10^{-5}$	$1.7 \cdot 10^{-10}$
Trypsin	RAE*PTI	$8.2 \cdot 10^4$	$7.5 \cdot 10^{-4}$	$9.1 \cdot 10^{-9}$
Trypsin	RCOM*PTI	0	—	—
Pseudo-trypsin	PTI	$7.0 \cdot 10^4$	$6.3 \cdot 10^{-4}$	$9.0 \cdot 10^{-9}$
RCOM*trypsin	PTI	$2.0 \cdot 10^4$	$1.2 \cdot 10^{-4}$	$6.0 \cdot 10^{-9}$
Trypsinogen	PTI	$4.4 \cdot 10^3$	$8.8 \cdot 10^{-4} < k_{-1} < 8.8 \cdot 10^{-2}$	$2 \cdot 10^{-7} < K < 2 \cdot 10^{-5}$
Chymotrypsin	PTI	$1.1 \cdot 10^5$	$9 \cdot 10^{-4}$	$9 \cdot 10^{-9}$
Chymotrypsin	R*PTI	$8 \cdot 10^4$	$1.4 \cdot 10^{-3}$	$18 \cdot 10^{-9}$

k_1 and k_{-1} are the rate constants for association and dissociation, respectively. K is the dissociation constant of the complex at 25°, pH 8.0. Description of the determination of k_1, k_{-1} and K is given elsewhere (VINCENT and LAZDUNSKI, 1972).

substrate complexes (10^{-2} M to 10^{-7} M) or for macromolecular complexes such as those formed between tRNAs and activating enzymes (10^{-8} M). The first-order rate constant for the dissociation of the complex, $6.6 \cdot 10^{-8}$ sec^{-1}, corresponds to a half-life of more than 17 weeks. Such data favour considerably the existence of a covalent attachment of the inhibitor to the enzyme. Very interesting models of interaction have recently been proposed by crystallographers. They have been presented by Dr. HUBER. These models were built by assembling the structures of trypsin (STROUD et al., 1970) or chymotrypsin (BLOW et al., in press; HUBER et al., 1971) with that of the pancreatic trypsin inhibitor [HUBER et al., 1970 (1) and (2)]. In these models, the recognition of lysine 15 (PTI) by aspartic acid 177 (trypsin) is followed by a split of the Lys_{15}-Ala_{16} bond concurrent with an acylation of Ser_{183} in the active site of trypsin. Deacylation is prevented by the absence of water molecules in the area of contact between trypsin and PTI. The covalent bond is then an ester bond between the carboxylate of lysine 15 and the alcohol function of serine 183 (trypsin). This type of interaction was suggested by the elegant analysis done by LASKOWSKI Jr. and his group on the association of trypsin with the soybean trypsin inhibitor (LASKOWSKI, Jr. et al., 1970). Acylation of the enzyme with the inhibitor implies that the rate limiting step for the dissociation of the complex is the deacylation. If that were the case, the dissociated pancreatic inhibitor would be a 2-chain molecule because of the splitting of the Lys_{15}-Ala_{16} bond. In fact, a difficulty of the acylation mechanism is that no digestion of the pancreatic trypsin inhibitor occurs near neutral pH (KRESS and LASKOWSKI, Sen., 1968). This is in contrast with the situation for the soybean trypsin inhibitor. In that case, splitting occurs at the strategic Arg_{64}-Ile_{65} bond (OZAWA and LASKOWSKI, Jr., 1966).

It appeared first necessary to evaluate the contribution of lysine 15 to the inhibitor-enzyme association. Lysine 15 is thus far the only chemically identifiable element of the inhibitor active site (CHAUVET and ACHER, 1967; KRESS and LASKOWSKI, Sen., 1968; FRITZ et al., 1969). The α-amino side chain of this residue mimics the structure of natural substrates of the enzyme and is believed to form an ion-pair interaction with the side chain of Asp_{177}, the anionic part of the specificity site in trypsin.

A quantitative evaluation of the importance of the ion-pair interaction can be obtained by modifying either one of the two partners. We have chosen to act at the level of Asp_{177}. The first possibility is to disconnect Asp_{177} from the specificity site of trypsin. There is a lysine residue in position 176; the Lys_{176}-Asp_{177} bond can be selectively cleaved by trypsin itself; the product of this limited autolysis is called pseudo-trypsin; the active site is not destroyed but pseudo-trypsin has lost its high specificity for the basic substrates of trypsin (SMITH and SHAW, 1969). The second possibility is to replace the trypsin partner by chymotrypsin. Chymotrypsin and trypsin present a high degree of homology which explains their cross-specificity. However, Asp_{177} in the specificity site of trypsin is replaced by a serine residue in the specificity site of chymotrypsin (SHOTTON, 1971). As indicated in the table, pseudo-trypsin and chymotrypsin still form stoichiometric complexes with PTI but the stability of these complexes is considerably decreased as compared to the stability of the trypsin-PTI complex. The dissociation constant of the complexes formed by PTI with pseudo-trypsin and chymotrypsin is about 1.5×10^4 times higher than that of the complex with trypsin. The important difference is not on k_1, the second-order rate constant of association, it is on k_{-1} which is increased by a factor of 10,000 by the removal of the Asp_{177} partner. The half-life of these complexes with pseudo-trypsin and chymotrypsin is only 18 and 12 min respectively.

These results indicate that the salt bridge Lys_{15} (PTI)-Asp_{177} (trypsin) is very important but not absolutely essential for the trypsin-PTI interaction. Complementary evidence for this conclusion was recently obtained by RIGBI (1970) through modification of the Lys_{15} component of the ion pair.

Considerable attention has been given to the role of disulfide bridges in the association of trypsin with the inhibitor. It is of interest to point out that all the essential components of the active sites of trypsin and PTI are in close proximity to disulfide bridges. The Cys_{14}-Cys_{38} bond of PTI is near the lysine 15 residue of the active site. The Cys_{31}-Cys_{47} bond (trypsin) is near the essential histidine 46 residue of the catalytic site of trypsin. The Cys_{179}-Cys_{203} bond (trypsin) is near both the active serine 183 and aspartic acid 177, the essential element of the specificity site of the enzyme. Such proximity of critical amino-acids to disulfide bridges appears to be more than a simple coincidence.

The Cys_{14}-Cys_{38} bridge connects the two strands of the β-structure in PTI and is in a very exposed position at the top of the pear-shaped molecule (HUBER et al., 1970). This disulfide bridge can be very selectively reduced in the free inhibitor (KRESS and LASKOWSKI, Sen., 1967; KRESS et al., 1968). However, we have found that it was completely protected against reduction not only in the trypsin-PTI complex (see also LIU et al., 1971) but also in the chymotrypsin-PTI complex.

Reduction of the Cys_{14}-Cys_{38} bridge (and subsequent chemical modifications) does not appear to modify the conformation of the inhibitor at all as far as we can judge from a classical analysis of its physico-chemical properties (VINCENT et al., 1971). This is not unexpected, since it is well known that disulfide bridges do not determine protein conformations; they only stabilize them.

Reduction of the Cys_{14}-Cys_{38} bridge of PTI (R*PTI) has a considerable effect on the characteristics of the association of the inhibitor with trypsin, as indicated in the table 1. Although the rate of association of R*PTI is decreased by a factor of only 3.5 as compared to native PTI, the dissociation constant of the complex is increased by a factor of 3×10^4 by selective reduction of the inhibitor. This is mainly due to a considerable change of k_{-1} which increases by a factor of about 10^4.

Chemical modifications of -SH groups produced by reduction of the Cys_{14}-Cys_{38} bridge may have different effects. Alkylation by iodoacetamide (RCAM*PTI) improved the binding (as compared to R*PTI) but treatment with ethyleneimine (RAE*PTI) decreased it slightly. Alkylation with iodoacetic acid (RCOM*PTI) completely prevented the association, as previously observed by KRESS et al. (1968). These data taken together indicate that the Cys_{14}-Cys_{38} bridge plays a very important role in the association of PTI with trypsin. Interestingly enough, reduction of the Cys_{14}-Cys_{38} bridge has practically no influence on the association (k_1, k_{-1}, K) of the inhibitor with chymotrypsin.

The Cys_{179}-Cys_{203} bridge of trypsin can also be selectively reduced and alkylated (RCOM*trypsin) without destruction of the active site (LIGHT et al., 1969; HATFIELD et al., 1971). Again, stoichiometric association with PTI persists but the strength of the association decreases considerably. The dissociation constant increases by a factor of 10^5, while the rate of dissociation increases by a factor of about 2×10^3.

For all these reasons a possible functional role of disulfide bridges has been considered. A tentative model of interaction involving multiple disulfide interchange is presented in Fig. 1. Such a model would certainly explain the data which have been presented.

Although no information is available at the moment concerning the Cys_{31}-Cys_{47} bond, this bond has been incorporated in the interchange mechanism because of its proximity to His_{46} and also because proteolytic digestion of the complex gave a high molecular

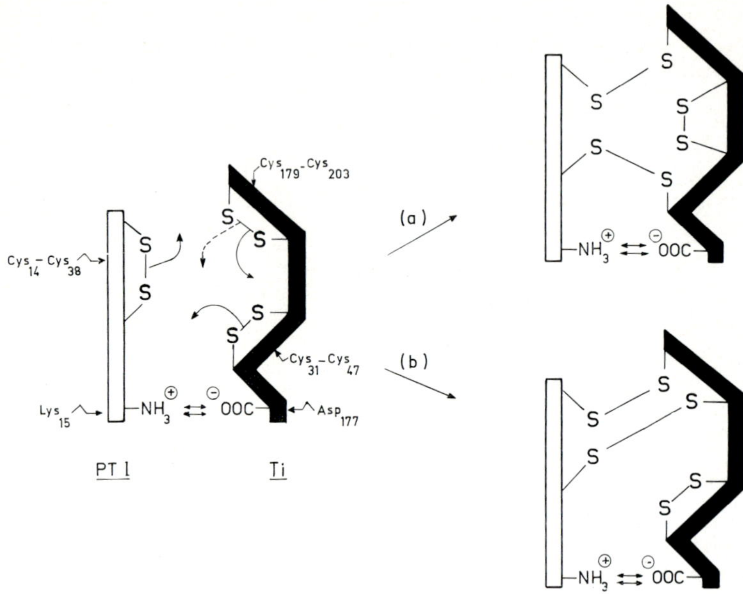

Fig. 1. Disulfide interchange, a plausible explanation of the "quasi-irreversible" trypsin-PTI association. The "primary event" in this hypothetical mechanism is the recognition, through ion-pair formation, of the ε-amino group of Lys_{15} (PTI) by the β-carboxylate of Asp_{177} (trypsin). Disulfide bonds are then correctly positioned for the interchange. Two possible mechanisms were considered: (pathway [a]) a multiple interchange (solid arrows) involving Cys_{14}-Cys_{38} (PTI) and both Cys_{179}-Cys_{203} and Cys_{31}-Cys_{47} of trypsin; (pathway [b]) a simple interchange involving only Cys_{14}-Cys_{38} (PTI) and Cys_{179}-Cys_{203} (trypsin) (broken arrow). These two disulfide bridges have been found to be unusually reactive as compared to others (KRESS and LASKOWSKI, Sen., 1967; LIGHT et al., 1969). The final mechanism will have to await further X-ray crystallographic data on the trypsin-PTI complex

weight fragment, comprising the undigested inhibitor apparently covalently bound to a fragment of trypsin which contained the sequence Ala_{146}-Ser_{198}, that is serine 183 and aspartic acid 177 of the active site, as well as the two histidines His_{29} and His_{46} (DLOUHA et al., 1968).

Although the three disulfide bridges involved in the interchange mechanism are all in the area of contact between the inhibitor and the enzyme (HUBER et al., 1971) they are not necessarily close enough for a direct interchange and a conformational rearrangement may be needed.

The disulfide bridges play an important role in the interaction. However, the argument that these bonds participate in an interchange rests largely on indirect evidence and a definite mechanism will probably be obtained only through the X-ray analysis of the crystal structure of the trypsin-PTI complex (HUBER et al., 1971). Because it does not make use of the special catalytic properties of the enzymatic receptor, a disulfide interchange mechanism could occur in a number of effector-receptor systems and in the first place in other associations involving protein inhibitors and degradative (non proteolytic) enzymes, such as phosphodiesterases or phospholipases, for example.

Work is presently being carried out in this laboratory on neurotoxins. Snake neurotoxins have structural properties similar to those of PTI. They are small proteins of 60 to 71 amino acids cross-linked internally by 4 or 5 disulfide bridges. Studies of NAJA HAJE neurotoxin I indicated two lysines in the active site — lysines 26 and 46 (CHICHEPORTICHE et al., 1972). Lysine 26 is very near the disulfide bridge Cys_3-Cys_{23} (BOTES and STRYDOM, 1969). Snake neurotoxins form "quasi-irreversible complexes" with the acetylcholine receptor of excitable membranes (CHANGEUX et al., 1970; MILEDI and POTTER, 1971). This is an indication that a covalent bond is formed between the neurotoxin and the receptor. Here again, we would be tempted to postulate that the formation of an ion pair between the side chains of neurotoxin active-site lysines and the anionic site of the receptor which normally recognizes the positive charge of acetylcholine, is followed by disulfide interchange involving critical disulfide bonds of the neurotoxin and of the receptor. In this respect, it is extremely interesting that the receptor protein of acetylcholine has a super-reactive disulfide bond in the vicinity of the acetylcholine binding site. This disulfide bond can be selectively reduced by dithiothreitol. Such a reduction of the receptor inhibits the response to carbamylcholine (KARLIN and WINNIK, 1968). The reactive disulfide bond may well undergo a disulfide interchange with the neurotoxin molecule. It was recently shown in this laboratory that one of the five disulfide bridges of neurotoxin III of NAJE HAJE was selectively reduced and that alkylation with iodoacetic acid, although it did not change the conformational properties, suppressed the toxic activity (CHICHEPORTICHE, SCHWEITZ and LAZDUNSKI, unpublished results).

Most protein hormones such as insulin (SANGER et al., 1955), growth hormone (LI et al., 1966), nerve growth factor (ANGELETTI and BRADSHAW, 1971) etc. contain one or more disulfide bridges; this is also true for a number of polypeptide hormones such as oxytocin or vasopressin, for example. Although no detailed information concerning protein hormone receptors is available at the moment, one can also consider the possibility of complex formation involving disulfide interchange.

References

ANDERER, F. A., HÖRNLE, S.: J. biol. Chem. **241**, 1568 (1966).

ANGELETTI, H. R., BRADSHAW, R. A.: Proc. nat. Acad. Sci. (Wash.) **68**, 2417 (1971).

BOTES, D. P., STRYDOM, D. J.: J. biol. Chem. **244**, 4147 (1969).

CHANGEUX, J. P., KASAI, M., LEE, C. Y.: Proc. nat. Acad. Sci. (Wash.) **67**, 1241 (1970).

CHAUVET, J., NOUVEL, G., ACHER, R.: Biochim. biophys. Acta (Amst.) **92**, 200 (1964).

— ACHER, R.: J. biol. Chem. **242**, 4274 (1967).

CHICHEPORTICHE, R., ROCHAT, C., SAMPIERI, F., LAZDUNSKI, M.: Biochemistry **11**, 1681 (1972).

CUATRECASAS, P., DESBUQUOIS, B., KRUG, F.: Biochem. biophys. Res. Commun. **44**, 333 (1971).

DLOUHA, V., KEIL, B., ŠORM, F.: Biochem. biophys. Res. Commun. **31**, 66 (1968).

FRITZ, H., SCHULT, H., MEISTER, R., WERLE, E.: Hoppe-Seylers Z. physiol. Chem. **350**, 1531 (1969).

GREEN, N. M., WORK, E.: Biochem. J. **54**, 347 (1953).

HATFIELD, L. M., BANERJEE, S. K., LIGHT, A.: J. biol. Chem. **246**, 6303 (1971).

HUBER, R., KUKLA, D., RÜHLMANN, A., EPP, O., FORMANEK, H.: (1) Naturwissenschaften **57**, 389 (1970).

— — — STEIGEMANN, W.: (2) Proc. Int. Res. Conf. on Proteinase Inhibitors, Munich, p. 56 (1970).

— — — — Cold Spr. Harb. Symp. quant. Biol. **36**, 141 (1971).

KARLIN, A., WINNIK, M.: Proc. nat. Acad. Sci. (Wash.) **60**, 668 (1968).

KASSELL, B., LASKOWSKI, M., Sen.: Biochem. biophys. Res. Commun. **20**, 463 (1965).

KRESS, L. F., LASKOWSKI, M., Sen.: J. biol. Chem. **242**, 4925 (1967); **243**, 3548 (1968).

— WILSON, D. A., LASKOWSKI, M., Sen.: J. biol. Chem. **243**, 1758 (1968).

LASKOWSKI, M., Jr., DURAN, W. R., FINKENSTADT, W. R., HERBERT, S., HIXSON, H. F., Jr., KOWALSKI, D., LUTHY, J. A., MATTIS, J. A., MC KEE, R. E., NIEKAMP, C. W.: Proc. Int. Conf. on Proteinase Inhibitors, Munich, p. 117 (1970).
LI, C. H., LIU, W. K., DIXON, J. S.: J. Amer. chem. Soc. **88**, 2050 (1966).
LIGHT, A., HARDWICK, B. C., HATFIELD, L. M., SONDACK, D. L.: J. biol. Chem. **244**, 6289 (1969).
LIU, W., TRZECIAK, H., SCHUSSLER, H., MEIENHOFER, J.: Biochemistry **10**, 2849 (1971).
MILEDI, R., POTTER, L. T.: Nature (Lond.) **233**, 599 (1971).
OZAWA, K., LASKOWSKI, M., Jr.: J. biol. Chem. **241**, 3955 (1966).
RIGBI, M.: Proc. Int. Conf. on Proteinase Inhibitors, Munich, p. 74 (1970).
ROUGET, F., CHAPEVILLE, F.: Europ. J. Biochem. **23**, 443 (1971).
SANGER, F., THOMPSON, E. O. P., KITAI, R.: Biochem. J. **59**, 509 (1955).
SHOTTON, D.: Proc. Int. Res. Conf. on Proteinase Inhibitors, Munich, p. 47 (1970).
SMITH, R. L., SHAW, E.: J. biol. Chem. **244**, 4704 (1969).
STROUD, R. M., KAY, L. M., DICKERSON, R. E.: Cold Spr. Harb. Symp. quant. Biol. **36**, 125 (1971).
VINCENT, J. P., CHICHEPORTICHE, R., LAZDUNSKI, M.: Europ. J. Biochem. **23**, 401 (1971).
— LAZDUNSKI, M.: Biochemistry **11**, 2967 (1972).
WALSH, K. A., NEURATH, H.: Proc. nat. Acad. Sci. (Wash.) **52**, 889 (1964).

Discussion

E. HELMREICH (Würzburg): I wonder how you can measure the dissociation of such a tightly bound complex? Dissociation must take hours, doesn't it!

M. LAZDUNSKI (Marseille): Dissociation constants of the order of 10^{-8} M, 10^{-9} M are easily determined either by a direct estimation of K or by measuring k_1 and k_{-1} (for example t 1/2 for dissociation is of the order of 15 min for the trypsin-R*PTI complex, $K = k_{-1}/k_1$).

To evaluate the dissociation constant of the trypsin-PTI complex, we proceeded as follows:

a. Determination of the dissociation constant of the trypsin-RCAM*PTI complex from kinetic data. k_1 was measured by following the loss of activity. k_{-1} was measured by following the displacement of the radioactive RCAM* PTI by the native inhibitor (t 1/2 = 500 min).

b. When the dissociation constant for the trypsin-RCAM*PTI complex is known, the dissociation constant of the trypsin-PTI complex is determined by measuring the competition between PTI and RCAM*PTI. In a typical experiment, the trypsin-RCAM*PTI complex is incubated in the presence of a 100-fold excess of labelled RCAM*PTI and a stoichiometric amount of native PTI. When equilibrium is attained, 15% of trypsin is associated in the form of the trypsin-RCAM*PTI and 85% in the form of the trypsin-PTI complex.

Knowing K, the equilibrium constant for the dissociation of the trypsin-PTI complex, and k_1, the second-order rate of association, k_{-1} can be easily determined. The half-life of this complex is more than 17 weeks.

R. Huber (München): Dr. Lazdunski's kinetic data of the interaction of different chemically modified forms of PTI, trypsin and chymotrypsin are extremely interesting for the study of structure and function relationships of the inhibitor-proteinase complexes. His model of "disulfide exchange" as an important contribution to the interaction of PTI with proteinases is however incompatible with our knowledge of the three-dimensional structures of PTI, the proteinases and the tentative model of the complex. The mutual distances between the three cystines are in the range of 9 to 13 Å units (see Figures in [1], Fig. 5 in [2] and Fig. 14 in [3]). "Disulfide exchange" would require a complete refolding of both molecules. This is highly improbable in view of the apparent rigidity of PTI and the absence of induced fit in various proteinase inhibitor complexes [4].

1. Blow, D. M., Wright, C. S., Kukla, D., Rühlmann, A., Steigemann, W., Huber, R.: J. molec. Biol. (1972) (in press).
2. Huber, R., Kukla, D., Rühlmann, A., Steigemann, W.: Cold Spr. Harbor Symp. quant. Biol. **36**, 141 (1971).
3. Stroud, R. M., Kay, L. M., Dickerson, R. E.: Cold Spr. Harbor Symp. quant. Biol. **36**, 125 (1971).
4. Henderson, R., Wright, C. S., Hess, G. P., Blow, D. M.: Cold Spr. Harbor Symp. quant. Biol. **36**, 63 (1971).

M. Lazdunski (Marseille): I think it would take too much time to discuss that point. Naturally there could be a rearrangement; furthermore the problem with the acyl – enzyme data is that you never isolate two-chain inhibitors and that, of course, is another point against the model.

B. J. G. Butler (Cambridge): Dr. Lazdunski suggests the bonding of the inhibitor to the enzyme by disulphide bonds. Has he looked for these directly, by acidifying and then digesting the complex with pepsin? This should show the appropriate disulphide bridged peptides, and thus either confirm or disprove his hypothesis simply and directly.

M. Lazdunski (Marseille): Proteolytic degradation of the complex was carried out some years ago by Dlouha and collaborators. They found a high molecular weight fragment containing parts of both the inhibitor and the enzyme. The fact that peptides belonging to the two partners remained linked together after extensive degradation does suggest disulfide cross-linking.

Subunit Interactions in Lactate Dehydrogenase

M. J. Adams, M. Buehner, K. Chandrasekhar, G. C. Ford, M. L. Hackert, A. Liljas, P. Lentz Jr., S. T. Rao, M. G. Rossmann, I. E. Smiley, J. L. White

Department of Biological Sciences, Purdue University, Lafayette, Indiana 47907, USA

With 9 Figures

I. Introduction

The NAD$^+$ dependent tetrameric enzyme lactate dehydrogenase (EC 1.1.1.27), of molecular weight 140,000 Daltons, catalyzes the interconversion of L (+) lactate and pyruvate. Five isoenzymes result from the two subunit types abundant in tissues. The H subunit type is predominant in heart tissue and the M type in skeletal muscle. The M_4 isoenzyme (LDH-5) from dogfish *(squalus acanthius)* has been most studied in this laboratory, while investigations are now proceeding on the H_4 (LDH-1) and M_4 (LDH-5) isoenzyme of pig.

An ordered binding of coenzyme followed by substrate has been shown for LDH (Novoa, Schwert, 1961; Gutfreund et al., 1968). The reaction is stereospecific for the exchange of the A hydrogen at C 4 of the nicotinamide ring of NADH (Levy, Vennesland, 1957; Cornforth et al., 1962). An abortive ternary complex of LDH, NAD$^+$ and pyruvate has been studied in solution (Fromm, 1961, 1963; Kaplan et al., 1968; DiSabato, 1968; Wuntch et al., 1969; Everse et al., 1971) and crystallographically (Smiley et al., 1971). Complexes of the dogfish muscle enzyme with substrate inhibitors, LDH-NAD$^+$-oxalate and LDH-NADH-oxamate, isomorphous with the abortive ternary complex have also been grown.

Drs. S. Taylor and W. Allison at University of California (San Diego) are determining the sequence of M_4 dogfish LDH and we are indebted to them for information on unpublished sequences

Table 1. *Crystalline*

Species	Isoenzyme	Complex	Spacegroup
Dogfish	M_4 (LDH-5)	Apo	F422
Dogfish	M_4 (LDH-5)	With NAD^+ diffused	$C4_22_12$
Dogfish	M_4 (LDH-5)	With NAD^+ co-crystallized	I2
Dogfish	M_4 (LDH-5)	With NAD^+-pyruvate, or NAD^+-oxalate, or NADH-oxamate	$C42_12$
Pig	M_4 (LDH-5)	With NAD^+-pyruvate, or NADH-oxamate	$P22_12_1$
Pig	H_4 (LDH-1)	With NAD^+-oxalate, or NADH-oxamate	C2
Pig	M_4 (LDH-5)	With NAD^+ co-crystallized	C2
Chicken	M_4 (LDH-5)	Apo	F222

of this enzyme. We are grateful to Prof. G. Pfleiderer and Dr. K. Mella for sequences of peptides of pig H_4 and M_4 LDH.

It could be expected that the NAD^+ dependent dehydrogenases may represent a class of proteins with isostructural subunits, particularly if they have the same stereospecificity for the exchanging hydrogen atom of NADH. Recently the first major result to

Forms of LDH

Cell dimensions	Subunits per asymm. unit	Symmetry of Tetramer	Directions of molecular two-fold axes
a = 146.8 Å c = 155.35 Å	1	222	Molecular center at 1/4, 1/4, — 1/4 P ‖ —x, Q ‖ z, R ‖ y
a = 146.9 Å c = 155.8 Å	2	2	Molecular center at 1/4, 1/4, —1/4 Q ‖ z
a = 149.0 Å b = 149.0 Å c = 146.0 Å β = 98°	4	1	Packing Similar to M_4 apo dogfish
a = 134.5 Å c = 85.9 Å	1	222	Molecular center at 1/2, 1/2, 0 P ‖ z, Q ‖ x, R ‖ y
a = 86.0 Å b = 60.0 Å c = 136.2 Å	2	2	P ‖ x Q in yz plane 13° from z non-crystallographic
a = 162.9 Å b = 60.8 Å c = 138.7 Å β = 93.2°	4	1	P approx ‖ x, Q in yz plane 13° from z non-crystallographic
a = 143.6 Å b = 148.0 Å c = 84.3 Å β = 97°	4	1	—
a = 190 Å b = 108 Å c = 620 Å	8	1	—

confirm this suggestion has been obtained by Dr. L. Banaszak and coworkers (private communication, 1972) who have calculated a 3.0 Å resolution electron density map of soluble malate dehydrogenase (s-MDH) from pig, a dimer of molecular weight 70,000 Daltons. This shows two subunits with tertiary structure extremely similar to that of LDH [Adams et al., 1970 (2)]. The MDH subunits are related to each other by a (non-crystallographic) two-fold axis in a direction very similar to that of one of the molecular two-fold axes of LDH.

II. LDH Complexes Studied Crystallographically

The various crystalline forms of LDH which have been studied in this laboratory are shown in Table 1. Of these, a 2.0 Å resolution electron density map has been calculated for dogfish M_4 apoenzyme and a 3.0 Å resolution map for the abortive ternary complex of the dogfish M_4 enzyme with NAD^+ and pyruvate. The form obtained by soaking apoenzyme crystals in oxidized coenzyme has been most studied to 5.0 Å resolution only, permitting the difference between its space group and that of the apoenzyme to be ignored [ADAMS et al., 1970 (1)]. Difference electron density maps have been calculated for the dogfish M_4 LDH-NAD^+-oxalate complex at 3.0 Å resolution and for the dogfish M_4 LDH-NADH-oxamate complex at 5.0 Å resolution.

A low resolution study of the pig M_4 LDH ternary complex using the rotation function (ROSSMANN and BLOW, 1962) has revealed the orientation of the tetramer in the unit cell. The results are given in Table 1. Crystals suitable for a low resolution investigation of the complex of pig H_4 LDH with NAD^+ and oxalate or with NADH and oxamate have now been obtained. The molecular orientation in these crystals is closely related to that of the pig M_4 LDH ternary complex.

III. The Subunit of Lactate Dehydrogenase

Schematic diagrams of the LDH subunit are shown in Fig. 1 a) and b). Letter symbols have been used for α-helices and strands of β-structure with the prefixes α or β respectively. The defined areas of secondary structure are indicated in these diagrams as well as the directions of the molecular axes P, Q, and R. These conventions have now been adopted for both MDH and LDH. The main chain-main chain hydrogen bonds are shown diagrammatically in Fig. 2. The large amount of ordered secondary structure is evident in this figure. About 40% of the residues are in α-helical conformation which is on some occasions entered by a turn of 3_{10} helix.

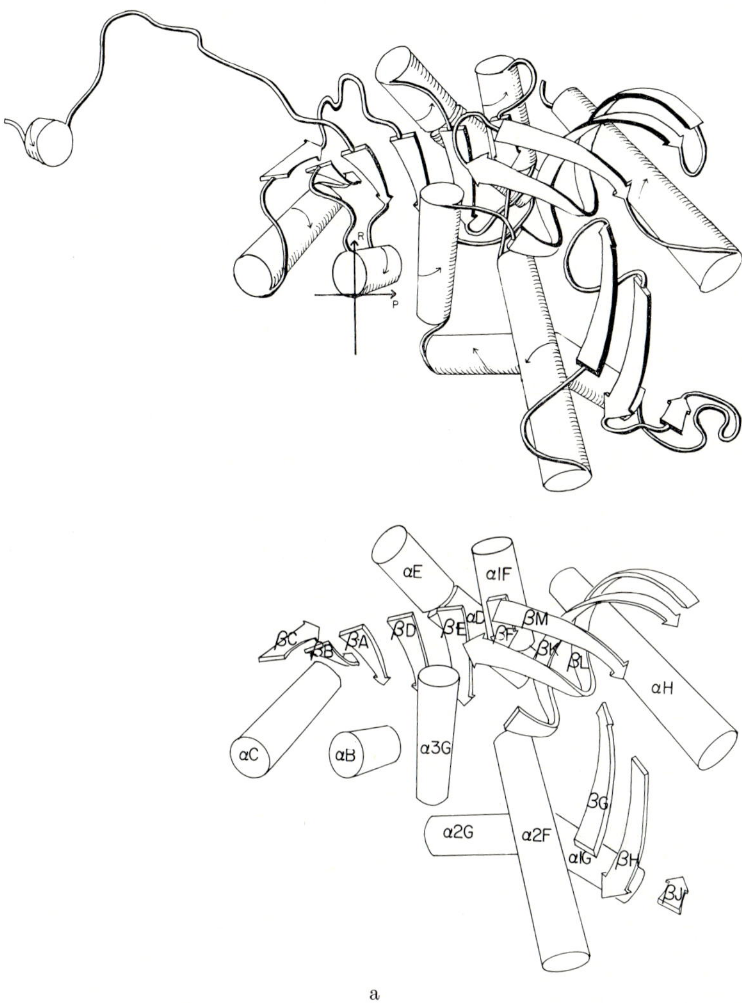

Fig. 1 a and b. Schematic drawing of LDH subunit when viewed along molecular Q axis; a) looking from inside the molecule outwards and b) (see next page) looking from outside the molecule inwards. Both diagrams show the named helical and β sheet regions

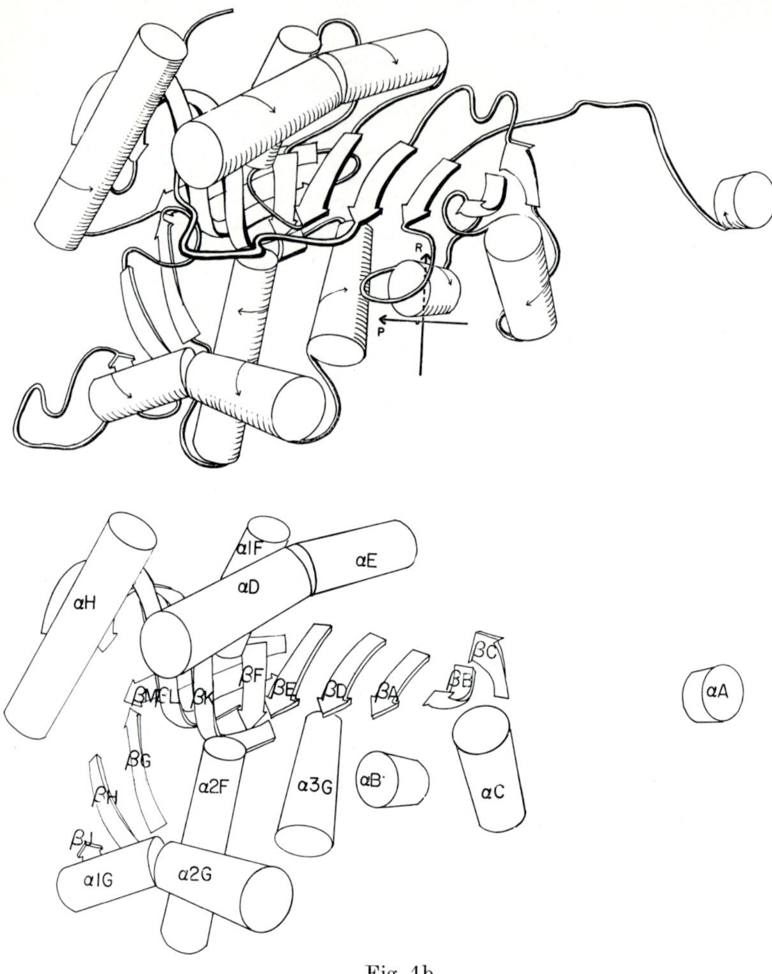

Fig. 1b

The most surprising feature of the conformation of the subunit is the extended arm consisting of the first 22 residues from the N terminus. These residues are important in the subunit interactions and will be described later. The subunit is folded so that from residue 23 it would build onto a core beginning with the central

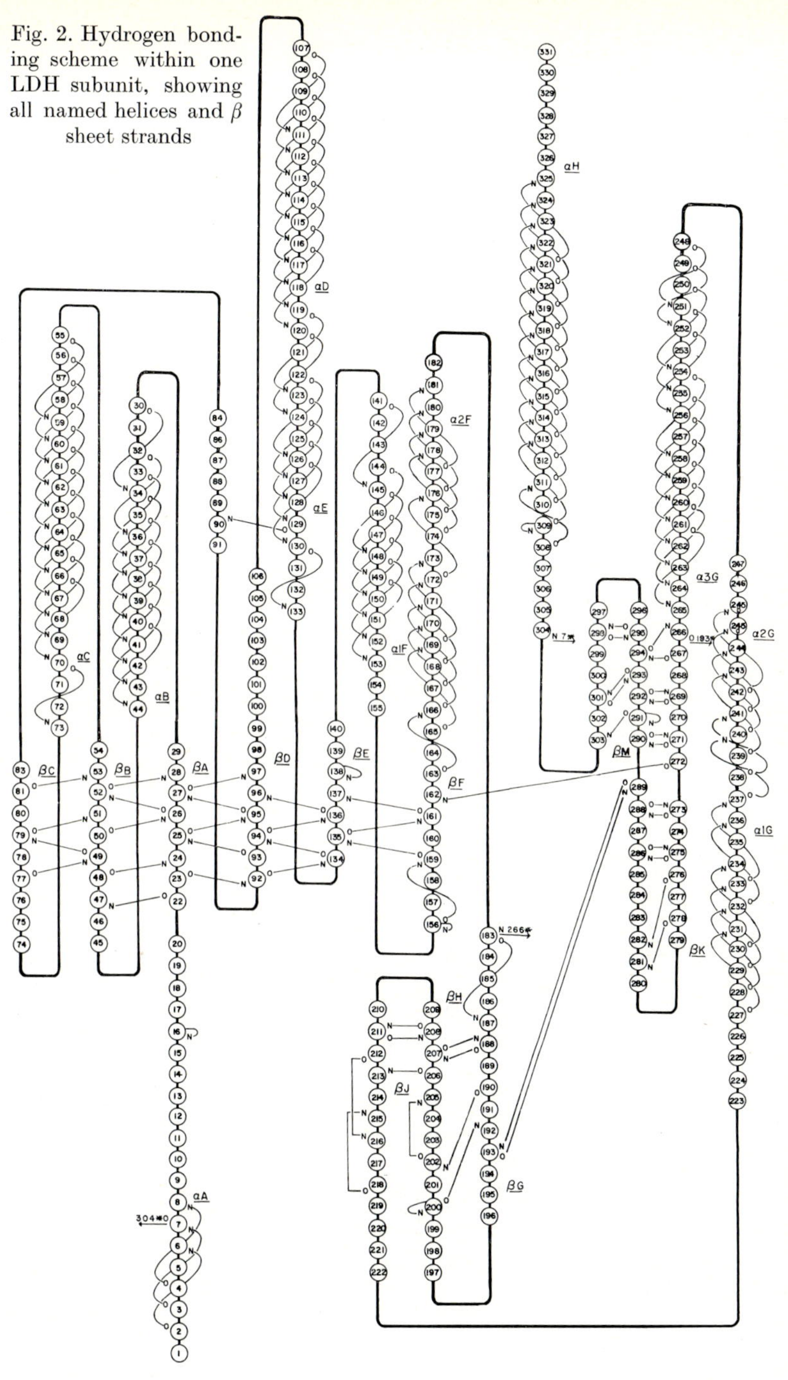

Fig. 2. Hydrogen bonding scheme within one LDH subunit, showing all named helices and β sheet strands

(and first) strand of the parallel pleated sheet βA. This sheet (βC, βB, βA, βD, βE, βF) has an anti-clockwise twist when viewed perpendicular to the successive strands, a feature which has also been observed in other sheets as in subtilisin (Wright et al., 1969), in carboxypeptidase (Lipscomb et al., 1969) and carbonic anhydrase (Kannan et al., 1971). The two antiparallel sheets (or ribbons) in the subunit, formed by the strands βH, βG, βJ and βK, βL, βM are

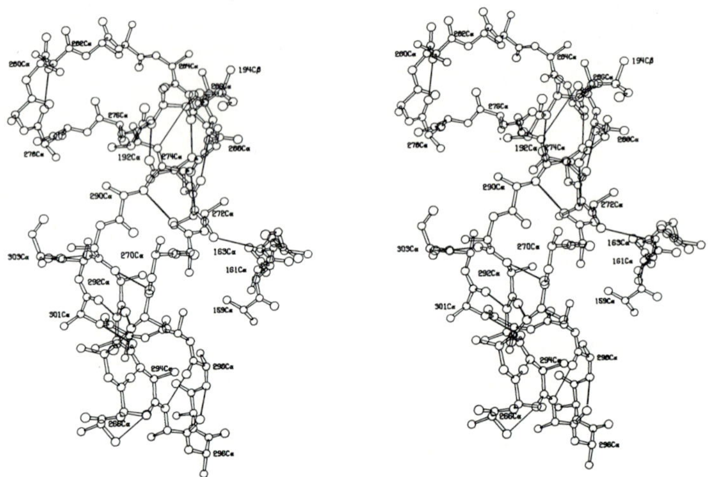

Fig. 3. Stereo drawing of the anti-parallel β pleated sheet and ribbon (βK, βL, βM)

connected to one another by two hydrogen bonds between residues 193 in βG and 289 in βM. One hydrogen bond between residue 162 in βF and 272 in βK connects the parallel sheet and one anti-parallel sheet. The antiparallel sheet and ribbon formed by strands βK, βL and βM is illustrated in Fig. 3. The approach of residues in different parts of the sequence which results from the chain folding can be seen in Fig. 4 where all α-carbon-α-carbon distances of less than 6 Å are indicated.

In addition to main chain-main chain hydrogen bonds, a fair number of main chain-side chain hydrogen bonds have been identified. These will not be considered further here since the sequence data are incomplete.

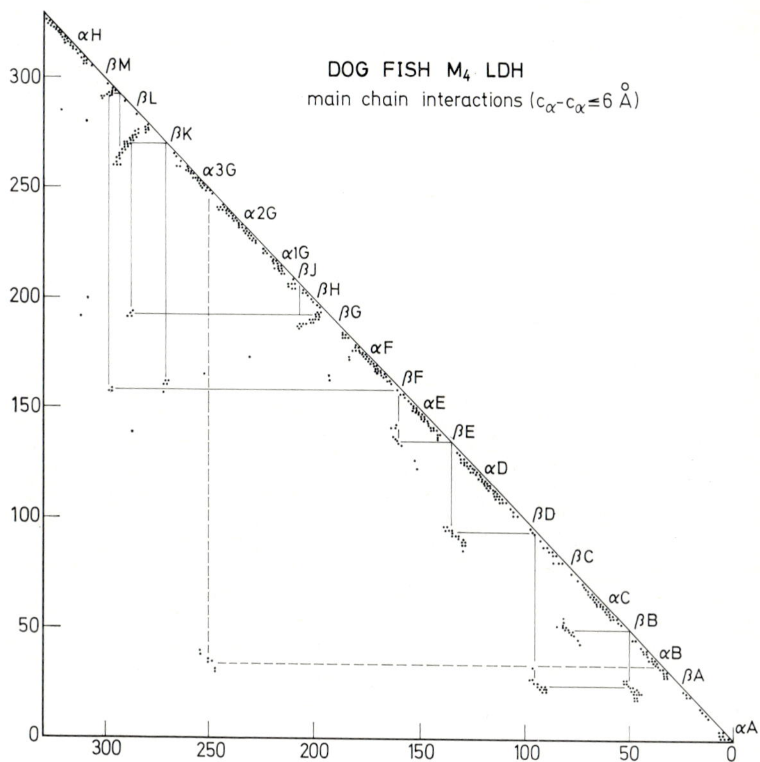

Fig. 4. Amino acids with α-carbons separated by less than 6.0 Å in one subunit

IV. Interactions between Subunits

Fig. 5 shows a stereoview of the LDH tetramer (α-carbons only) looking down the molecular two-fold R axis.

The molecular axes for a molecule (P, Q, R) with center at 1/4, 1/4, –1/4 in the apoenzyme cell (F422, right handed axis system X, Y, Z) can be related to the crystallographic axes by:

$$P = -X + (1/4)\,a$$
$$Q = Z + (1/4)\,c$$
$$R = Y - (1/4)\,b$$

where P, Q, R and X, Y, Z are in Ångstrom and a, b, c are the lattice translations of the F422 cell. For convenience, the subunits have been color coded. The standard subunit is red and is principally in the positive octant of the molecular axes PQR. The

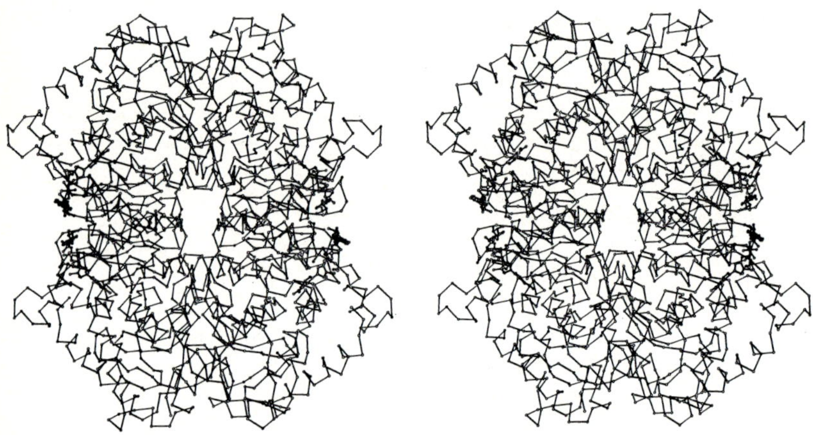

Fig. 5. Stereo view of LDH tetramer down molecular R axis in apo-enzyme conformation with the coenzyme placed in position

Table 2. *Molecular two-fold operations on each, color coded, subunit*

Rotation axis	Interconversions	
P	Red ⇄ Blue	Yellow ⇄ Green
Q	Red ⇄ Yellow	Blue ⇄ Green
R	Red ⇄ Green	Blue ⇄ Yellow

transformation about P gives the blue, about Q the yellow, and about R the green subunit (Table 2).

Interactions between residues in different subunits, with α-carbons separated by less than 6 Å, are shown diagrammatically in Fig. 6. Fig. 6 a) shows the three interfaces red-blue, red-yellow and red-green. The insert (6 b) indicates that a solid figure showing all possible contacts between subunits may be made by folding an

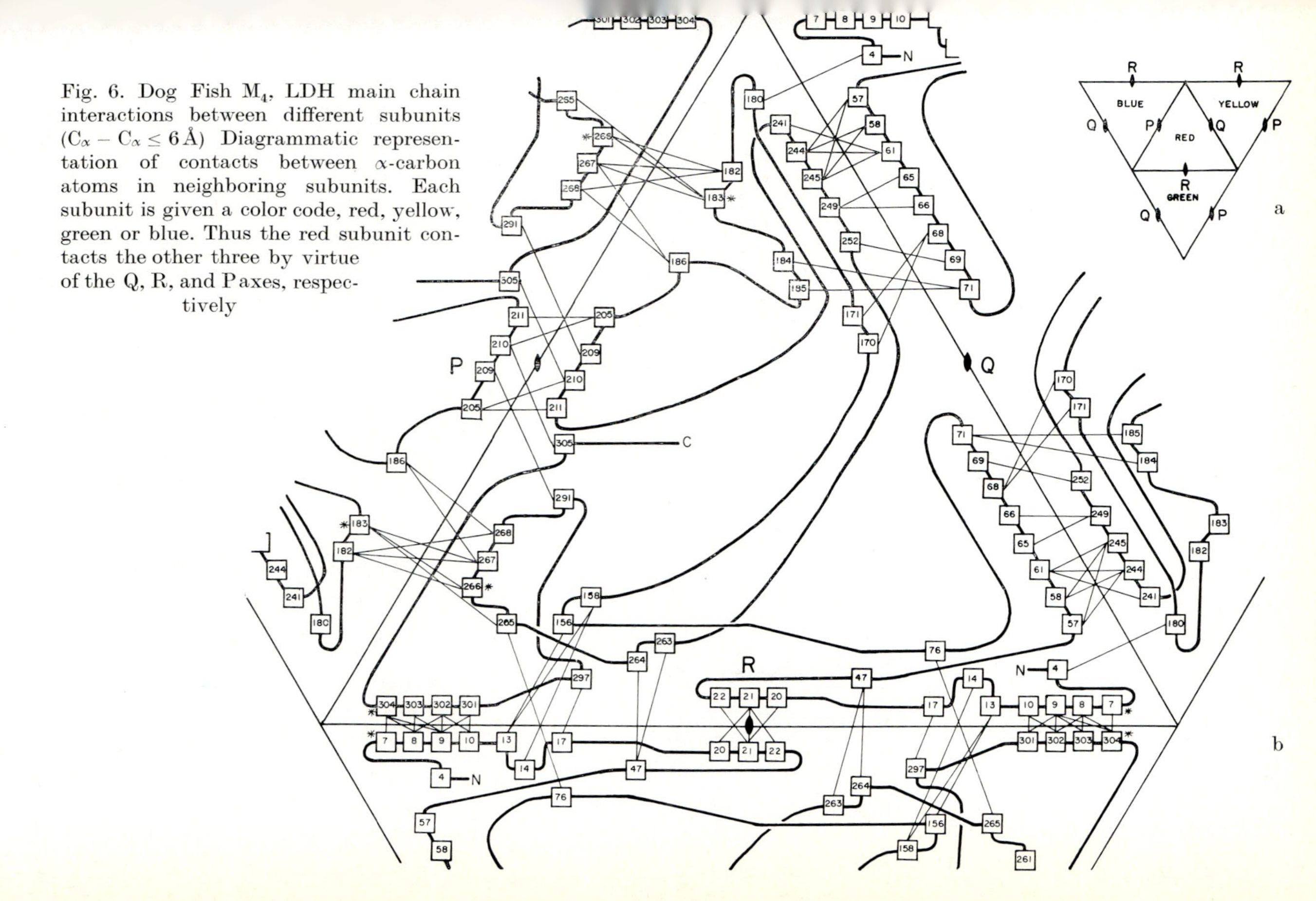

Fig. 6. Dog Fish M_4, LDH main chain interactions between different subunits ($C_\alpha - C_\alpha \leq 6$ Å) Diagrammatic representation of contacts between α-carbon atoms in neighboring subunits. Each subunit is given a color code, red, yellow, green or blue. Thus the red subunit contacts the other three by virtue of the Q, R, and P axes, respectively

extended version of Fig. 6 a). Fig. 6 also shows the different characters of the three interfaces. That formed by the two-fold axis P (red-blue) results primarily from contacts between the two anti-parallel sheet areas (βG red with βM blue and βH red with βH blue). The interface between the red and yellow subunits (formed by rotation about Q) mainly shows contacts between two helices, αC and α3G. Almost all the contacts of the red-green interface (formed by rotation about R) arise from the arm (residues 1 to 22) of one of the subunits. It is interesting to note that the molecular two-fold axis retained in malate dehydrogenase corresponds with the Q axis of LDH. The N terminal arm of LDH is not present in MDH. A tetramer of MDH subunits would thus be energetically less favorable since it would have very few interactions at the red-green interface.

Two of the close contacts between subunits are main chain-main chain hydrogen bonds. These are marked with asterisks in Fig. 6 a) and in Fig. 2. One hydrogen bond, a part of the red-green interface, links residue 7 carbonyl oxygen (in αA) to residue 304 nitrogen (the residue following βM). The second hydrogen bond is part of the red-blue interface and connects residue 183 nitrogen to 266 carbonyl oxygen. This hydrogen bond links the antiparallel sheet βK, βL, βM of the red subunit to a corner preceding the sheet βG, βH, βJ of the blue subunit and vice versa. The contact areas of which these hydrogen bonds form a part are close to one another and are within a stack, 50 Å high, containing the N terminal helix αA of the green subunit, the sheet βG, βH, βJ of the blue subunit, the sheet βG, βH, βJ of the red subunit and the N terminal helix αA of the yellow subunit.

There are also hydrogen bonds between the main chain of one subunit and the side chain of another. Only four of these occur where the side chains are known. Two of them take part in the stack just described. Those are from lysine 6 of the yellow subunit to the carbonyl oxygen of residue 216 of the red subunit and from carbonyl oxygen 209 of the red subunit to a tryptophan (190) of the blue subunit. The other main chain-side chain hydrogen bonds are in the red-green interface and link the arm of one subunit to the other.

V. Complexes of LDH with Coenzyme, Coenzyme Parts and Ternary Complexes — the Active Site

Various binary and ternary complexes of LDH have been studied in order to explore the active center of the enzyme. In addition to investigations of the complexes mentioned in section II, two 2.8 Å resolution difference electron density maps have been computed. These represent studies of adenosine and AMP when bound to the apo-enzyme at pH 7.8. The determination of the binding site for coenzyme at 5 Å resolution, using crystals into which NAD^+ had been diffused, has already been discussed [Adams et al., 1970 (1)].

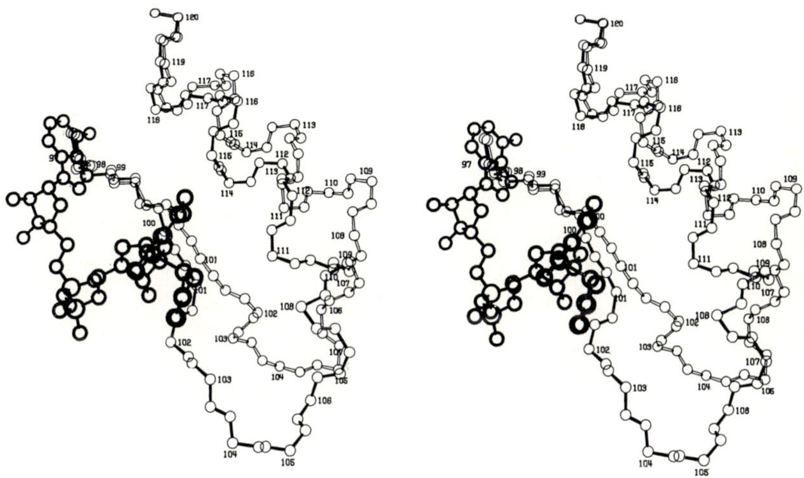

Fig. 7. The two different positions of the loop (βD-αD) in the apo-enzyme and in the ternary complex

The study of the ternary complex LDH-NAD-pyruvate for which phases had to be determined independently has also been described (Smiley et al., 1971; Adams et al., 1971). The major change of conformation seen in the ternary complex is that of the "loop" (residues 97 to 118) which in the ternary complex folds down over the coenzyme and substrate. The two different positions of this loop are shown in Fig. 7.

The comparison of the active center region in the various maps has been aided by a small half silvered mirror device with interchangeable electron density sheets and model cages (after RICHARDS, 1968). A large number of smaller conformation changes have been

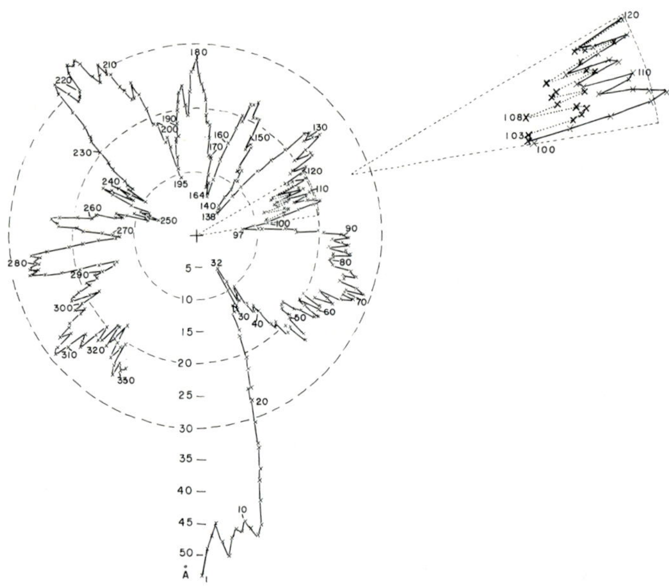

Fig. 8. Dog Fish M_4, LDH. Distances between the active site and C_4. Distance of each successive α-carbon (plotted arround a circle) from a point near the substrate binding site. This shows the residues in various parts of the sequence which approach the active center. Insert shows an enlarged version for the loop region. The position of the loop in the ternary complex is dotted, demonstrating the folding down and approach of the loop to the active center when coenzyme and substrate are bound

observed between the apo-enzyme and the different binary and ternary complexes. Movements of certain atoms of 1 or 2 Å are not uncommon for several residues in the active site cavity. These shifts result in a correspondingly modified position of the AMP portion of the coenzyme relative to the molecular center, when bound alone at pH 7.8 and when part of the ternary complex.

The active center of LDH is large. The coenzyme itself extends more than 15 Å from the adenine ring to the nicotinamide ring in the open conformation it adopts on binding. The different parts of the coenzyme and abortive substrate or inhibitor are held in position by residues far apart in the primary sequence. The approach of the different parts of the main chain to the active center is shown

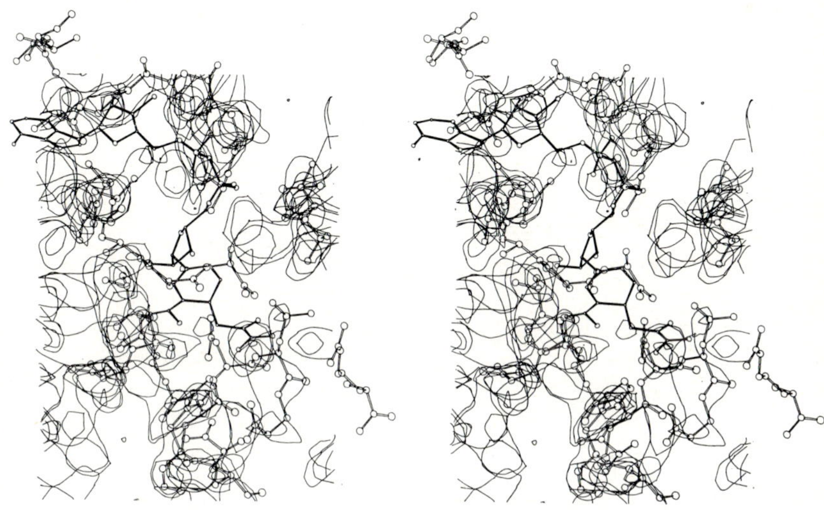

Fig. 9. View of apo-enzyme electron density in the active center region with molecular interpretation superimposed. The NAD-pyruvate adduct has been added. The electron density is represented by a single contour on successive layers one Ångstrom apart, and indicates positions of side chains which have not yet been sequenced

in Fig. 8. A point close to the nicotinamide ring has been chosen as center. The close approach to the backbone at residues 32, 97, and 138 to 140 at the ends of strands βA, βD and βE of the parallel sheet, at residue 195 between βG and βH and at 246 to 252 at the end of helix α2G can be seen. This figure also shows (Fig. 8 insert) that the conformation change in the ternary complex moves the loop residues closer to the active center.

Fig 9. shows a superposition of the model onto the apoenzyme electron density in the active site region. A model of the NAD-pyruvate adduct, at the position it occupies in the abortive ternary complex, is included. Residues which interact with it are discernable in the electron density. A large number of residues are involved in contacts to parts of the coenzyme. The side chain of residue 86 between βC and βD may interact with N 1 of the adenine. Residue 53 and the main chain at 55, at the end of the βD strand of the sheet, form hydrogen bonds to oxygens of the adenine ribose while residue 28 is almost certainly a glycine as anything larger would interfere with the binding of this ribose. The adenine phosphate is close to residue 101 of the folded down loop. Residues 247 at the end of the helix α2G and 97 at the end of βD of the parallel sheet contact the second (nicotinamide) phosphate. One of the sheet hydrogen bonds is broken to enable this contact with 97 to be made in the ternary complex. The main chain at residue 140 (the end of βE of the sheet) hydrogen bonds to the nicotinamide ribose as it does at the loop residue 100.

The position of the abortive substrate or inhibitor can be identified in maps of the ternary complexes. The pyruvate is expected to form a covalent adduct at the C 4 position of the nicotinamide in the abortive ternary complex (EVERSE et al., 1971). The electron density supports this view and also shows it to be hydrogen bonded to histidine 195. WOENCKHAUS and co-workers [WOENCKHAUS et al., 1969; BERGHÄUSER et al., 1971 (1)] have suggested that a histidine is essential to the catalysis. A trough and peak in this area in the difference map between the LDH-NAD$^+$-oxalate and LDH-NAD-pyruvate complexes confirm the substrate position and show that the oxalate is not covalently bound to the nicotinamide. A second contact to the abortive substrate or inhibitor molecule is made by a long residue from the folded down loop at 109. This might be an arginine [BERGHÄUSER and FALDERBAUM, 1971 (2)] or possibly a tyrosine (PFLEIDERER and JECKEL, 1967; JECKEL et al., 1971). A sulfate ion occupies the substrate position in the apo-enzyme structure.

The orientation of the substrate during catalysis becomes clear once his 195 has been recognized as the proton acceptor and the nicotinamide, which receives the hydride in the 4 (A) position, has been oriented. The carboxyl group is apparently held in position by arg 171 and by residue 109. The methyl group of the substrate thus points out into solution. It is noteworthy that phenylpyruvate can act as a substrate (W. J. RAY, J. BURGNER, private communication, 1972). If the substrate had been oriented any other acceptable way the methyl would approach leucine 167 closely. The necessary orientation of the four groups (H, OH, CH_3 and COO^-) of the substrate, when bound to the enzyme, give rise to the specificity for L rather than D-lactate.

The "essential thiol" at 165 is not directly involved in binding of either coenzyme or substrate. The thiol group is 10 to 11 Å from the substrate site and about 6 Å from the histidine ring. The histidine and main chain move towards the abortive substrate in the ternary complex by 1.5 Å. It is interesting to note that the site of mercury substitution is between cysteine 165 and histidine 195, and may thus interfere in the conformational change required in the formation of the ternary complex.

Acknowledgements. This work was supported by NIH grant No. GM 10704 and NSF grant No. GB-29596x. MARVIN HACKERT and MANFRED BUEHNER were recipients of NIH and DFG postdoctoral fellowships, respectively. Discussion of nucleotide conformations was greatly assisted by Drs. D. HUKINS and S. ARNOTT. We appreciate many helpful discussions with Dr. WM. J. RAY, Jr., and Mr. J. BURGNER, and we are grateful for the technical assistance of Mrs. J. BARRETT and Mrs. A. ROSSMANN.

References

ADAMS, M. J., MCPHERSON, A., Jr., ROSSMANN, M. G., SCHEVITZ, R. W., WONACOTT, A. J.: (1) J. molec. Biol. **51**, 31 (1970).

— FORD, G. C., KOEKOEK, R., LENTZ, P. J., Jr., MCPHERSON, A., Jr., ROSSMANN, M. G., SMILEY, I. E., SCHEVITZ, R. W., WONACOTT, A. J.: (2) Nature (Lond.) **227**, 1098 (1970).

— BUEHNER, M., FORD, G. C., HACKERT, M. L., LENTZ, P. J., Jr., MCPHERSON, A., Jr., ROSSMANN, M. G., SCHEVITZ, R. W., SMILEY, I. E.: Cold Spr. Harb. Symp. quant. Biol. **36**, 179 (1971).

BERGHÄUSER, J., FALDERBAUM, I.: (1) Hoppe-Seylers Z. physiol. Chem. **352**, 1189 (1971).

— — WOENCKHAUS, C.: (1) Hoppe-Seylers Z. physiol. Chem. **352**, 52 (1971).

CORNFORTH, J. W., RYBACK, G., POPJAK, G., DONNINGER, G., SCHROEPFER, G., Jr.: Biochem. biophys. Res. Commun. **9**, 371 (1962).

DISABATO, G.: Biochem. biophys. Res. Commun. **33**, 688 (1968).

EVERSE, J., BARNETT, R. E., THORNE, C. J. R., KAPLAN, N. O.: Arch. Biochem. **143**, 444 (1971).

FROMM, H. J.: Biochem. biophys. Acta (Amst.) **52**, 199 (1961).

— J. biol. Chem. **238**, 2938 (1963).

Gutfreund, H., Cantwell, R., McMurray, C. H., Criddle, R. S., Hathaway, G.: Biochem. J. **106**, 683 (1968).

Jeckel, D., Anders, R., Pfleiderer, G.: Hoppe-Seylers Z. physiol. Chem. **352**, 769 (1971).

Kannan, K. K., Liljas, A., Waara, I., Bergsten, F. L., Lovgren, S., Strandberg, B., Bengtsson, U., Carlbom, U., Fridborg, K., Jarup, L., Petef, M.: Cold Spr. Harb. Symp. quant. Biol. **36**, 221 (1971).

Kaplan, N. O., Everse, J., Admiraal, J.: Ann. N.Y. Acad. Sci. **151**, 400 (1968).

Levy, M. R., Vennesland, B.: J. biol. Chem. **228**, 85 (1957).

Lipscomb, W. N., Hartsuck, J. A., Reeke, G. N., Quiocho, F. A., Bethge, P. H., Ludwig, M. L., Steitz, T. A., Muirhead, M., Coppola, J. C.: Brookhaven Symp. Biol. **21**, 24 (1969).

Novoa, N. B., Schwert, G. W.: J. biol. Chem. **236**, 2150 (1961).

Pfleiderer, G., Jeckel, D.: Europ. J. Biochem. **2**, 171 (1967).

Richards, F. M.: J. molec. Biol. **37**, 225 (1968).

Rossmann, M. G., Blow, D. M.: Acta Cryst. **15**, 24 (1962).

Smiley, I. E., Koekoek, R., Adams, M. J., Rossmann, M. G.: J. molec. Biol. **55**, 467 (1971).

Woenckhaus, C., Berghäuser, J., Pfleiderer, G.: Hoppe-Seylers Z. physiol. Chem. **350**, 473 (1969).

Wright, C. S., Alden, R. A., Kraut, J.: Nature (Lond.) **221**, 235 (1969).

Wunch, T., Vesell, E. S., Chen, R. F.: J. biol. Chem. **244**, 6100 (1969).

Discussion

K. Holmes (Heidelberg): What is the nature of the interaction between the two subunits across the boundary involving the Woenckhaus peptide? Is it a continuous β-pleated sheet?

M. Rossmann (Lafayette): The hydrogen bonding pattern within one subunit is shown in Fig. 2, while the main chain—main chain hydrogen bonds between subunits are shown in Fig. 6 (the interactions). The 266 (red) to 183 (blue) main chain interaction creates a small but continuous sheet between the anti-parallel ribbon (βK) and the end of the essential thiol peptide (βG) across the subunit boundary. This sheet is part of the stack of interactions caused by the P 2-fold axis at the center of which is the Woenckhaus peptide. A tryptophan at position 207 in the red subunit (X-ray identification) gives rise to a strong hydrophobic interaction with its symmetry-related trp 207 in the blue subunit.

C. Woenckhaus (Frankfurt): Lactate dehydrogenase is inactivated by 3-(4-bromoacetyl-pyridino) propyl-adenosine-pyrophosphate, while the isomeric 3-bromoacetyl-compound does not affect the enzyme. The inactivation is a consequence of the alkylation of histidine. On the other hand bromopyruvate inactivates the enzyme, again as a consequence of a modification of histidine residues. In this case the holoenzyme reacts faster than the apoenzyme. The reaction can be explained by the shift of a positively charged amino acid residue, when the coenzyme is bound to the enzyme. Dr. Berghäuser found that a guanidinium residue of arginine is essential for the catalytic activity of the enzyme. The substrate pyruvate or lactate is bound and activated by a hydrogen bridge between the imidazole residue of a histidine. The carboxyl group of the substrate is oriented in the active center by a guanidinium moiety:

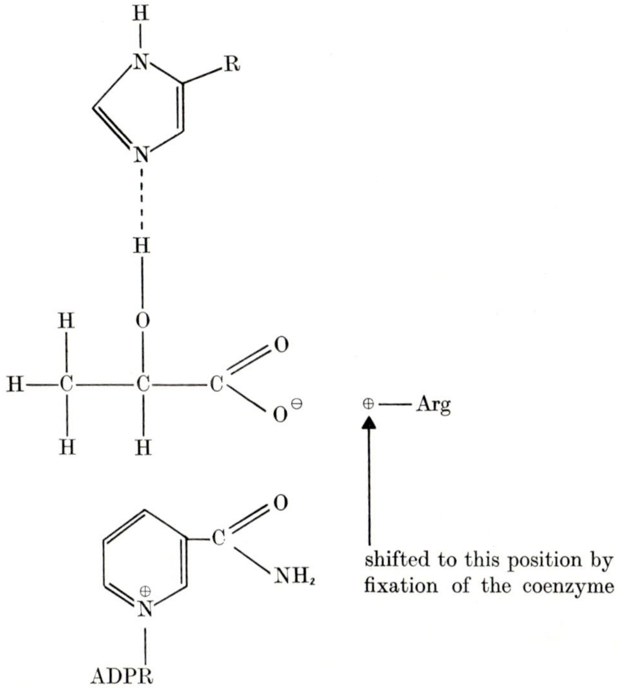

⊕— Arg

shifted to this position by fixation of the coenzyme

ADPR

My question is: Did you observe the migration of a positively charged amino acid residue during NAD^+ fixation by X-ray investigations?

M. Rossmann (Lafayette): Yes, indeed, we did and it is a pleasure to hear of your and Dr. Berghäuser's results. The residue in position 109 is a long residue, most probably an arginine. In the apoenzyme structure it is not easily visible as it interacts with solvent on the outside of the molecule. However, in the LDH-NAD^+-pyruvate complex the "loop" (roughly residues 97 to 118) has folded down and the large residue at 109 comes forward into the active center pocket interacting with the substrate. Residues 109 and arg 171 would appear to be associated with the binding of the substrate carboxyl group. It is noteworthy that this position is occupied by an SO_4^{--}-ion in the apoenzyme structure.

C. Woenckhaus (Frankfurt): We believe that the arginine comes from the positively charged pyridinium part and is shifted towards the position of the substrate, since efficient inactivation by bromopyruvate needs the coenzyme to be present in the incubation mixture.

K. Holmes (Heidelberg)—Chairman's remark—: I should like to summarize this part of the meeting by saying we have seen two very good examples of strong interactions between different subunits or different polypeptide chains. In both cases these were mediated by organized and extensive hydrogen bonding, coupled with close van der Waals contacts of a type more usually thought of as existing within a subunit.

The trypsin inhibitor is obviously one such case and parts of the LDH structure might also show this with the relatively strong interactions.

Cooperative Association

J. ENGEL and D. WINKLMAIR

Biozentrum der Universität Basel, Abteilung für Biophysikalische Chemie, Basel, Switzerland
and Fachhochschule, Fachbereich Physik, München, Germany

With 9 Figures

Introduction

Cooperative processes may be defined in a very general sense as processes in which the individual elementary steps are not independent of each other. According to this definition cooperativity can show up only in complex processes, which consist of a number of elementary steps. Usually the individual sites of reaction are connected in some steric arrangement. In general the equilibrium constant and the rate constants of the elementary steps will depend on the state of the neighbouring segments: reacted or unreacted. Since spatial arrangement is an important prerequisite for cooperativity we can classify cooperative processes according to the arrangement of the segments at which the elementary processes take place. The most simple case which can be treated with relative ease theoretically is a linear arrangement of segments (for a recent review see Ref. [1]). A well known example is the conformational transition of segments in a chain from state A to state B

$$-\mathrm{A}-\mathrm{A}-\mathrm{A}- \rightleftharpoons -\mathrm{B}-\mathrm{B}-\mathrm{B}- . \tag{1}$$

State A can stand for an amino acid residue in a coiled state and B for a residue in an α-helical conformation. The equilibrium and the kinetics of the α-helix $\rightleftharpoons$ coil transition [2, 3] and of the helix I $\rightleftharpoons$ $\rightleftharpoons$ helix II transition of poly-L-proline [4, 5] have been successfully described assuming that the first step (nucleation)

$$-\mathrm{A}-\mathrm{A}-\mathrm{A}- \overset{\sigma\,\mathrm{K}}{\rightleftharpoons} -\mathrm{A}-\mathrm{B}-\mathrm{A}- \tag{2}$$

has a much smaller equilibrium constant than all following steps (propagation), for example

$$-A-A-B- \overset{K_P}{\rightleftharpoons} -A-B-B- . \qquad (3)$$

Experimentally, it was found that σ may be as small as 10^{-5} which means that nucleation is 100,000 times more difficult than propagation. This implies that junctions between different states (A and B) have a very small probability. At high cooperativity (small σ) there is a large tendency to avoid these unfavourable junctions and to form long uninterrupted sequences of identical states. This explains the word cooperativity: a segment in a given state helps its neighbouring segment to assume the same state. As a consequence, even relatively small variations in the external parameters (e.g. temperature, pH, solvent composition) can lead to practically complete conformational transitions. For small chain length and high cooperativity the probability that reacted and unreacted segments occur in the same chain is so small that we are faced with an all-or-nothing process. For example at chain length $n = 6$ such a process may be described by:

$$A-A-A-A-A-A \overset{K}{\rightleftharpoons} B-B-B-B-B-B \qquad (4)$$

with an equilibrium constant

$$K = \sigma K_P^n . \qquad (5)$$

Another example is the binding of small molecules B to a linear array of binding sites A

$$-A-A-A- + n\,B = \begin{matrix} -A-A-A- \\ \dot{B}\quad \dot{B}\quad \dot{B} \end{matrix} . \qquad (6)$$

Here again for simple model systems a high cooperativity was found and the process was successfully described by a model [6] with a nucleation step:

$$-A-A-A- + B \overset{\sigma K_P}{\rightleftharpoons} \begin{matrix} -A-A-A- \\ \dot{B} \end{matrix} \qquad (7)$$

and a propagation step:

$$\begin{matrix} -A-A-A- \\ \dot{B} \qquad\quad \end{matrix} + B \overset{K_P}{\rightleftharpoons} \begin{matrix} -A-A-A- \\ \qquad \dot{B}\quad \dot{B} \end{matrix} . \qquad (8)$$

Similarly PÖRSCHKE and EIGEN [7, 8] assumed an unfavourable nucleation

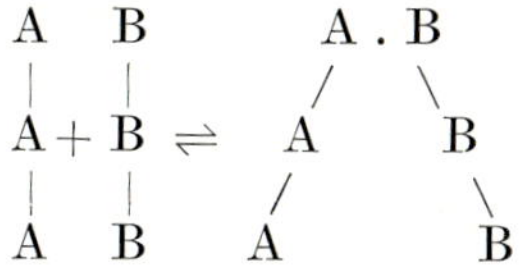

$$\begin{array}{ccc} \mathrm{A} & & \mathrm{B} \\ | & & | \\ \mathrm{A} & + & \mathrm{B} \\ | & & | \\ \mathrm{A} & & \mathrm{B} \end{array} \rightleftharpoons \begin{array}{ccccccc} & & & \mathrm{A}\,.\,\mathrm{B} & & & \\ & & / & & \backslash & & \\ & \mathrm{A} & & & & \mathrm{B} & \\ / & & & & & & \backslash \\ \mathrm{A} & & & & & & \mathrm{B} \end{array} \tag{9}$$

followed by easier propagation steps

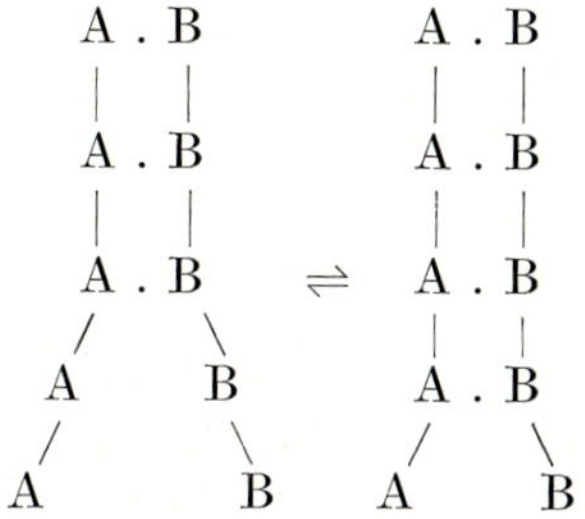

$$\begin{array}{ccccc} & \mathrm{A} & . & \mathrm{B} & \\ & | & & | & \\ & \mathrm{A} & . & \mathrm{B} & \\ & | & & | & \\ & \mathrm{A} & . & \mathrm{B} & \\ & / & & \backslash & \\ \mathrm{A} & & & & \mathrm{B} \\ / & & & & \backslash \\ \mathrm{A} & & & & \mathrm{B} \end{array} \rightleftharpoons \begin{array}{ccccc} & \mathrm{A} & . & \mathrm{B} & \\ & | & & | & \\ & \mathrm{A} & . & \mathrm{B} & \\ & | & & | & \\ & \mathrm{A} & . & \mathrm{B} & \\ & | & & | & \\ & \mathrm{A} & . & \mathrm{B} & \\ & / & & \backslash & \\ \mathrm{A} & & & & \mathrm{B} \end{array} \tag{10}$$

in the formation of a double helix from two coiled stands of oligoriboadenylic acid and oligoribouridylic acid.

It is shown in the following paragraph that a similar model may be applied to the binding of a peptide to a linear array of binding sites at a protein. Another type of linear cooperative process is the linear association of molecules A

$$\mathrm{i\,A} \rightleftharpoons -\mathrm{A}-\mathrm{A}-\mathrm{A}- \tag{11}$$

with an unfavourable nucleation step

$$\mathrm{A} + \mathrm{A} \overset{\sigma\,\mathrm{K_P}}{\rightleftharpoons} \mathrm{A}_2 \tag{12}$$

and propagation steps

$$\mathrm{A}_{i-1} + \mathrm{A} \overset{\mathrm{K_P}}{\rightleftharpoons} \mathrm{A}_i, \quad i \geq 3\,. \tag{13}$$

This simple model will be discussed in connection with association processes in which proteins assemble to form linear structures.

1. Binding of a Chain-Like Molecule to a Linear Array of Binding Sites

Well known examples for this type of process are the binding of a peptide (substrate or inhibitor) to a proteolytic enzyme, the binding of polysaccharides to lysozyme and the binding of polynucleotides to nucleases. In all these examples the specific binding is an essential step of the catalytic process. The ideas developed in the following paragraph were stimulated by the work of BERGER and SCHECHTER [9], who investigated the binding of small peptides to papain. On the basis of systematic variations in the sequence of the peptides they were able to show that the active site of papain may be conveniently pictured by seven sub-binding sites, four of which are located at one side and three at the other side of the active center where the splitting of a substrate chain takes place. It was found that the contributions of the various subsite — segment interactions to the total binding energy are additive. This suggests that the process is of the all-or-nothing type e.g. the peptide binds via all subsites or not at all.

1.1 The Model

Schematically the binding of a randomly coiled peptide chain with n segments: 1′ 2′ 3′ ... i′ ... n′ to a rigid arrangement of a series of n binding sites 1, 2, 3 ... i ... n can be represented by the scheme shown in Fig. 1.

In order to derive the general properties of this complex binding mechanism it is useful to discuss the standard enthalpies ΔH^0 and standard entropies ΔS^0 which are connected with the individual steps. We shall see from very qualitative considerations that the arrangement of sub-binding sites in fixed positions at the enzyme surface together with the fact that the segments which bind are connected in a peptide chain is sufficient to account for a very high positive cooperativity of the process. The standard free energy of any step (binding of segment i′ to subsite i) is given by the relation

$$\Delta G^0_{i,i'} = \Delta H^0_{i,i'} - T\,\Delta S^0_{i,i'}. \tag{14}$$

The corresponding equilibrium constant can be obtained from the relation

$$\Delta G^0_{i,i'} = -RT \ln K_{i,i'}\,. \tag{15}$$

We shall first concentrate on the entropy contribution. The entropy of the step in which E and S are connected the first time via one subsite certainly contains two entropy terms which do not appear in all successive elementary steps. The first one is a loss of translational entropy ΔS^0_{tr} due to the loss of translational freedom of the entire molecule S. In addition a large fraction of conformations in space which otherwise would be accessible to the non-bonded segments of the S-chain (in our scheme 2′ to 6′) are excluded because of the impossibility for the chain to penetrate the enzyme. The loss of entropy which will result may be called ΔS^0_{ex}. Another entropy term ΔS^0_{co} shall be distinguished which contrib-

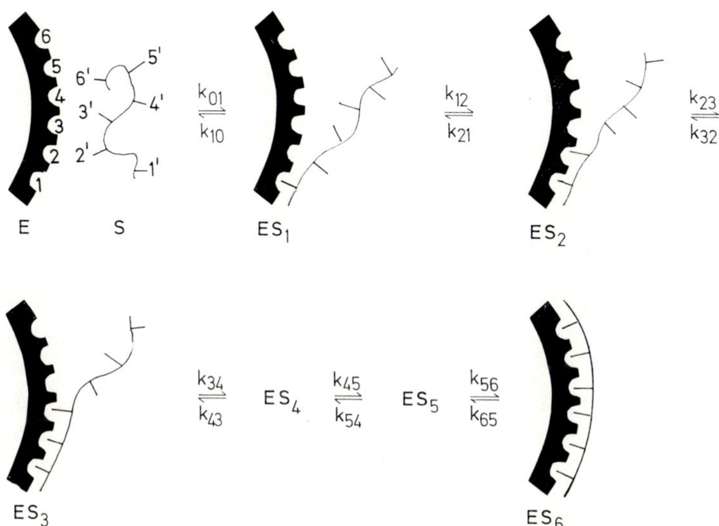

Fig. 1. Schematic representation of the binding of six segments (1′ ... 6′) in a randomly coiled peptide chain S to a rigid arrangement of six binding sites 1 ... 6 on an enzyme E

utes to each binding step, independently whether it is the first or a following. Each residue looses some or all of its conformational freedom when it gets fixed on a binding site. (N-C_α and C_α-CO bonds loose their free rotation.)

For our discussion it is not necessary to know exact values of ΔS^0_{tr}, ΔS^0_{ex} and ΔS^0_{co} since we are not dealing with a specific system. Cooperativity will arise as long as these entropy terms remain negative. It is however instructive to select reasonable numerical values. Translational entropies have been estimated for some association processes between proteins [10, 11]. The values depend on the size and nature of the molecules but also on a number of assumptions concerning the conformational freedom of the partners in the complex, the role of the solvent and other effects. For the dimerisation of insulin values between $\Delta S^0_{tr} = -10$ e.u. [11] and -120 e.u. [10] have been calculated. The latter figure represents an upper limit of negative entropy which is valid when two partners of the molecular weight of insulin (M. W. = 12,000) loose all their freedom in the complex and when solvent effects are excluded. A value of $\Delta S^0_{tr} = -10$ e.u. will be chosen for the following discussion which according to Eq. (14) contributes an unfavourable free energy of $-T\Delta S^0_{tr} \approx 3$ kcal/mol.

LAIKEN and NEMETHY [12, 13] have calculated ΔS^0_{ex} for a model in which the first segment of the chain binds to a point at the surface of a sphere. The ΔS_{ex} value per non-bonded segment of the chain is about -1 e.u. (For the relation between the number of non-bonded segments linearity is a fair approximation [12]). In our example with 5 non-bonded segments after the first binding step $\Delta S_{ex} \approx -5$ e.u. which is equivalent to an unfavourable free energy contribution of $-T\Delta S^0_{ex} \approx 1.5$ kcal/mole. Conformational entropies between $\Delta S^0_{ex} = -5$ and $-8{,}5$ e.u. have been calculated for different amino acid residues [14] and a value of $\Delta S^0_{co} = -5$ e.u. will be used. If the binding step under consideration is a propagation step, its conformational freedom has already been diminished during the first binding of the peptide chain (see ΔS_{ex}). We may account for this by subtracting $\frac{1}{n-1}\Delta S^0_{ex}$ from S^0_{co}. All other entropy contributions of an individual binding step can be summed up in a term ΔS^0_{in} "entropy of interaction". This contains for example the gain of entropy in the surrounding water which is the main driving force for hydrophobic interactions. With all that in mind we can write the entropy balance of a nucleation (1st step) and of a propagation step (i[th] step)

$$\Delta S^0_N = \Delta S^0_{in,1} + \Delta S^0_{co,1} + \Delta S^0_{ex} + \Delta S^0_{tr} \tag{16}$$

$$\Delta S^0_P = \Delta S^0_{in,i} + \Delta S^0_{co,i} - \frac{1}{n-1}\Delta S^0_{ex} \tag{17}$$

$$\Delta S^0_N - \Delta S^0_P = (\Delta S^0_{in,1} - \Delta S^0_{in,i}) + (\Delta S^0_{co,1} - \Delta S^0_{co,i}) + \frac{n}{n-1}\Delta S^0_{ex} + \Delta S^0_{tr}\,. \tag{18}$$

The terms in brackets reflect individual differences between different binding processes (hydrophobic interactions of different strength, hydrogen bonds etc.). They do not reflect characteristic differences between a propagation step and the first step. Even if we assume that these differences are zero we arrive at

$$\Delta S_N^0 - \Delta S_P^0 = \frac{n}{n-1} \Delta S_{ex} + \Delta S_{tr} \approx -16 \text{ e.u.} \quad (19)$$
$$(\text{for } n = 6)$$

Assuming also for simplicity that the enthalpies of the individual steps ΔH_i^0 are the same (for a detailed consideration of a real system this assumption must of course be dropped) one obtains

$$\Delta G_N^0 - \Delta G_P^0 \approx 5 \text{ kcal/mole.}$$

From this a ratio σ between the two equilibrium constants can be estimated

$$\sigma = \frac{K_N}{K_P} = \exp\left(-\frac{\Delta G^0{}_N - \Delta G^0{}_P}{RT}\right) \approx 10^{-4} \text{ m}^{-1} \cdot \quad (20)$$

One has to keep in mind that the above estimate is very approximate and the values are based on entropy estimates which are not very certain. Apart from the quantitative aspects of the calculation it is a notable result that nucleation is much more difficult than propagation for simple entropic reasons. The difficulty of nucleation expressed by a small σ-factor is expected to increase if the binding process takes place in a deep cleft in the enzyme. In this case ΔS_{ex} is expected to exceed the value for the binding at a surface considerably because the non-bonded segments are forced to assume a very restricted conformation when the first residue binds. The maximum value of ΔS_{ex}^0 is $(n-1)\,\Delta S_{co}^0 \approx -25$ e.u. in our example. This leeds to

$$\Delta G_N^0 - \Delta G_P^0 \approx 12 \text{ kcal/mole}$$
$$\text{and } \sigma \approx 10^{-9} \text{ m}^{-1}.$$

These values would be valid for a process in which all the conformational freedom of the peptide chain is destroyed when the first segment binds.

1.2 Equilibrium Properties

According to the scheme in Fig. 1 ES may consist of six different species

$$ES_1 \ldots ES_r \ldots ES_6 \quad (21)$$

with different numbers r of bonds formed. In some experimental techniques like equilibrium dialysis or ultracentrifugation these are not distinguished and we are only interested in the sum of their concentrations

$$c_{ES} = \sum_{r=1}^{n} c_{ES_r} \qquad (n = 6 \text{ in our example}) . \tag{22}$$

For other ends like spectroscopic measurements, enzymatic activity or kinetics (see below), we are however interested to calculate the concentrations c_{ES_r} in which the individual states occur (population analysis).

We can write with c_E = concentration of free E and c_S = concentration of free S

$$\begin{aligned} c_{ES_1} &= 6\, c_E\, c_S\, \sigma\, K_P \\ c_{ES_2} &= 5\, c_E\, c_S\, \sigma\, K_P^2 \\ c_{ES_6} &= 1\, c_E\, c_S\, \sigma\, K_P^6 . \end{aligned} \tag{23}$$

The factors 6, 5, ... 1 account for the number of possibilities to place r bonds in ES. (For calculation of the factors it was assumed that the subsites on E and their corresponding segments in the peptide chain always stay in proper register: non-staggering zipper model. It is further assumed that only uninterrupted sequences of bound and non-bound segments and subsites exist. Then there are 6 possibilities to form the first bond; 5 possibilities to get ES_2 etc.). The total concentration of E is:

$$c_E^0 = c_E + \sum_{r=1}^{n} c_{ES_r} = c_E + c_E c_S \sigma \sum_{r=1}^{n} (n + 1 - r)\, K_P^r . \tag{24}$$

The fraction of a state with r bonds is:

$$\frac{c_{ES_r}}{c^0_E} = fr = \frac{c_S\sigma\, (n + 1 - r)\, K^r{}_P}{1 + c_S\sigma \sum_{r=1}^{n} (n + 1 - r)\, K^r{}_P} . \tag{25}$$

The distribution of all states can be easily calculated for different values of σc_S and K_P. Such a population analysis is shown for $\sigma\, c_S = 10^{-9}$ in Fig. 2 a.

K_P values have been selected at which E is about 90, 50 and 10% saturated. It can be seen that only two states exist in equilibrium: E and ES_6 whereas the concentrations of all intermediates

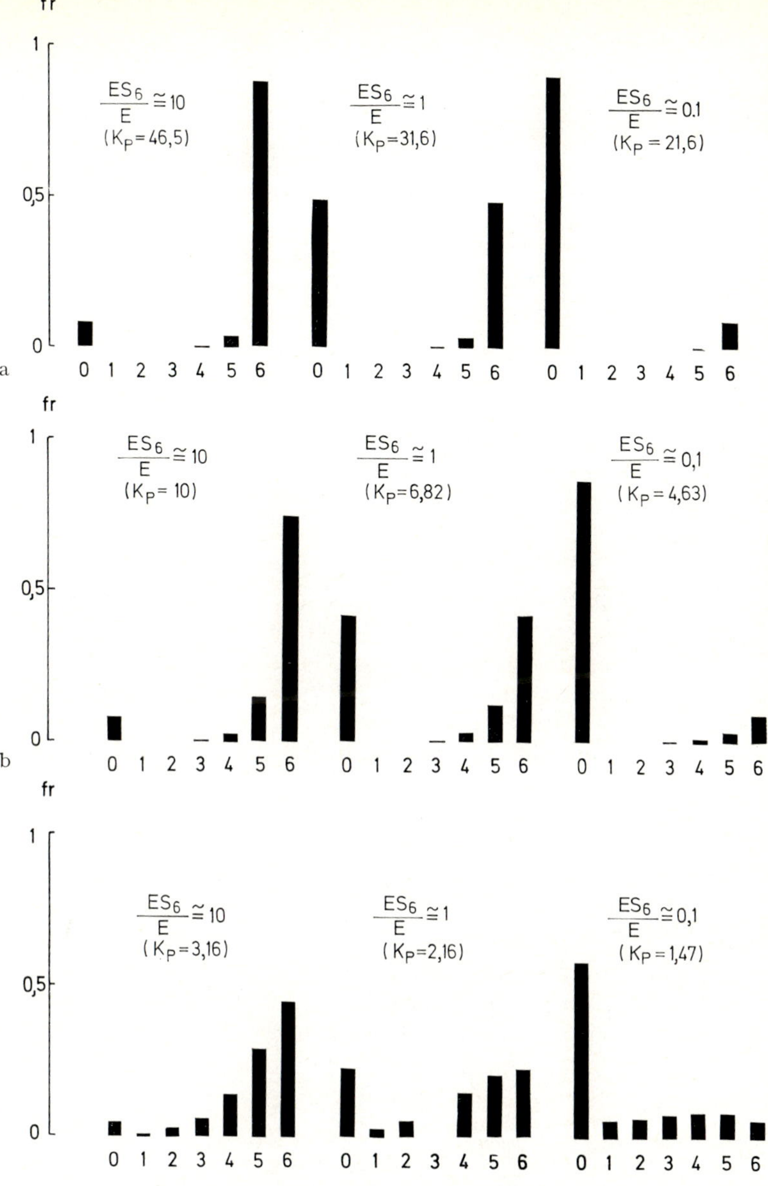

Fig. 2. Population analysis for a) σ $c_S = 10^{-9}$, b) σ $c_S = 10^{-5}$, and c) σ $c_S = 10^{-2}$ (fr = fraction of E with r subsites saturated; r = number of subsites by which E and S are connected)

are negligible. This all-or-nothing behavior reflects the high cooperativity of the process. For $\sigma c_S = 10^{-5}$ the all-or-nothing case is still maintained as a reasonable approximation (Fig. 2 b), but at $\sigma c_S = 10^{-2}$ intermediate states are already heavily populated (Fig. 2c).

It is interesting to note that cooperativity is determined by the product of σ and c_S. The smaller σ and the smaller c_S the more pronounced is the all-or-nothing behavior. The same distribution of states at half saturation is expected for $\sigma = 10^{-5}$ m^{-1} with $c_S = 1$ m and for $\sigma = 1$ m^{-1} with $c_S = 10^{-5}$ m. Since the overall binding constant K of a binding process is usually quite high — under favourable external conditions — and since at half saturation $c_S = 1/K$ the all-or-nothing situation will be very common even for those systems for which σ is not too small.

It is interesting to know the dependence of the degree of saturation Θ on K_P because K_P as any equilibrium constant is sensitive to changes of external conditions. It contains the enthalpy and entropy terms of hydrogen bonding, hydrophobic interactions etc. The σ-parameter is certainly less sensitive because the entropic arguments given in the preceding paragraph do not depend on external conditions (a small temperature dependence is expected due to $T\Delta S$).

Fig. 3 shows the degree of saturation

$$\Theta = \frac{c_{ES}}{c^0_E} \frac{c_S\sigma \sum\limits_{r=1}^{n} (n+1-r) K^r_P}{1 + c_S\sigma \sum\limits_{r=1}^{n} (n+1-r) K^r_P} \tag{26}$$

as a function of $\ln K_P$ for various values of σ c_S. The dependences were also calculated with σ c^0_S as the parameter, where c^0_S is the total concentration of S (dotted lines). For simplicity these calculations were done for the special case $c^0_S = c^0_E$ where

$$\Theta = 1 - \frac{\left(1 + 4 c_S^0\sigma \sum\limits_{r=1}^{n} (n+1-r) K^r_P\right)^{1/2} - 1}{2 c_S^0\sigma \sum\limits_{r=1}^{n} (n+1-r) K^r_P} . \tag{27}$$

The dotted curves are more important for practical processes because it is difficult to maintain a constant c_S under most experimental conditions.

The steepness of the curves first increases with decreasing $\sigma\, c_S$ but it is almost the same for $\sigma\, c_S = 10^{-5}$ and 10^{-9}. This results from the fact that at $\sigma\, c_S \leq 10^{-5}$ the process can be well described by an equilibrium between E and ES_6 with

$$\frac{c_{ES}}{c_E} = K\, c_S \approx \frac{c_{ES_6}}{c_E} = \sigma\, c_S\, K_P^6\ . \tag{28}$$

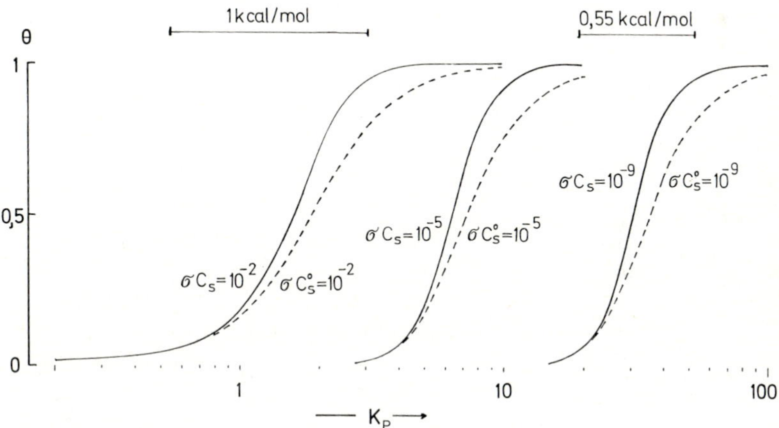

Fig. 3. Dependence of the degree of saturation $\Theta = c_{ES}/c_E{}^0$ on the binding constant of propagation K_P. The parameter is $\sigma\, c_S$ (drawn out lines) or $\sigma\, c_S{}^0$ (dotted lines) where c_S and $c_S{}^0$ are the free and total concentrations of S respectively

For example to change Θ from 0.05 (at K_1) to 0.95 (at K_2) a ratio $K_2/K_1 = 380$ is needed (at constant c_S). According to

$$\frac{K_2}{K_1} = \frac{(K_{P,2})^6}{(K_{P,1})^6} \tag{29}$$

the ratio $K_{P,2}/K_{P,1}$ is 2.7. Expressed in free energy this is a change of about 0.55 kcal/mole subsites which is represented by a bar in Fig. 3. (This is possibly due to the logarithmic scale of the abscissa.) The example illustrates that the equilibrium of the cooperative process is very sensitive to small changes of free energy which occur at all sub-binding sites (for example a weakening of hydrophobic interactions caused by the addition of organic solvent). Also it may be seen from the figure that at small $\sigma\, c_S$ no binding occurs inspite of the fact that the free energy of binding is very favourable in the propagation step. For example at $\sigma\, c_S = 10^{-9}$ very little binding occurs at $K_P = 10$ inspite of the fact that the free energy of the propagation step is by 1.3 kcal/mole subsites in favour of binding.

At this point it should be noted that in our model if ΔS^0_{ex}, K_P and σ are not independent (see Eqs. 16 and 17). K_P increases, when ΔS_{ex} becomes more negative according to

$$K_P = K'_P \exp\left(-\frac{\Delta S_{ex}}{(n-1)T}\right) \tag{30}$$

$$\text{with } K'_P = \exp\left(-\frac{H_{in} - T\Delta S_{in} - T\Delta S_{co}}{RT}\right). \tag{31}$$

This reflects the fact that some (and in the extreme all) of the conformational entropy of a segment which binds in a propagation step has been lost already during nucleation. In order to study the influence of external conditions it is therefore useful to plot Θ versus K'_P instead of K_P. In such plots a decrease in σ caused by a more negative ΔS_{ex} will not shift the saturation curves. Also the release of products after cleavage can be understood on the basis of the estimated entropy values. A hexapeptide which is split into two halves looses $-T\Delta S_{tr} = 3$ kcal/mole of its binding energy due to the gain in translational freedom of the products. The binding constant of the tripeptides is by a factor $(K'_P)^3$ smaller than the binding constant of the hexapeptide.

1.3 Kinetics

Under all-or-nothing conditions ($\sigma\, c_S \leq 10^{-5}$) we can savely assume steady state conditions for intermediates:

$$\frac{dc_{ES_1}}{dt} = \frac{dc_{ES_2}}{dt} = \frac{dc_{ES_3}}{dt} = \frac{dc_{ES_4}}{dt} = \frac{dc_{ES_5}}{dt} = 0\,. \tag{32}$$

The process will follow second order kinetics in one and first order kinetics in the other direction

$$\frac{dc_{ES}}{dt} \approx \frac{dc_{ES_6}}{dt} = \vec{k}\, c_E\, c_S - \overleftarrow{k}\, c_{ES_6} \tag{33}$$

$\vec{k}$ and $\overleftarrow{k}$ are no true rate constants of elementary steps but will in general depend on all elementary rate constants of scheme 1. In order to derive a simple model which describes the main features of the kinetics but does not yield too complicated expressions, we shall introduce some assumptions. First, one can assume for the propagation steps that not only their equilibrium constants but also their rate constants are identical.

$$\begin{aligned} k_{12} = k_{23} = k_{34} = k_{45} = k_{56} = k_P \\ k_{21} = k_{32} = k_{43} = k_{54} = k_{65} = k_P' \, . \end{aligned} \tag{34}$$

Concerning the first step our molecular picture predicts that the formation of the nucleus is much slower than all following steps. The first segment of the chain can only bind when all other residues are in a conformation which does not interfere with the binding. For example 1 1′ can not form when the chain is in a conformation which would somewhere penetrate the enzyme. Therefore, ΔS_{ex} enters the activation entropy of nucleus formation. If the rate constant for dissociation of the nucleus is assumed to be identical with the other dissociation constants — and there is no obvious reason why the dissociation rates should be different — it follows from

$$\frac{k_{01}}{k_{10}} = \sigma \frac{k_P}{k_P'} \tag{35}$$

that

$$k_{01} = \sigma \, k_P .$$

With these assumptions and the steady state condition it is easy to derive [5, 8]

$$\vec{k} = \sigma \, k_P \frac{1 - K_P}{1 - K^n{}_P} K_P^{n-1} \tag{36}$$

$$\overleftarrow{k} = k_P \frac{1}{K_P} \, \frac{1 - K_P}{1 - K^n{}_P} \; . \tag{37}$$

For $K_P \gg 1$ these equations read

$$k \approx \sigma \, k_P \tag{36a}$$

$$\overleftarrow{k} \approx \frac{1}{K^n{}_P} k_P \, . \tag{37a}$$

From Eqs. 36a and 37a we can obtain an estimate of the rate constants of the process. The rate of the elementary step of propagation is certainly very high and may approach that of a diffusion controlled process [15]. A value of $k_P = 10^8$ sec^{-1} can be estimated as a lower limit. With $\sigma = 10^{-4}$ m^{-1} we obtain $\overrightarrow{k} = 10^5$ m^{-1} sec^{-1}. The value of $\overleftarrow{k}$ may be very small even under conditions at which the complex completely dissociates. For example at $c_S^0 = 10^{-5}$ m and $\sigma = 10^{-4}$ the complex is dissociated in spite of the fact that K_P is still as high as 10. When we start off from conditions under which ES_6 is stable and induce the kinetics by performing a sudden change of conditions to $K_P = 10$, the chain of binding sites will start to zipper up. Because of $K_P = 10$ when ES_5 is formed there is a 10 times higher probability that it reacts back to ES_6 than that it dissociates to ES_4. Since this applies to each step the opening of the entire site is very much slowed down. From Eq. 37a $\overleftarrow{k} = 10^2$ sec^{-1} is obtained for our example.

A few words should be said on the situation when the all-or-nothing case does not hold true. Of course then we cannot assume steady state conditions and we have to consider all intermediate species. Since the rate with which these react will depend on the number of open and closed sub-binding sites a spectrum of relaxation times is to be expected. This is perhaps a sensitive experimental tool for recognition of intermediates.

2. Linear Aggregation

Processes in which proteins assemble to form linear structures are very common. Examples are the formation of actin filaments from actin (for a recent review see Ref. [16]) the formation of bacterial flagella from flagellin [17] the assembly of tobacco mosaic virus [18] and the assembly of the tail tube of phage T 4 (for a review see Ref. [19]).

The idea of an unfavourable nucleation step was introduced in order to explain certain experimentally observable features of the aggregation process [20]. For example in the reversible aggregation of actin a critical monomer concentration is observed [21, 22] below which no association takes place. This is clearly shown when the reduced viscosity $(\eta)_{red}$ measured after establishment of equilibrium is plotted versus the total concentration of actin (Fig. 4).

In general the aggregation process can be described by an equilibrium in which aggregates with $i-1$ protomers (A_{i-1}) form i-mers A_i by addition of a monomer A

$$A_{i-1} + A \overset{K}{\rightleftharpoons} A_i \,. \qquad (38)$$

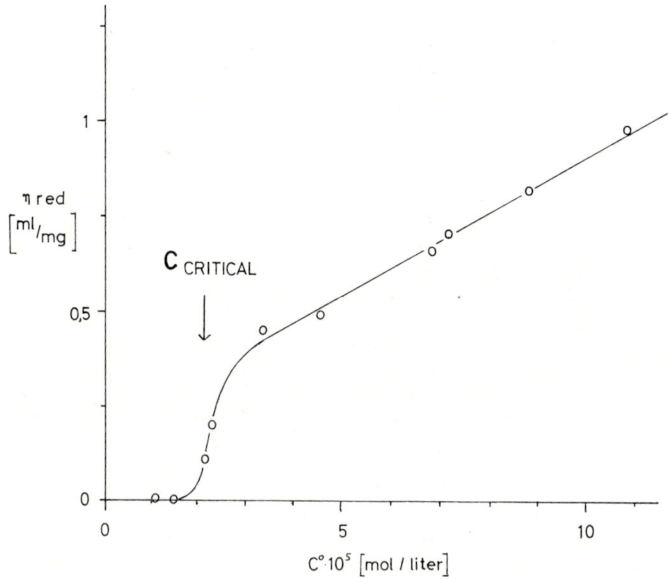

Fig. 4. Dependence of the association of actin on the total actin concentration as followed by reduced viscosity $\eta_{red} = (\eta_{solution} - \eta_{solvent}/c_w\eta_{solvent}$ where c_w is the concentration of actin in mg/ml). The magnesium and calcium concentrations were 2.10^{-4} m and $1.5\ 10^{-4}$ m respectively, pH = 7.5

If the molar concentrations are designated c_i the equilibrium constant of the elementary step (Eq. 38) reads

$$K_i = \frac{c_1}{c_{i-1}\, c_1} (i \gg 2) \cdot \qquad (39)$$

2.1 The Model

A positive cooperativity of the association process is obtained if K_i is smaller for small values of i than for large values of i. The reason of such a behavior may be different for different cases. For instance, for the aggregation of actin the first step is expected to be less probable than all successive ones because within a nucleus the protomers are in contact with one or two neighbours only whereas in the following propagation steps the reacting protomer can form a larger number of contacts (see Fig. 5).

If one bears in mind that the contact between two protein molecules may result in a large binding enthalpy the propagation binding constant K_P is expected to be much higher than the nucleation binding constant.

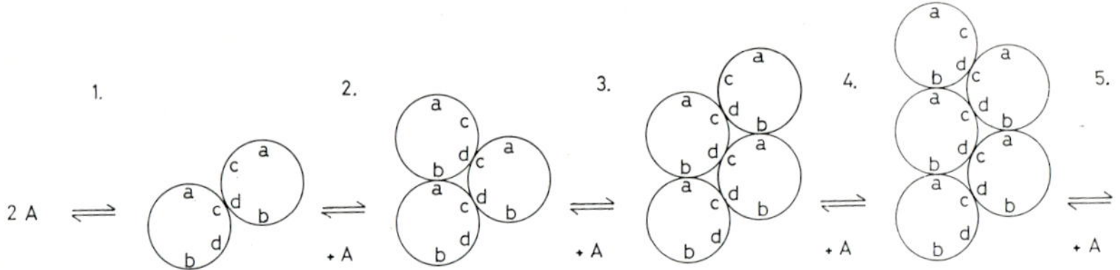

Fig. 5. Schematic drawing of contacts formed in the double helix of actin (designation of contact pattern according to OOSAWA and KASAI, Ref. [20]). In the formation of a dimer (step 1) only a contact of the c d type is formed whereas in all successive steps (2, 3, 4, 5) an additional a b contact is formed

For actin aggregation K_P can be estimated to be about $5 \cdot 10^4$ m^{-1} from the critical concentration, (see section 2.2) under the conditions given in the legend of Fig. 4. This corresponds to a free energy of binding of $\Delta G^0 \approx -6$ kcal/mole which stems from the formation of two contacts (a b and c d in Fig. 5). If one assumes that both contacts contribute equally, nucleation (dimerization step in Fig. 5) would be by about 3 kcal/mole less favourable which would correspond to a σ-factor of about 5.10^{-3}.

Another molecular source of cooperativity can act in addition to the mechanism discussed above or may be the only source of cooperativity in cases in which the number of contacts are equal in nucleation and propagation. If a conformational transition within the protomers is associated with the binding process the dimerization constant should be different from the binding constants of the propagation steps. The situation that all the nucleation difficulty lies in the dimerization step is a very special one. For many practical systems one may expect that also the third and even higher steps are more difficult than propagation. It turns out, however, that even for those cases a mathematical model is realistic in which the total difficulty of nucleus formation is formally ascribed to the dimerization process. This holds true if cooperativity is high. Then the average length of aggregates is very high (see below) and trimers, tetramers etc. can be practically neglected. The very simple model with

$$K_2 = \sigma K_P \text{ and } K_i = K_P \text{ for } i \geqslant 3 \tag{40}$$

(Eqs. 12 and 13) should therefore be of much more general validity than perhaps expected at first glance.

2.2 Equilibrium Treatment

Since the total number of protomers remains constant

$$\sum_{i=1}^{\infty} i\, c_i = c^0 \tag{41}$$

where c^0 is the total concentration of protomers. From Eq. (39) and definitions (40) c_i can be determined

$$c_i = c_1^i \sigma K_P^{i-1} \, . \tag{42}$$

Insertion of c_i into Eq. (41) and introduction of

$$s = K_P c^0 \text{ and } x = s \frac{c_1}{c^0}$$

yields

$$x \left(1 + \sigma \sum_{i=2}^{\infty} i\, x^{i-1}\right) = s$$

or

$$x \left(1 + \sigma x \frac{2 - x}{(1 - x)^2}\right) = s \tag{43}$$

with $0 \leq x < 1$. For a given value of s this equation can be solved and the relative concentrations $\gamma_i = \frac{c_i}{c^0}$ are given by

$$\gamma_1 = \frac{x}{s} \text{ and } \gamma_i = \frac{\sigma}{s} x_i \quad (i \geq 2) \, . \tag{44}$$

The two parameters σ and s completely determine the thermodynamic equlibrium. Since $x < 1$, γ_i decreases monotonically with increasing i. The model (Eq. 40) can therefore not explain a maximum in the length distribution of aggregates as it is found for flagella [17] or actin filaments [16]. Such a maximum is however obtained if a monotonic decrease of the propagation binding constant with increasing i (which may result from the chain length dependence of configurational entropy) is incorporated into the model [23]. It is impossible on the basis of an equilibrium model to explain the formation of linear structures with an exactly defined size from one component only. This problem has been discussed by

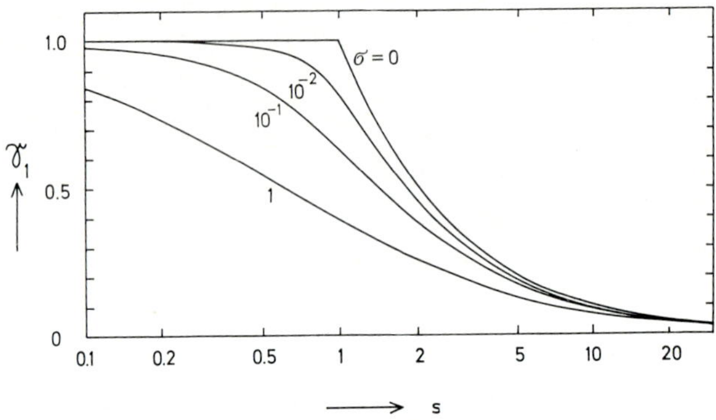

Fig. 6. Plot of the fraction monomers γ_1 versus $s = K_P c^0$ for different cooperativity parameters σ

KELLENBERGER [19]. The phenomenon of a critical concentration is however explained by the simple model defined by Eq. (40).

Fig. 6 shows that for very high cooperativity ($\sigma \to 0$) a critical point occurs at $s = 1$. For $s < 1$, all protomers exist as monomers, whereas for $s > 1$, monomers and aggregates coexist and the fraction of monomers decreases with increasing s. The occurrence of such a critical concentration is well known from the formation of micelles. If $\sigma > 0$ no sharp break is found at $s = 1$ but there is still a threshold concentration below which very little association takes place. The model allows the calculation of other observable quantities, for instance the number and weight averages of the particle size $\overline{N}_n$ and $\overline{N}_w$, which are defined by

$$\overline{N}_n = \frac{\sum_{i=1}^{\infty} i\,\gamma_i}{\sum_{i=1}^{\infty} \gamma_i} \qquad \text{and } \overline{N} = \frac{\sum_{i=1}^{\infty} i^2\,\gamma_i}{\sum_{i=1}^{\infty} i\,\gamma_i} \tag{45}$$

Insertion of Eq. (41) and Eq. (44) yields

$$\overline{N}_n = \frac{s}{x\left(1 + \frac{\sigma x}{1-x}\right)} \tag{46}$$

and

$$\overline{N}_w = 1 + \frac{\sigma}{s} \cdot \frac{2x^2}{(1-x)^3} . \tag{47}$$

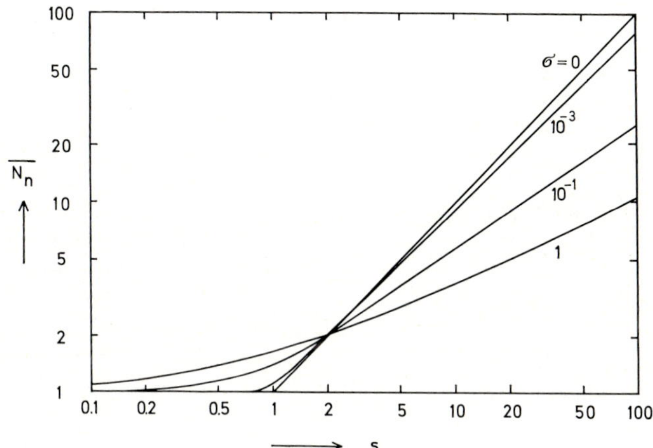

Fig. 7. Plot of the number average particle size $\overline{N}_n$ versus $s = K_P c^0$ for different values of σ

The plots of $\overline{N}_n$ and $\overline{N}_w$ versus s (Figs. 7 and 8) again show the existence of the critical point $s = 1$ for high cooperativity. For $\sigma \ll 1$ the weight average particle size increases very sharply at $s \approx 1$. It is interesting to note that the average particle sizes increase with increasing cooperativity at constant s. The high polydispersity of the system in equilibrium is expressed by the fact that $\overline{N}_w$ always is considerably larger than $\overline{N}_n$ for constant values of σ and s.

Because of $s = K_P c^0$ aggregation can be induced by variation of K_P (which as any binding constant is influenced by external conditions) or via variation of c^0. The concentration at which significant aggregation starts at constant K_P is called critical concentration $c^0_{critical}$. Since $s = 1$, at this point $K_P = 1/c_{critical}$. By this relation $K_P = 5.10^4\ m^{-1}$ has been determined from the data in Fig. 4.

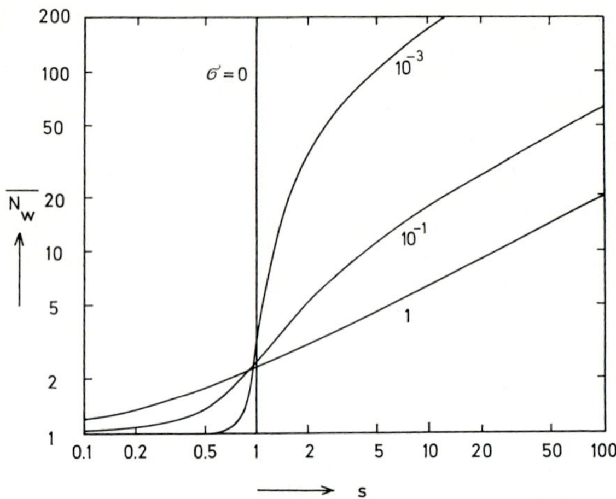

Fig. 8. Plot of the weight average particle size $\overline{N}_w$ versus $s = Kc^0$ for different values of σ

2.3 Kinetics

The description of the kinetics of a cooperatively aggregating system is possible with reasonable effort only for high cooperativity. Since the formation of nuclei then is extremely unfavourable, the growth of an existing chain will occur via subsequent binding of monomers and not by condensation of chains. In this instance it is not necessary to extend the reaction scheme used in the description of the equilibrium (Eqs. 12 and 13). In terms of our simple model the rate constants of the propagation step are denoted by k_P and k_P':

$$A_{i-1} + A \underset{k_P'}{\overset{k_P}{\rightleftharpoons}} A_i \qquad \text{for } i \geq 3 .$$

The suitable choice of the dimerization rate constants now depends on the aggregating system under consideration. Two limiting cases should be distinguished:

Case 1: $$A + A \underset{k_P'}{\overset{\sigma k_P}{\rightleftharpoons}} A_2 \,. \tag{48}$$

Case 1 possibly is a realistic choice as long as conformational changes within the protomers are the reason for the cooperativity. The forward rate constant then should be very small whereas the dissociation rate constant should be about the same for propagation and nucleation.

Case 2: $$A + A \underset{1/\sigma\, k_P'}{\overset{k_P}{\rightleftharpoons}} A_2 \,. \tag{49}$$

Case 2 should be an approach to the situation where the cooperativity results from the fact that the number of steric contacts is different for protomers in a nucleus or in larger aggregates. The bimolecular association rate constant should be affected little by the number of contacts in contrast to the dissociation constant which may be higher by orders of magnitude for the nucleation step if less contacts have to be disrupted.

The differential equations of such a system can be forwarded in a straight forward manner. One of them can be replaced by Eq. (41). For small perturbations of an equilibrium state these equations can be linearized and the mean relaxations times τ^* [24] for different observable quantities (for instance γ_1, $\overline{N}_n$ and $\overline{N}_w$) are obtained. For instance, the mean relaxation time $\tau^*_{\gamma_1}$ which is observed if the change of γ_1 is followed with time after a very small perturbation of an equilibrium state follows from

$$\frac{1}{\tau^*_{\gamma_1}} = k_P' \, L \, \frac{2 - x}{2\,(1 - x)} \qquad \text{(case 1)} \tag{50}$$

and

$$\frac{1}{\tau^*_{\gamma_1}} = \frac{k_P'}{\sigma} L \left(1 + \frac{\sigma\, x}{2\,(1 + x)}\right) \qquad \text{(case 2)} \tag{51}$$

where $L = (1 - x)^3 + \sigma\, x\, (x - 3\, x + 4)$ and x is the solution of Eq. (43) for that value of s which defines the equilibrium state before perturbation.

The functions (50) and (51) are plotted in Figs. 9 a and 9 b respectively.

In case 1 $\tau^*_{\gamma_1}$ exhibits a very sharp maximum at $s = 1$ for high cooperativity (Fig. 9 a) whereas in case 2, $\tau^*_{\gamma_1}$ sharply increases at $s = 1$ and remains nearly constant for $s > 1$ (Fig. 9 b). Similar curves are obtained for the relaxation times of N_n and N_w.

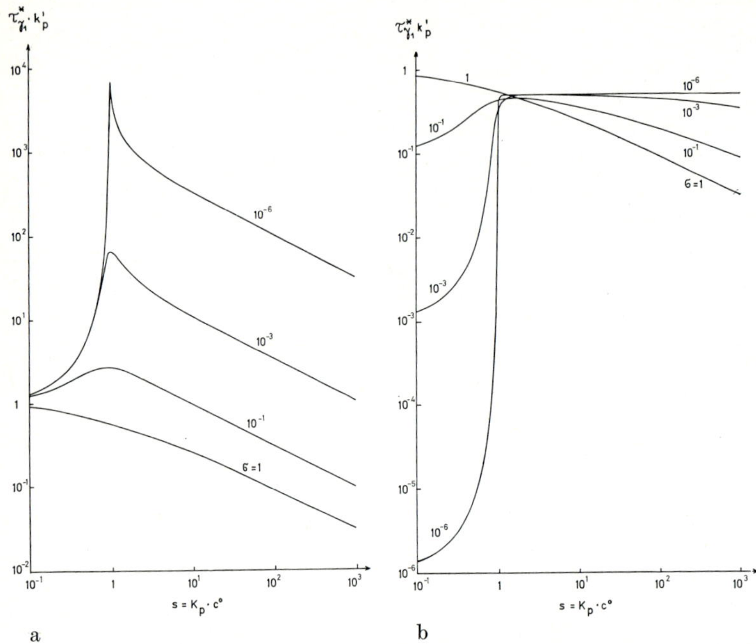

Fig. 9. Plot of the mean relaxation time $\tau^*_{\gamma_1}$ times the elementary rate constant of propagation k_P' versus $s = k_P c^0$ for different σ. a) for case 1 (Eqs. 48 and 50) and b) for case 2 (Eqs. 49 and 51)

Summary

Cooperative associations are binding processes in which the individual binding steps are not independent of each other. For example in the linear association of actin monomers (formation of actin filaments) the first nucleation step is more difficult than the following propagation steps. Another example is the binding of a peptide (substrate or inhibitor) to an enzyme via a number of sub-binding-sites. Additivity of the binding energies of the sub-sites (as has been demonstrated for the binding of inhibitors to papain) indicates very high cooperativity. Therefore the process is of the all-or-nothing type e.g. the peptide binds via all sub-sites or not at all.

The unique thermodynamic and kinetic properties of such systems are discussed in general terms and some experimental systems are presented.

Acknowledgement. The authors wish to express their thanks to the Schweizerische Nationalfonds for financial support of this work and to Drs. G. NEMETHY and G. SCHWARZ for stimulating discussions.

References

1. ENGEL, J., SCHWARZ, G.: Angew. Chem. **82**, 468 (1970); Angew. Chem. Internat. Ed. **9**, 389 (1970).
2. ZIMM, B. H., BRAGG, J. K.: J. chem. Physics **28**, 1246 (1958); **31**, 526 (1959).
3. WINKLMAIR, D.: Ber. Bunsen-Ges. physikal. Chem. **75**, 815 (1971).
4. GANSER, V., ENGEL, J., WINKLMAIR, D., KRAUSE, G.: Biopolymers **9**, 329 (1970).
5. WINKLMAIR, D., ENGEL, J., GANSER, V.: Biopolymers **10**, 721 (1971).
6. SCHWARZ, G.: Europ. J. Biochem. **12**, 442 (1970).
7. EIGEN, M., PÖRSCHKE, D. J. molec. Biol. **53**, 123 (1970).
8. PÖRSCHKE, D., EIGEN, M. J. molec. Biol. **62**, 361 (1971).
9. BERGER, A., SCHECHTER, I.: Phil. Trans. B **257**, 249 (1970).
10. DOTY, P., MYERS, G. E.: Disc. Faraday Soc. **13**, 51 (1953).
11. STEINBERG, I. Z., SCHERAGA, H. A.: J. biol. Chem. **238**, 172 (1963).
12. LAIKEN, N., NEMETHY, G.: J. physiol. Chem. **74**, 4421 (1970).
13. — — Biochemistry **10**, 2101 (1971).
14. GO, M., GO, N., SCHERAGA, H. A.: J. chem. Physics. **52**, 2060 (1970).
15. EIGEN, M., DE MAEYER, L.: In: Technique of organic chemistry, 2nd ed., vol. 8, no. 2, (WEISSBERGER, A., Ed.) (1963).
16. OOSAWA, F., KASAI, M.: In: Subunits in biological systems, Part A. (TIMASHEFF, S. N., FASMAN, G. D., Eds.). New York: Marcel Dekker 1971.
17. ASAKURA, S.: In: Adv. in Biophys., Vol. 1, p. 99. (KOTANI, M., Ed.). Manchester: Univ. Park Press 1970.
18. DURHAM, A. C. H., KLUG, A.: Nature (Lond.) New Biology **229**, 42 (1971).
19. KELLENBERGER, E.: In: Polymerization reactions in biological systems. Ciba Foundation Publication. Amsterdam: Elsevier 1972.
20. OOSAWA, F., KASAI, M.: J. molec. Biol. **4**, 10 (1962).
21. — ASAKURA, S., HOTTA, K., IMAI, N., OOI, T.: J. polymer. Sci. **37**, 323 (1959).
22. KASAI, M., KAWASHIMA, H., OOSAWA, F.: J. polymer. Sci. **44**, 51 (1960).
23. WINKLMAIR, D.: Arch. Biochem. **147**, 509 (1971).
24. SCHWARZ, G.: Rev. modern Physics **40**, 206 (1968).

Analysis of the Structure of Complex Proteins by Electron Microscopy

N. M. Green

National Institute for Medical Research, Mill Hill, London NW7 1AA, Great Britain

With 10 Figures

Most multisubunit proteins and simple viruses are self-assembling systems, in that their structure is completely specified by that of their subunits. On thermodynamic grounds we would expect such limited structures to show point group symmetry, since this ensures that the maximum number of links of a given type [1], in this case of the most stable type, are formed. However, since proteins are not completely rigid structures and can undergo mutual distortions when they interact it would not be surprising if they sometimes deviate from the simple rules. Sufficient structural information is now available to provide an approximate estimate of the frequency of different types of symmetry and of deviations from them. It is of particular interest to have some idea of the frequency of such deviations since this will tell us how effectively we can apply arguments based on symmetry in our attempts to determine the structure of multisubunit proteins.

I will commence with an empirical approach to a description of the point groups which owes much to previous accounts by other workers [2, 3, 4] and differs in emphasis rather than in essentials. It has the merit of being comprehensive and of providing an analogy with possible steps in the evolution of oligomeric proteins. I will consider how far known structures can be fitted into this classification, and then after a brief digression on the experimental problems of assigning symmetry from the results of electron microscopy I will conclude with an application of these principles and methods to an analysis of the structure of a complex enzyme.

In our model for the evolution of complex proteins we will start from an asymmetric unit which does not form any stable bonds and consider the consequences of mutations which lead to interacting sites. In a far sighted paper in 1953, PAULING [5] pointed out that the most general result of introducing a single 'site A'—'site B' interaction was an infinite helix in which the only symmetry element was a screw axis. He went on to discuss the possible functional significance of a variety of other helices and closed structures. If the pitch of the helix is very low then steric interference will halt growth after one turn. If the pitch is zero and the bond angles are convenient than a closed ring will result with n members related to each other by an n-fold rotation axis. If the strain involved is not great then these closed rings will always be favoured over infinite helices since their entropy is higher. Such a structure has cyclic symmetry, designated C_n, the value of n depending on the angle between the normals to the A and B sites. If these normals are approximately parallel and if A and B are close together than a pair of molecules can be joined by two A-B links and the structure can grow no further. We thus have a symmetric dimer, (C_2), in which the symmetry axis passes through the intersubunit bond, which is effectively a double bond. MONOD, WYMAN and CHANGEUX [2] referred to this type of bond as isologous, as opposed to the heterologous (or asymmetric bond). It is unnecessary to introduce these additional technical terms and we will retain the more basic terms symmetric and asymmetric. We can express the above argument in a slightly different way. If we have a population of identical non-interacting subunits and introduce bonding sites the first links may be either symmetric or asymmetric. Asymmetric bonds lead to helices while a symmetric bond will always give a limited dimer. Stable dimeric proteins are therefore most probably symmetrical dimers.

Using a single intersubunit link we can thus generate only two types of symmetrical structure a simple helix and a cyclic ring. Before going on to see what happens on introducing the possibility of two more different intersubunit bonds, we will briefly summarize the description of symmetrical structures. Any symmetrical structure can be generated from the basic unit of that structure by application of an appropriate group of symmetry operators (rotations, reflections and translations). Fortunately, proteins are asymmetric and no mirror images exist, so that we can omit reflections from the groups to be considered. The resulting structures may be limited and symmetrical about a point or they may be infinite in one, two or three dimensions. The appropriate groups of symmetry operators which generate these structures are called point groups, line groups, plane groups and space groups. Here we shall consider only the first two, of which we have already met the simplest members: 1. the cyclic structures generated by the action of a single rotational operator and 2. the simple helix. In these two cases the group contains a single operator only, apart from that of identity.

Since these structures utilise all the potential bonding sites they will not grow further without introducing new classes of stable bond (i.e. a second mutation would be required). The consequences of introducing a new symmetric (S) or asymmetric (A) bond can be visualised with the aid of Table 1. If both are asymmetric each subunit in the original helix will form part of a set of new helices intersecting at an arbitrary angle with the first. Each member of the new helix will act as a nucleus for further helices so that in most cases the result will be a tangle and it will be impossible to form the maximum number of stable bonds. Such tangles have presumably been eliminated long ago by natural selection. Only if the original asymmetrically linked structure was cyclic and the new helix axis coincides with the original rotation axis will a unique structure result, which can be described either as a stack of rings or as a multiple helix (symmetry s_r [6]).

Limited structures with point group symmetry arise only if at least one of the bonds is symmetrical. For example, we can take two cyclic structures formed from 'A' bonds and join them by 'S' bonds parallel to the symmetry axis to form a double ring with two faces. Such a structure is called dihedral and its symmetry is designated D_n. In addition to the original n-fold axis it will have n-twofold axes at right angles to the main axis and passing through the new S bonds. All the other structures will be collections of infinite helices, planar arrays or crystals. If both bonds are symmetric the most general result will be an antiparallel double helix (symmetry s2), with two fold axes at right angles to the helix axis. As with the simple helix, the pitch may be so low that no more than one turn can form, so that ring structures may result. These

Table 1. *Legend see opposite page*

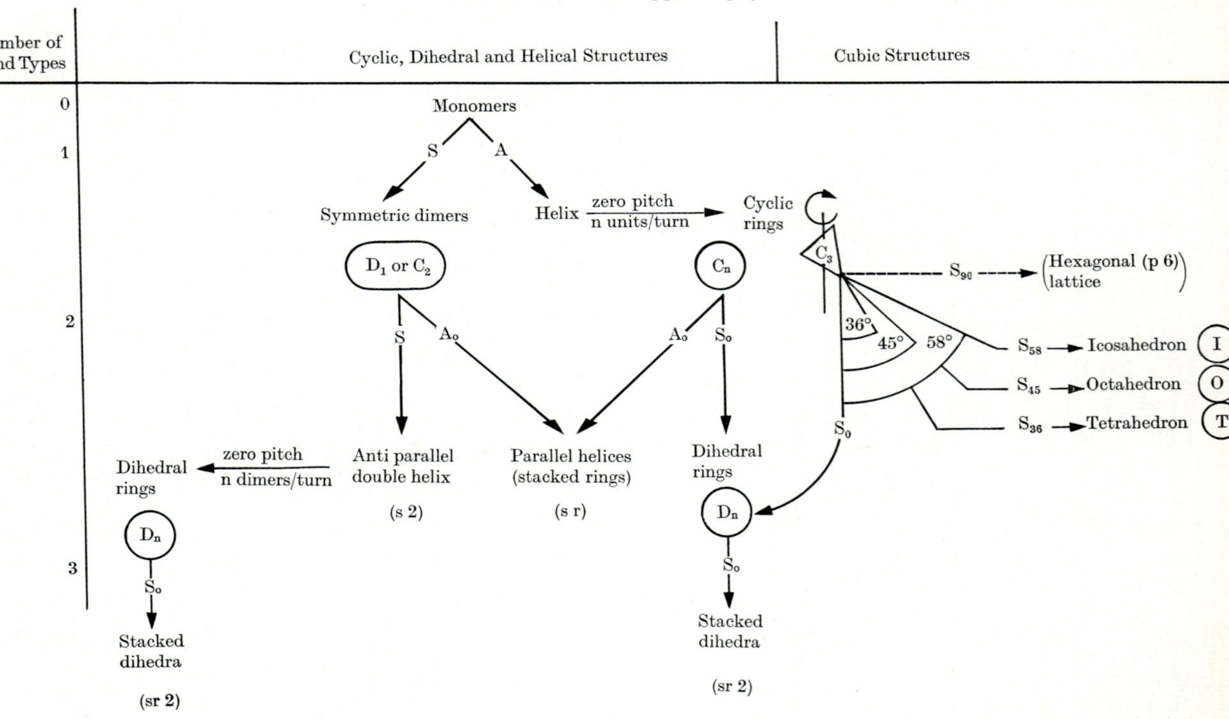

structures are, however, dihedral rather than simply cyclic since the twofold axes of the original dimers rotate one face into the other. Although the symmetry is the same as that of the A, S linked dihedral structures, the bonding pattern n is different. It is also possible to introduce an 'A' bond into this structure, thus combining the bonding patterns of the two types of dihedral structure so that the subunits are joined by one 'A' bond and two 'S' bonds. In general, however, introduction of a third bond type leads to infinite growth in one, two or three dimensions.

There is only one further class of structures with rotational point group symmetry, the cubic class, which can most simply be described as originating from cyclic trimers by joining them through a symmetrical bond at a specific angle to the threefold axis. Since three such bonds will be formed by each trimer, a planar network with hexagonal (p 6) symmetry will be generated if the bonds are at right angles to the threefold axis. Such a network can be curved in one direction, with only slight distortion of the bonding pattern, to give a cylinder with a hexagonal net of subunits on its surface, but totally closed structures require more drastic modification of the bonding pattern [8]. If the new symmetric bonds make suitable angles of less than 90° to the threefold axis then closed structures

---

Table 1. *Hierarchy of structures which result when asymmetric monomers are united by different combinations of symmetric (S) and asymmetric (A) bonds. Only closed and helical structures are shown. They arise when the angle between the second type of bond and the symmetry axis generated by the first is that indicated by the subscript to A or S. (e.g. A_0 implies that the second bond is asymmetric and parallel to the existing symmetry axis). In the absence of a subscript this angle is arbitrary. Bond angles other than those indicated lead to infinite planar structures or more usually to three-dimensional tangles. The cubic structures originate by uniting cyclic trimers, as shown, through appropriately angled symmetric bonds between subunits. The Schoenflies symbols for the point-group symmetry of the closed structures are encircled. Apart from 'p 6' (hexagonal lattice), the symbols in parentheses are those introduced by* KLUG, CRICK *and* WYCKOFF [6] *to describe helical structures. Many of the structures can be formed in more than one way (e.g. an octahedral structure can be assembled from dimers or tetramers as well as, as indicated, from trimers) but in general only one has been shown. Two modes are shown for dihedral structures since different bond types are involved. The sequence of events on the left of the table is illustrated by models in Fig. 1*

will be generated with symmetry characteristic of one of the three cubic point groups. If the angle is about 58° the original hexagonal net will become pentagonal and the product will be a dodecahedron (icosahedral symmetry, I). An angle of 45° will give a cube (octahedral symmetry, O) while an angle of 36° will unite the trimers in a triangular pattern (tetrahedral symmetry, T).

This completes the enumeration of all the types of closed symmetrical structures that can be formed from identical subunits, identically linked.

The appreciation of the relationship between bonding patterns and structure is facilitated by use of suitable asymmetric building blocks. Plastic monkeys are very suitable for this purpose [7] and an illustration of their capabilities is given in Fig. 1. We will now consider some general properties of these closed structures and then see how far actual proteins assemble themselves according to these simple, thermodynamically based rules.

1. Each structure can accommodate a specific number of subunits in equivalent positions (Table 2). Therefore, if we know the number of subunits in a protein, the symmetry possibilities are limited. The number of subunits can be increased beyond the nominal values by bending the rules slightly and putting subunits in quasi-equivalent positions. This has been done in an ingenious and plausible manner by Caspar and Klug [8] in their interpretation of the structure of spherical viruses which possess far more than the nominal maximum of 60 identical subunits.

2. The actual quaternary structure generated by a particular subunit is determined only by the number of stable bonds it can form and by the angles between them.

3. Protein structures held together by more than one different bond type will in general show stable intermediates when they dissociate, since conditions which break one type of bond will not necessarily affect the others. However, it is unlikely that more than one well defined intermediate stage will be observed since structures with more than two types of bonding will almost always be infinite. Observation of more than one stable intermediate suggests more than one type of subunit in the structure, or some allosteric change following breakage of one bond of a given type which leads to a stabilization of neighbouring bonds of the same type.

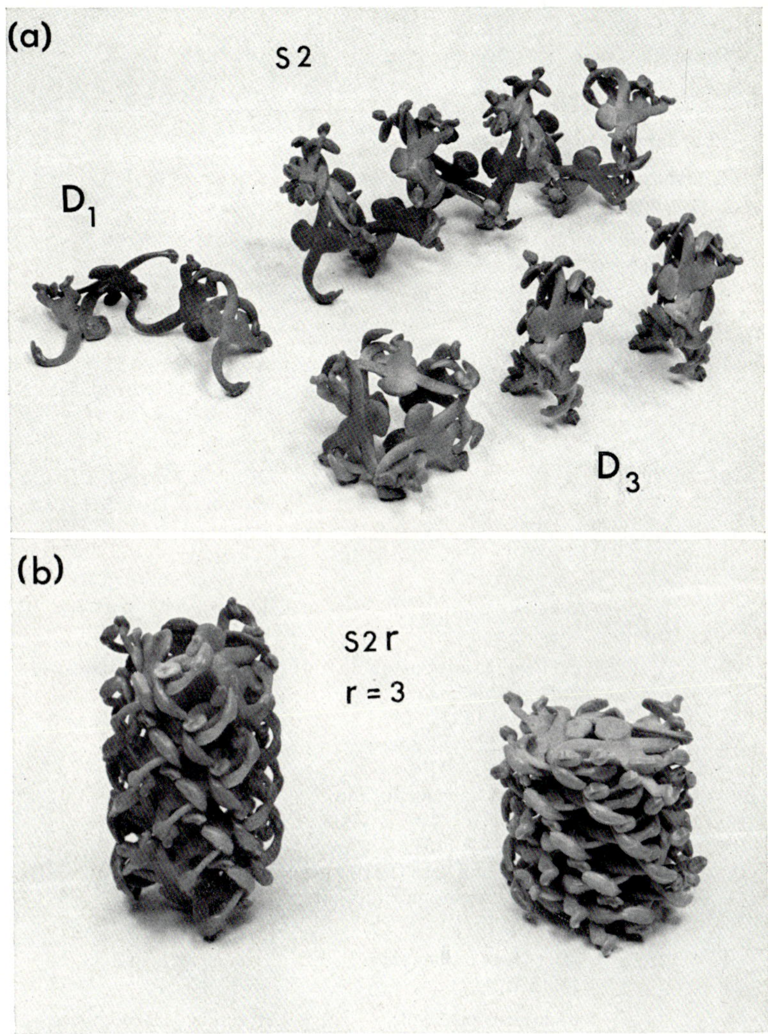

Fig. 1 a and b. a) Illustration of the formation of an anti parallel double helix (symmetry s 2) from a symmetrical dimer (D_1). When the pitch is lowered sufficiently it dissociates into hexamers (D_3) without change in the bonding pattern. b) An interlocking stack of such hexamers can be induced to undergo a cooperative transition, reminiscent of the contraction of the tail sheath of a bacteriophage

Table 2. *Symmetry Classes*

Symmetry Class	Cyclic C_n	Dihedral D_n
Rotational axes	n	n, 2
Equivalent positions =protomer number	n	2 n
Intersubunit bond types Symmetric (S) or Asymmetric (A)	A	A + S, $S_1 + S_2$ or A + $S_1 + S_2$
Examples with n =		
1	Monomers	Most dimers. e.g. Alcohol dehydrogenase (X) [9] *β*-lactoglobulin (X) [10]
2		Aldolase (X [19], EM [20]) Avidin (X [20], EM [22]) Phosphorylase (X, EM) [23], [24] Tryptophanase (EM) [25] Lactic Dehydrogenase (X) [26]
3	ATCase (catalytic subunit)[a] [11] Adenovirus hexon (X, EM)[a] [12] Fatty acid synthetase (EM) [13], [39] Influenza haemagglutinin (EM) [14] Insulin (X)[b] [15] Transcarboxylase (EM)[b] [16] Ketodeoxyphosphogluconate aldolase [67]	Aspartic transcarbamylase (X) [27] Glutamic dehydrogenase (EM) [28]
4	(Pyruvate carboxylase ? D_2 (EM) [17]	(Haemerythrin ? C_8) [29]
5	(Celovirus penton ?) (EM)[a] [18]	Arginine decarboxylase (EM) [30]
6	—	Glutamine synthetase (EM) [31] Phycocyanin (EM) [32]
9	—	Limulus haemagglutinin (EM) [33]

of Oligomeric Proteins

Cubic		
T	O	I
2, 3	4, 3, 2	5, 3, 2
12	24	60
A + S	A + S	A + S
Aspartic decarboxylase (EM) [34]	Ferritin (X, EM) [35], [36], [37] E. coli lipoate[a] transsuccinylase (X, EM) [38]	Mammalian lipoate Transacetylase (EM)[a] [39] Spherical viruses (EM)[b] [8]

[a] Breakdown product of larger molecule.
[b] Protomer contains two or more similar chains.

This table is not exhaustive and references to further examples will be found in other reviews [44, 45]. Dimers can be either symmetric or asymmetric. Asymmetric dimers which do not polymerize are likely to be rare and would only be detectable by X-ray crystallography. The only known example is the insulin subunit [15] (M. W. 12,000). Symmetric dimers have been classified as D_1, rather than C_2, since the symmetric bond is characteristic of dihedral rather than of cyclic structures.

Trimers and other odd numbered oligomers are rare [2, 4], but by no means non – existent. Most of the currently acceptable examples are listed and most of them show unusual features. Insulin and transcarboxylase have six subunits but they are organized with alternate chains in non equivalent positions. The fatty acid synthetase is complex, each protomer containing seven different enzymes. The adenovirus hexon and the catalytic subunit of aspartic transcarbamylase are breakdown products of larger structures. The hexon shows three-fold symmetry, rather than the six-fold symmetry required by the theory of Caspar and Klug [8]. It seems likely that in the adenovirus the triangular faces have p 3 rather than p 6 symmetry and that the inter-hexon bonds along the edges (i.e. between faces) are different from those within the face. This would imply that the departure from equivalent bonding is greater than that envisaged by Caspar and Klug.

Legend to Table 2

All tetramers that have been examined in detail show dihedral symmetry. Although pyruvate carboxylase shows superficial C_4 symmetry in the electron microscope, it is impossible to distinguish this from the D_2 class since the orientation of the individual subunits has not been determined (Fig. 2 a).

There are no well established pentamers. The penton of the celovirus shows five-fold symmetry in the electron microscope, but the number of peptide chains is not known. Hexamers like tetramers all show dihedral symmetry. No octamers have been examined. Decamers and dodecamers all show five or six-fold symmetry and are therefore dihedral structures. It was suggested that the haemagglutinin of Limulus (18 subunits) was a hexamer of trimers [33]. This conflicts both with the nine-fold symmetry seen in the electron micrographs and with the general symmetry considerations outlined here. The more probable D_9 structure is now supported by the original authors (personal communications from Dr. J. J. Marchalonis).

The apparent cubic symmetry of ferritin presented a problem as long as it was thought to contain 20 subunits [40]. The more recent estimate of 24 subunits [36, 37] would fit an octahedral structure, but there is still some conflict with X-ray crystallographic evidence for five-fold symmetry [35].

The only well characterized oligomeric proteins which cannot be accounted for as self assembling structures with point group symmetry are the chlorocruorin [41] and haemocyanin [42] respiratory carriers with 60 to 120 subunits which appear as stacked dihedral structures in the electron microscope. They appear to be potentially infinite structures whose growth has been limited by some special mechanism involving built in strain or extra peptide components, such as has been suggested in models for assembly of complex viruses [43].

The symmetry of a multisubunit structure can be unequivocally determined only by X-ray crystallographic analysis. Electron microscopy provides information about the number of subunits and the overall geometry of the structure, but does not define the orientation of subunits within the structure (e.g. Fig. 2). Assignment of symmetry from electron micrographs is therefore usually the result of an educated guess, making use of the general considerations outlined above and extraneous experimental information concerning numbers and kinds of subunits, binding sites and peptide chains in the molecule. We shall consider later additional information which can be obtained from the electron micrographs by specific labelling of subunits with suitable reagents.

Classification of Oligomers by Their Symmetry

Examples of the different types of symmetry which have been observed are summarized in Table 2, together with comments about specific points of interest. The main conclusions that can be drawn are that representatives of most of the symmetry classes with sixfold rotation axes or lower are now known, though simple cyclic structures are rare, and that most proteins belong to the class appropriate to the number of protomers that they contain. There are a few examples 'b' of proteins that contain double the expected number of subunits. When these have been examined closely it has usually been possible to establish chemical or conformational differences between apparently identical subunits, so that the asymmetric unit is a pair of subunits. Insulin [15] is a clear example of this, alternate subunits (M. W. 6000) having slightly different conformations. It was at one time thought that symmetry of phosphorylase [40], ferritin [40], and tryptophanase [25] might be similarly anomalous but more recent evidence [36, 37, 65, 66] has shown them to be normal. Authentic examples of this anomaly are provided by (1) the polar double disc formed by subunits of TMV [68] and (2) by the shells of icosahedral viruses. The reverse possibility of a protein which contains fewer peptide chains than would be expected from its morphology is less common, though there are a few examples which will be mentioned later. It is reasonable to conclude from the evidence summarized in Table 1 that the subunit number, if accurately known, is often a reliable guide to the symmetry of oligomeric proteins.

The predominance of symmetrically bonded structures is striking. Reasons for this have been discussed elsewhere [2] but I will add one further plausible rationalisation. Most intersubunit bonds contain a hydrophobic nucleus [46] and probably have their origin in mutations which lead to the appearance of a hydrophobic residue on the surface. These will tend to produce symmetrical dimers since a hydrophobic residue will tend to interact symmetrically with the homologous residue on another molecule. This symmetry will be reinforced, since for thermodynamic reasons the most stable combinations of hydrogen bonds and other supplementary interactions will tend to be arranged in pairs about a twofold axis passing through the hydrophobic nucleus. (By doubling the interactions in this way the stability will be increased). The monomer-dimer-tetramer series would thus be the result of two such mutations. This argument is based on our original assumption that the most primitive proteins were monomeric and that multisubunit structures arose from these. It is equally plausible to assume that primitive proteins were already aggregated into helices. The most stable helices would have a low pitch, since this allows more interactions, and by mutations and distortions these could generate cyclic and dihedral structures. However, this pathway would not necessarily favour symmetrical bonding. The transition between helical and cyclic structures of tobacco mosaic virus protein [47] provides a contemporary example of the process.

Relations between Symmetry and Function

The functional advantage of multisubunit structures have been discussed extensively elsewhere [2, 48] and can be briefly summarised:

1. The most general advantage is the increased stability due to multiplication of weak stabilizing interactions.

2. Such structures introduce the possibility of combining subunits with different preformed specificities so that either their mutual catalytic efficiency (e.g. in two stages of a complex reaction) is enhanced, or one ligand can regulate allosterically the metabolism of another ligand of different structure.

3. More subtle advantages follow from cooperative effects which are a natural consequence of a symmetrical structure and which have sometimes been made use of to increase the efficiency of respiratory carriers and of feedback control systems.

It can be seen from an examination of the examples listed in Table 2 that there is no direct connection between the actual symmetry of a particular protein and its function. Such relationships appear only with much larger structures, such as viral capsids, flagella, microtubules and muscle fibres, which we are not concerned with here. For enzymes, symmetry is related to structural stability rather than to function and from the point of view of the biochemist it is useful mainly because it simplifies the analysis of complex structures.

Labelling of Subunits for Electron Microscopy

Although electron microscopy often provides an accurate picture of the overall geometry of a multisubunit protein it is usually impossible to determine unequivocally the symmetry of the structure since the orientation of subunits remains in doubt (Fig. 2). A suitable label linked to a specific binding site could help to answer such questions. Positive staining, using a label of sufficient electron density (5 or 10 heavy atoms) should be applicable. However, it would not be easy to combine with negative staining and it has not often been used successfully at the molecular level. We have employed bifunctional ligands to unite proteins through their binding sites and these have provided useful information about the orientation of subunits in avidin [22] and in antibodies [49, 50]. The reagents used were symmetrical and produced polymers the symmetry of which revealed that of the subunit arrangement. The results obtained with avidin are shown in Figs. 2 b and 3. Each tetrameric molecule was doubly linked to its neighbours by the bis-biotinyl diamine, from which it could be concluded that the subunits were arranged with twofold symmetry.

It is also possible to use an asymmetric bifunctional reagent to link univalent labels to subunits and avoid the complications of polymerization. The Fab fragment of anti DNP antibody provides one suitable marker since it approaches the smallest size which is readily visible in negative stain. Even so the resolving power is limited to about 40 Å by the smallest dimension of the Fab fragment. We have used it in conjunction with 1-dinitrophenylamino-12-biotinamido-dodecane to label the biotin binding sites of avidin (Fig. 3 b) and thus confirm the conclusions obtained with the symmetrical bis-biotinamido reagent. Iodoacetyl DNP lysine would be a more generally useful reagent since it could provide a specific anchor for an anti-DNP Fab fragment at the binding site of any enzyme with a reactive -SH group, but it has not yet been used in this way. At a slightly lower level of resolution antibodies have been used to locate products of specific bacteriophage genes in the complete phage [51].

Another version of the technique which can be very useful in specific cases is to employ proteins with a natural affinity for each other. The use of heavy meromyosin to identify unknown protein

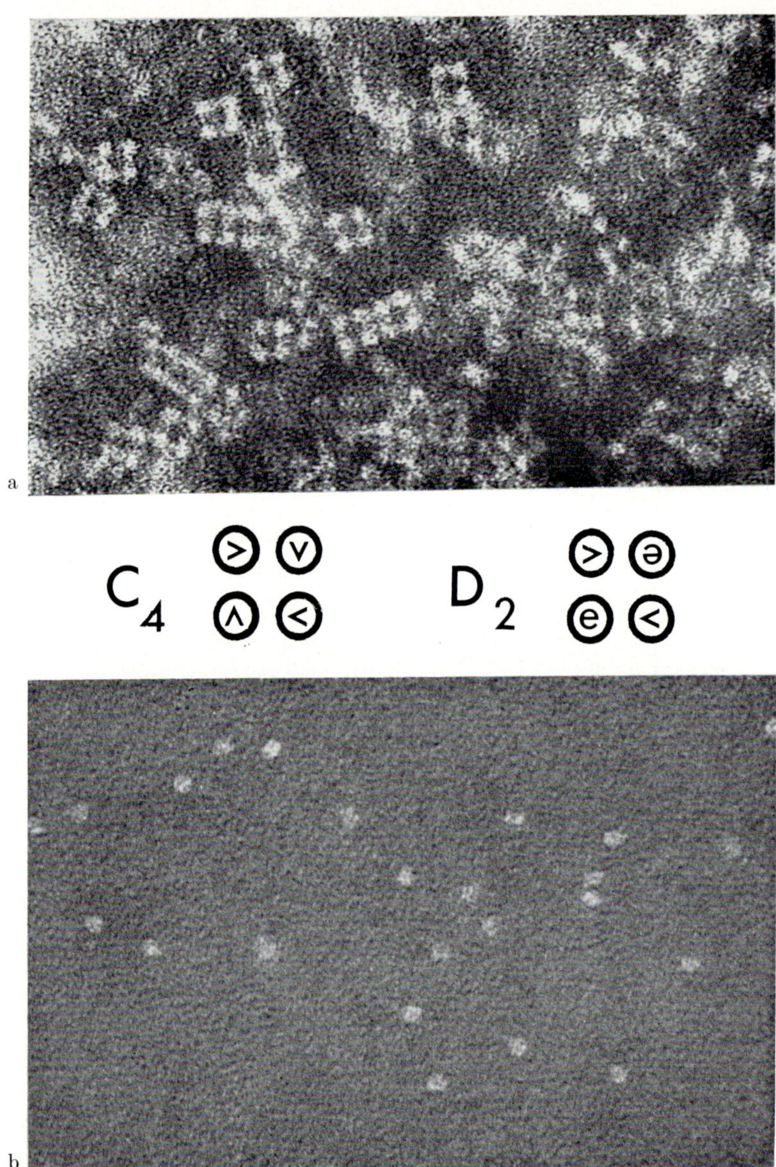

Fig. 2a and b

filaments as actin provides one example of this approach [52]. Here I will describe the use of avidin to determine the location of covalently bound biotin in transcarboxylase [16]. This enzyme was discovered and its subunit structure analyzed by H. G. Wood and his colleagues, and their results have been summarized in recent reviews [53, 64]. Our collaborative electron microscopic analysis of the enzyme illustrates many of the techniques and arguments that we have been discussing and shows how the structure of a complex enzyme with several unexpected features can nevertheless be reconciled with the simple basic principles of self assembly.

Transcarboxylase catalyses a two stage transfer of CO_2 from oxalacetate to propionyl coenzyme A, using covalently bound biotin as a carrier. The constituent reactions: —

1. Oxalacetate + TC-biotin = pyruvate + TC-biotin -CO_2,
2. TC-biotin-CO_2 + propionylCoA = TC-biotin + methylmalonylCoA

are catalyzed by two different subunits and the biotin is carried on a third, low molecular weight subunit, the biotin carrier protein [54]. The complete enzyme (18 S, M. W. 790,000) contains six of each type of subunit, and its dissociation products, named in accordance with their sedimentation coefficients are shown in Fig. 4. The enzyme is labile and at low salt concentration, pH 8.0, it loses three 6 S_E subunits in a stepwise manner, leaving a residual 12 S_H structure. Each 6 S_E subunit contains two molecules of biotin carrier protein (1.3 S_E, M. W. 12,000) and two Co^{++}/Zn^{++} containing subunits (2.5 S_E, M. W. 60,000). The 12 S_H subunit (M. W. 322,000) dissociates in two stages, to 3×6 S_E and finally in SDS, to 6×2.5 S_E subunits. Electron micrographs of the whole enzyme show a variety of profiles, most with two (Fig. 5 b) and three (Fig. 5 a) small subunits on one side of a large subunit. The small sub-

---

Fig 2 a and b. Electron micrographs of two tetrameric proteins. a) Pyruvate carboxylase. (M. W. 4 × 165,000). Four distinct subunits are arranged with apparent four fold symmetry. However, the symmetry could be either C_4 or D_2, as indicated in the diagram, since the orientation of the subunits is not defined. × 500,000. b) Avidin (M. W. 4 × 16,000), like other small proteins, shows no resolvable structure. [The smallest tetramer to show distinct subunits is aldolase [20] (M. W. 4 × 40,000)] × 500,000. Fig. 2a is taken from [17] and Fig. 2b from [22]

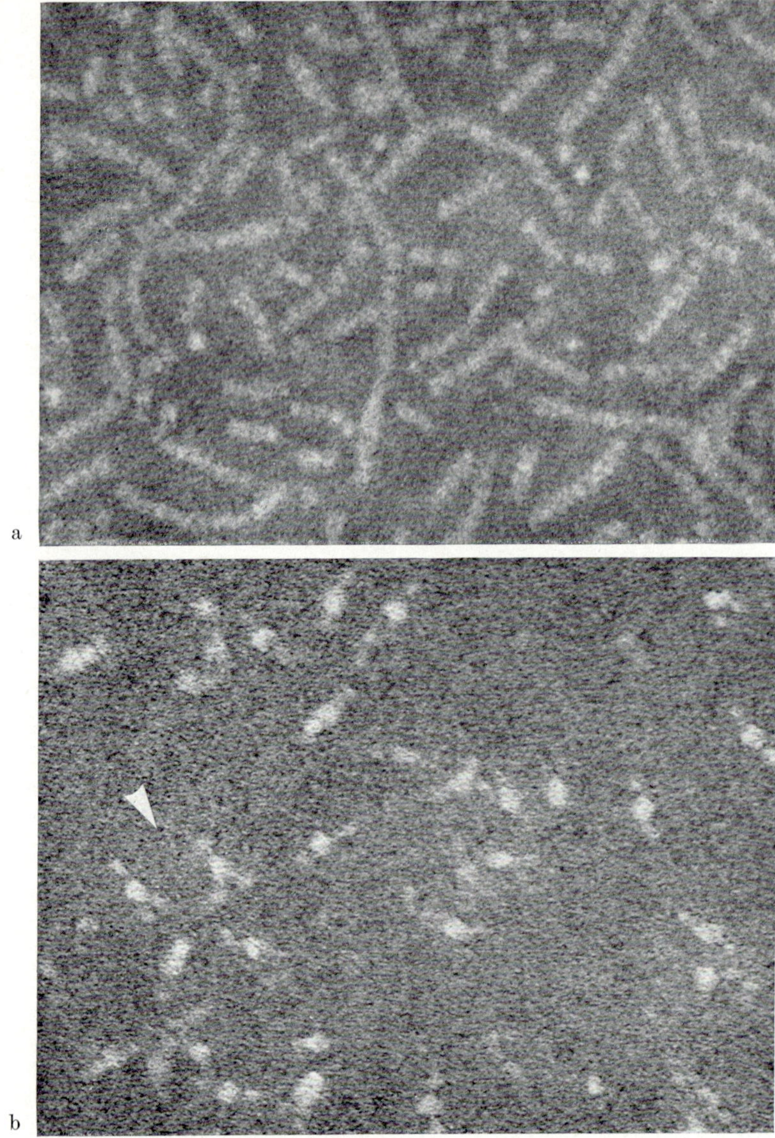

Fig. 3a and b

units appear sometimes as discrete, highly contrasted discs, sometimes as a single diffuse area of low contrast (Fig. 5 f). They were often well separted from the large subunit. One particularly striking profile (Fig. 5 b) suggested the head of Mickey Mouse and we have therefore added the subscripts H and E to the names of the subunits to indicate their origin from the head or the ear of this structure. There was no direct indication how this structure could be assembled from six of each type of subunit. However, one can make a tentative correlation between the two or three small subunits and the 6 S_E component of the enzyme, which is lost in a stepwise manner. In these terms the Mickey Mouse-like molecules would correspond to the 16 S enzyme which contains only two 6 S_E subunits.

Electron micrographs of the separated 6 S_E (Fig. 5 g; separate EM not shown) and 12 S_H (Fig. 6 a) components, confirmed that the 6 S_E had an elongated structure, similar to the ears of the enzyme in their diffuse orientation. The 12 S_H subunit usually had an approximately rectangular appearance, (70 × 100 Å) similar to the head of the whole enzyme. It was apparently divided into four subunits by penetration of stain along both its axes. This picture was initially misleading. A more useful clue to the structure was

---

Fig. 3 a and b. Determination of the orientation of subunits in avidin using bifunctional reagents. a) Avidin combined with 1:12 bis-biotinamidododecane (2 moles/mole avidin). The molecules form linear polymers with a repeat distance of 41 Å. This is significantly less than the width of the polymers (55 Å) and shows that the molecule approximates to an oblate ellipsoid. The linearity of the polymers shows that the four binding sites are arranged in two pairs at either end of the short axis of a molecule with D_2 symmetry. b) Avidin combined with Fab fragments of anti-DNP antibody, using the reagent 1-biotinamido-12-dinitrophenylaminododecane. Both fluorescence quenching titration and electron microscopy showed that the product contains only two Fab fragments per avidin molecule. This shows that the members of each pair of binding sites were close together so that when one Fab was bound, it blocked access to the neighbouring DNP group. The two Fabs were bound at 180° to each other confirming the two-fold symmetry of avidin. Occasional molecules (arrow) did show a third Fab, suggesting that weak binding could occur at the hindered sites. × 500,000. Fig. 3a is taken from [22] and Fig. 3b from unpublished experiments by N. M. Green and N. G. Wrigley

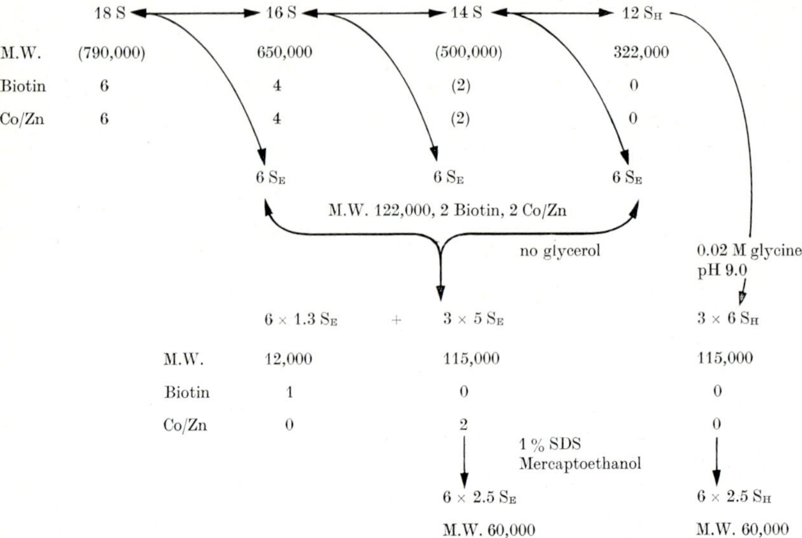

Fig. 4. Dissociation products of transcarboxylase. The subunits are named using their approximate sedimentation coefficients followed by subscripts H or E to indicate their origin from 'head' or 'ear' structures of the whole enzyme. The molecular weights in brackets have been estimated by summing those of the constituent subunits. The early stages of the dissociation are reversible

---

Fig. 5 a to g. Transcarboxylase (predominantly 18 S) [16]. The most common profile shows a large lozenge-shaped subunit (80 × 100 Å to 100 × 100 Å) with several smaller ones (50 × 50 Å to 50 × 80 Å) on one side of it. The large subunit sometimes shows stain penetration along either or both axes. The small subunits appear sometimes as a pair of high contrast 50 Å diameter discs [Mickey Mouse profiles b)]. More often an ill defined number are fused into a continuous region of low contrast f). Occasionally this region shows three clear subunits a), and this probably provides the most accurate picture of the whole enzyme. Another less common appearance e) shows the ears as elongated (50 × 70 Å) structures projecting from the head. In addition to profiles of the complete enzyme (2 or 3-eared) there are heads with only one ear c), isolated heads d) and ears g). × 400,000

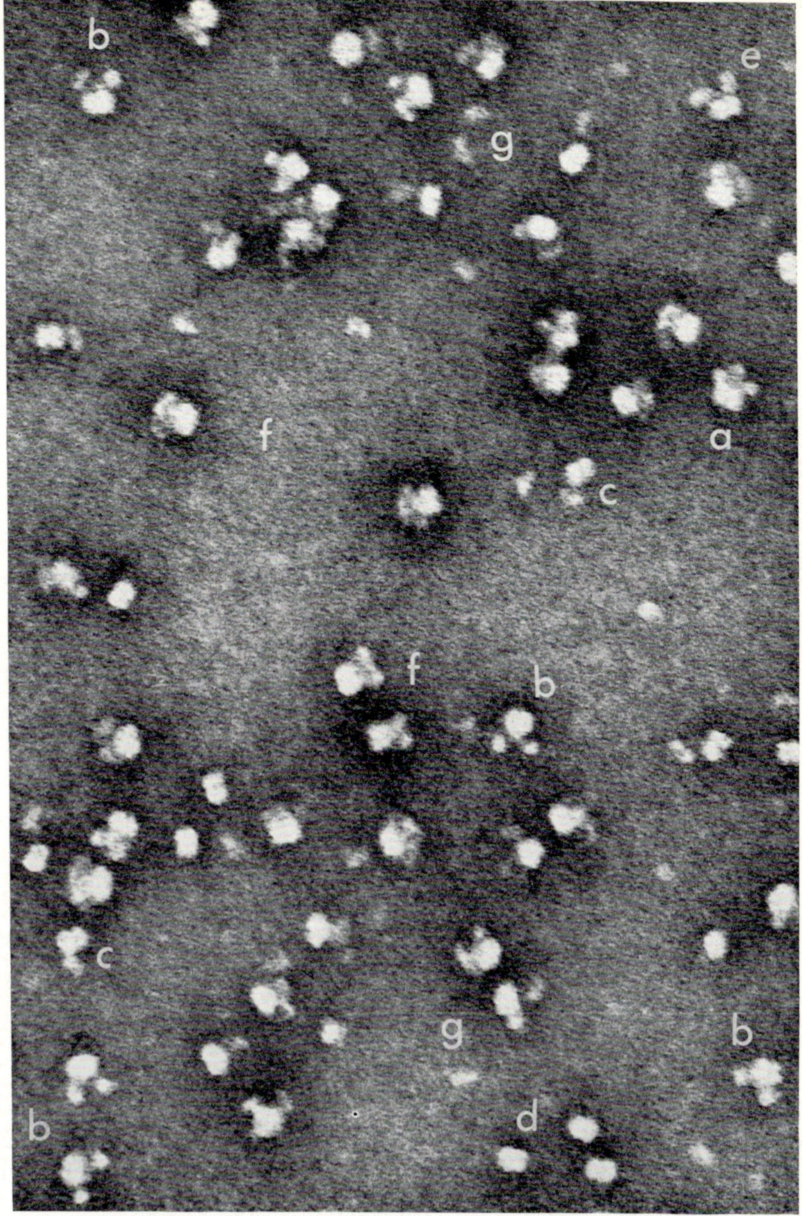

Fig. 5

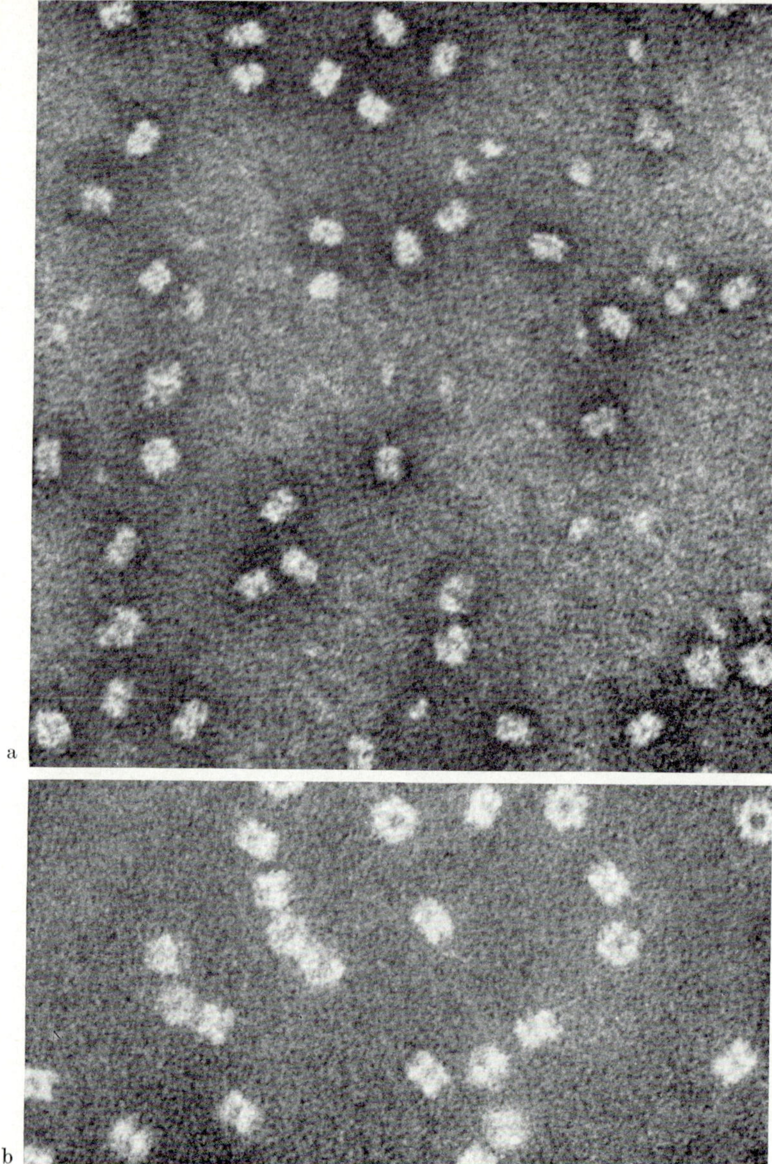

Fig. 6a and b

provided by the less common circular or polygonal profiles some of which showed a central hole. This suggested that the six 2.5 S_H subunits of the head were arranged in a ring and that the rectangular profile represents an edgewise view of this ring. In this respect the 12 S_H subunit resembles glutamine synthetase (Fig. 6 b) [31], which has a double ring of six subunits (D_6). We will return later to the question of the dihedral or cyclic symmetry of the head.

Accepting a ring structure of some type for the head, we can interpret the profiles of the whole enzyme as edgewise views of the ring with the 6 S_E subunits bound to one side only. The location of the 1.3 S_E subunit in this structure was deduced from electron micrographs of complexes of the enzyme with avidin (Fig. 7) which show, in addition to polymeric material, end on views of the whole enzyme in which the 12 S_H subunit is surrounded in alternating fashion by three 6 S_E subunits and three avidin molecules. The predominantly tangential orientation of the 6 S_E subunit suggests that the 1.3 S_E biotin units, which form part of it, are located at each end of the long axis, next to the avidin. The 20 to 30 Å gaps between avidin and ears and between ears and head suggested a flexible link between biotin and ear and between ear and head, probably supplied by the 1.3 S_E subunit. About 15 to 20 residues of unfolded peptide chain would suffice to account for the range of appearances of transcarboxylase. Supporting evidence for such a flexible link was provided by brief digestion with trypsin which removed the ears from the head and a biotinyl peptide from the ears.

Since the ears are dimeric they will be doubly linked to pairs of 2.5 S_H subunits in the head, as indicated in the model of the complete enzyme (Fig. 8). The occasional radially projecting ear (Figs. 5 e, and 7 d) could be result of dissociation of one of the two

---

Fig. 6 a and b. Comparison of electron micrographs of the purified 12 S_H subunit with those of glutamine synthetase. a) 12 S_H 'head' subunits of transcarboxylase [16] show two distinct profiles 1. rectangular or lozenge-shaped (70 × 100 Å) often with stain penetrating along both axes and less commonly 2. circular or polygonal (100 Å diameter) with a central hole. b) Glutamine synthetase [31] (M. W. 12 × 40,000). Both rectangular and polygonal profiles can be seen. The pattern of stain penetration is similar to that in 6 *a*, but is much more marked. × 500,000

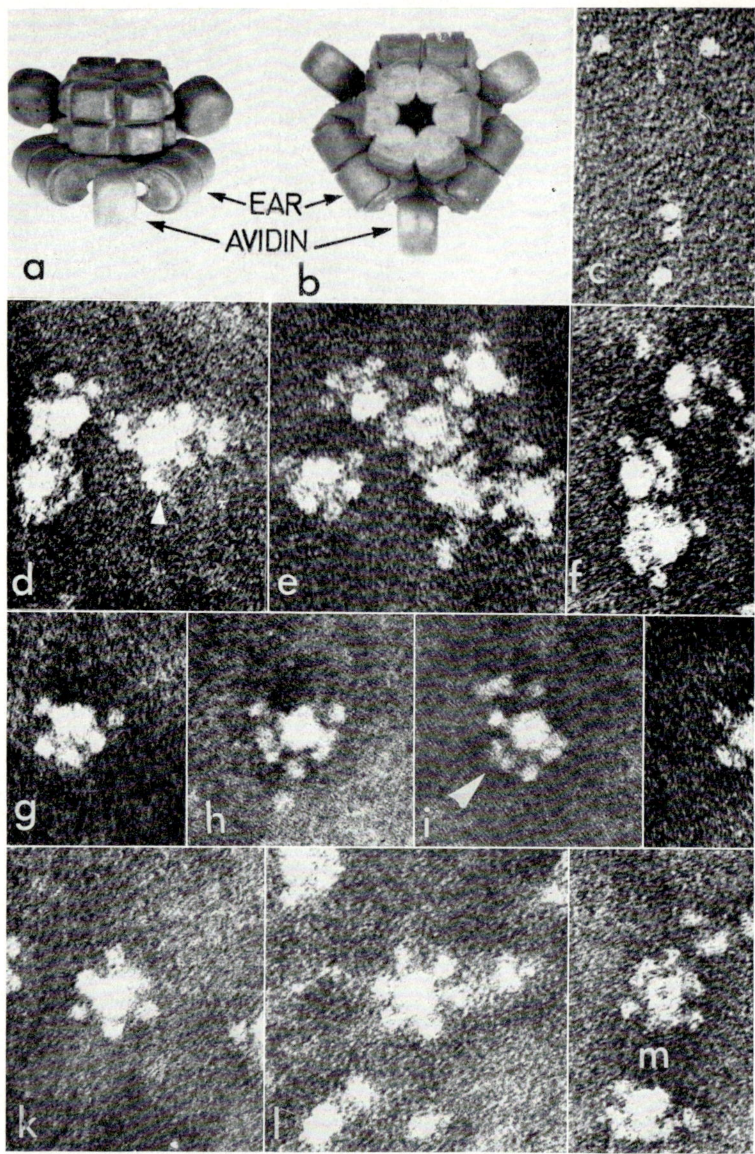

Fig. 7 a—m

links. In this model the 2.5 S_H subunit is shown as an elongated structure oriented parallel to the threefold axis of the molecule and divided into two separate regions of tertiary structure. We can produce the following plausible though not unequivocal arguments for this structure. The one sided location of the ears shows that the six binding sites on the head are all located on one surface, implying that the head has cyclic symmetry and belying its dihedral appearance. If each 2.5 S_H subunit possesses one ear binding site then each must have an axis running from one face to the other, approximately as indicated, and the equatorial stain penetration must divide each subunit in two.

Such a structure, although unusual has several precedents. Single peptide chains forming two or more regions of tertiary structure (domains) are found (1) among the immunoglobulins [50, 55], (2) in the pyruvate dehydrogenase subunit of the pyruvate dehydrogenase complex from E. coli [56, 57] and probably, (3) in transferrin, the single chain of which forms two binding sites and has duplicated peptide sequences [58]. Such structures will be formed whenever gene duplication followed by fusion has occurred, provided that the tertiary structures of the parent proteins persist in the new product [59]. More direct evidence is desirable and could be obtained from a study of peptide maps and of the number of catalytic sites of the 12 S_H subunit.

Further indirect evidence is provided by electron micrographs of the reconstituted enzyme. Dissociation of transcarboxylase in the presence of glycerol, followed by reconstitution at pH 7, gave normal 16 to 18 S enzymes, whereas more extensive dissociation in

---

Fig 7 a to m. [16] Complexes of transcarboxylase with an excess of avidin. Single avidin molecules are shown in c). Large numbers of polymeric complexes d), e), f) were seen, as well as single molecules showing three small, avidin like attachments alternating with two — g), h), i), or three k), l), m) — 'ear' subunits. In most molecules which showed clear three-fold symmetry in the arrangement of subunits around the head, h), l), m) the head itself showed the circular, end on, profile, sometimes with a central hole. Some of the two-eared forms g) showed the lozenge-shaped profile of the head, with the ears on a high contrast orientation. An interpretation of two different profiles in terms of the model discussed below (Fig. 8) is shown in a) and b)

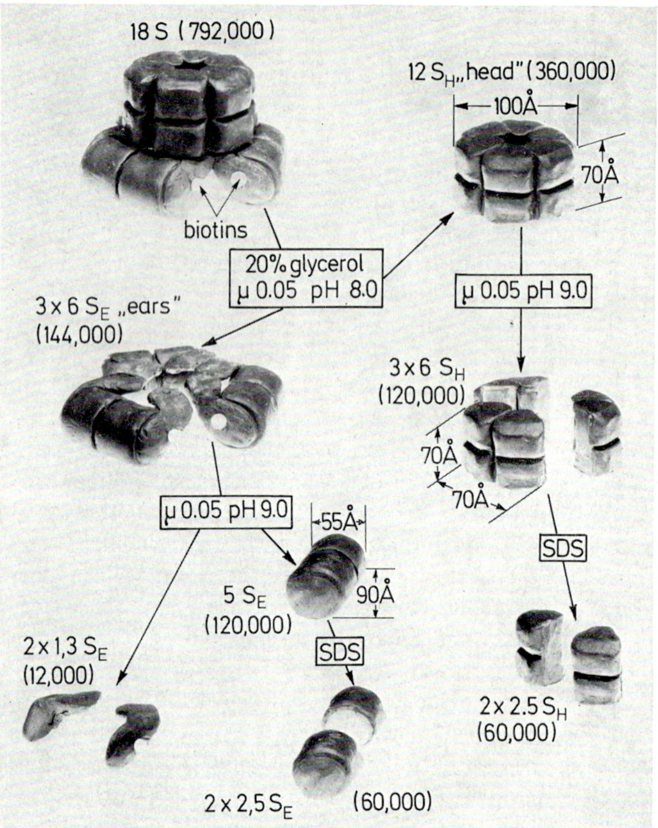

Fig. 8. [16] Dissected model of transcarboxylase showing sedimentation coefficients, molecular weights and dimensions of the various products of dissociation. The rationale for the subdivisions of the 12 S_H subunit is discussed in the text. The location of the division between the two halves of each ear is based on a few examples of stain penetration in this region (Fig. 7 i). The location of the 1.3 S_E subunit is deduced from the position of the avidin label (Fig. 2 b); its detailed shape is arbitrary. Approximately 40 Å (10 to 15 amino acid residues) of extended peptide chain which would be required to account for the variety of profiles of the whole enzyme have not been shown in the model

the absence of glycerol followed by reconstitution at pH 5 gave a 24 S product. In the electron microscope (Fig. 9) this appeared as a head with a double complement of ears showing that under these circumstances the two surfaces of the reconstituted head were

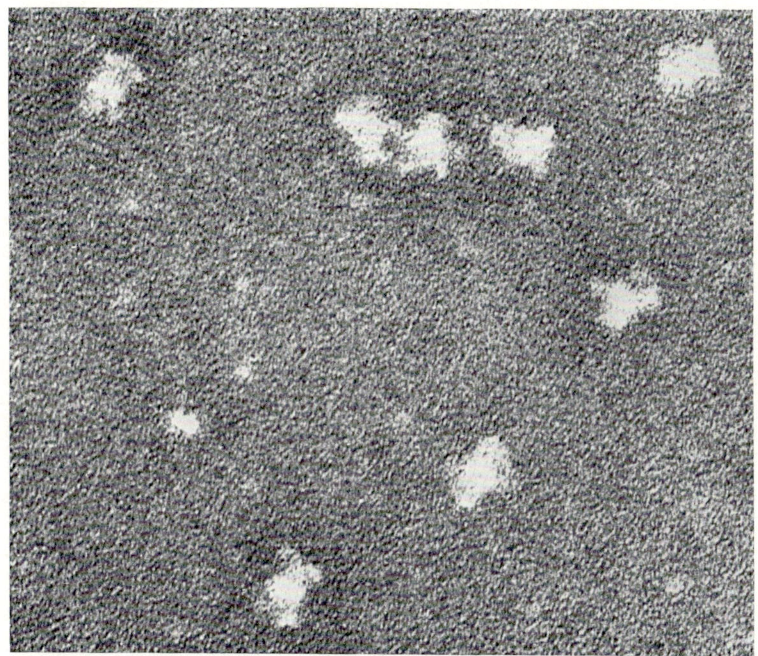

Fig. 9. Electron micrograph of reconstituted transcarboxylase [16]. Two blurred sets of ears can be seen on opposite faces of a single head. The molecular weight of such a structure ($12\ S_H + 6 \times 6\ S_E = 1{,}220{,}000$) is consistent with its sedimentation coefficient (~ 24 S)

similar and that the symmetry of the head could be described as pseudo-dihedral.

We can summarize our tentative conclusions about the symmetry relationships between 2.5 S_H subunits within the head in terms of the diagram (Fig. 10). We would expect the six subunits of a hexamer to be arranged with either C_6 or D_3 symmetry. However, a D_3 structure could not account for the presence of 3 doubly bonded ears on one surface and a C_6 structure is inconsistent with the existence of the dimeric 6 S_H intermediate dissociation product of the head. This intermediate dimer can be accounted for in terms of slight alternations in the bonding pattern between identical subunits, somewhat analogous to those observed in insulin [15]. In order to account for the equatorial stain penetration and the binding of an extra set of ears we have postulated a set of pseudo two-

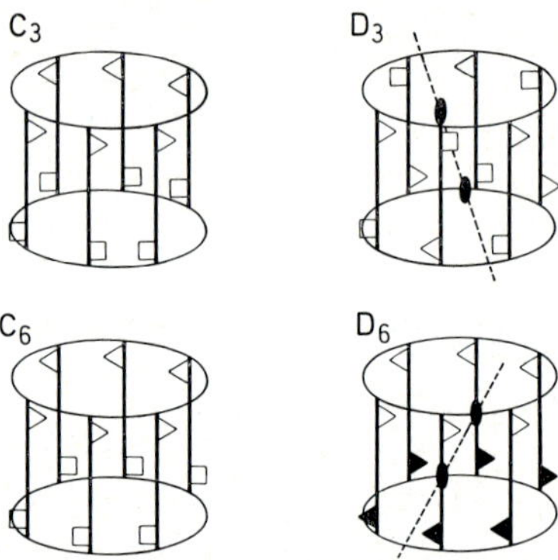

Fig. 10. Diagram of possible symmetry relations in a hexamer. C_3: Alternate subunits are in non-equivalent positions. C_6, D_3: All subunits are in equivalent positions. D_6: Additional two-fold axes are present within subunits, which are therefore dimeric

fold axes passing through each 2.5 S_H subunit, leading to a pseudo D_6 symmetry, analogous to the D_6 symmetry of glutamine synthetase. However, the true symmetry of the head remains no higher than C_3.

We can find other enzymes which show a related type of behaviour, in that they have potentially similar subunits which behave in a non identical fashion, although they are not covalently linked to each other, as suggested for the 12 S_H subunit. The enzymes threonine deaminase [60] and acetoacetate decarboxylase [61] have twice as many subunits as they have catalytic sites. The first has four very similar peptide chains but binds only one mole of pyridoxal phosphate for every two chains. Acetoacetate decarboxylase has twelve subunits but forms only six moles of Schiff's base with substrates and inhibitors. When it is heated its specific activity is increased [62], which provokes the intriguing but unsupported suggestion that the second set of subunits may have become activated.

The proposed structure relies fairly heavily on arguments based on considerations of symmetry since, as we saw in the introduction, most enzymes are assembled according to simple basic principles. The functional significance of the unusual features of transcarboxylase is only partially clear. The flexible link between biotin, head and ears has an obvious relation to the catalytic function of the biotin [63], but the significance of the structure of the 12 S_H subunit is obscure. It may be merely a reflection of the haphazard way in which that great opportunist, Natural Selection, arrives at an effective solution to a chemical problem. Nevertheless these features provide interesting clues to the protein chemist, which may throw light on the structure of other enzymes.

Acknowledgements. I would like to thank my colleagues Dr. N. G. Wrigley and the late Dr. R. C. Valentine for their extensive collaboration in all the electron microscopy, also Dr. H. G. Wood and his colleagues at Case-Western Reserve University, to whom most of the credit for elucidating the structure of transcarboxylase is due.

References

1. Caspar, D. L. D.: In: Principles of biomolecular organisation, p. 7. (Ciba Foundation Symposium, Wolstenholme, G. E., Ed.). London: J. and A. Churchill 1966.
2. Monod, J., Wyman, J., Changeux, J. P.: J. molec. Biol. **12**, 88 (1965).
3. Klug, A.: In: Symposium of the International Society for Cell Biology **6**, 1. Academic Press, New York (1967).
4. Hanson, K. R.: J. molec. Biol. **22**, 405 (1966).
5. Pauling, L.: Discussions of the Farad. Soc. **15**, 170 (1953).
6. Klug, A., Crick, F. H. C., Wyckoff, H.: Acta Cryst. **11**, 199 (1958).
7. Green, N. M.: Nature (Lond.) **219**, 413 (1968).
8. Caspar, D. L. D., Klug, A.: Cold Spr. Harb. Symp. quant. Biol. **27**, 1 (1962).
9. Branden, C.-I.: Arch. Biochem. **112**, 215 (1965).
10. Aschaffenburg, R., Green, D. W., Simmons, R. M.: J. molec. Biol. **13**, 194 (1965).
11. Rosenbusch, J. P., Weber, K.: J. biol. Chem. **246**, 1644 (1971).

12. Franklin, R. M., Pettersson, V., Akervall, K., Strandberg, B., Philipson, L.: J. molec. Biol. **57**, 383 (1971).
13. Oesterhelt, D., Bauer, H., Lynen, F.: Proc. nat. Acad. Sci. (Wash.) **63**, 1377 (1969).
14. Webster, R. G., Laver, W. G.: In: Progr. med. Virol. **13**, 275 (1971).
15. Adams, M. J., Blundell, I. L., Dodson, E. J., Dodson, G. G., Vijayan, M., Baker, E. N., Harding, M. M., Hodgkin, D. C., Rimmer, B., Sheat, S.: Nature (Lond.) **224**, 491 (1969).
16. Green, N. M., Valentine, R. C., Wrigley, N. G., Ahmad, F., Jacobsen, B., Wood, H. G.: J. biol. Chem. **246**, 6284 (1972).
17. Valentine, R. C., Wrigley, N. G., Scrutton, M. C., Irias, J. J., Utter, M. F.: Biochemistry **5**, 3111 (1966).
18. Laver, W. G., Younghusband, H. B., Wrigley, N. G.: Virology **45**, 598 (1971).
19. Eagles, P. A. M., Johnson, L. N., Joynson, M. A., McMurray, C. H., Gutfreund, H.: J. molec. Biol. **45**, 533 (1969).
20. Penhoet, E., Kochman, M., Valentine, R. C., Rutter, W. J.: Biochemistry **6**, 2940 (1967).
21. Green, N. M., Joynson, M. A.: Biochem. J. **118**, 71 (1970).
22. Green, N. M., Konieczny, L., Toms, E. J., Valentine, R. C.: Biochem. J. **125**, 781 (1971).
23. Eagles, P. A. M., Johnson, L. N.: J. molec. Biol. **64**, 693 (1972).
24. Kiselev, N. A., Lerner, F. Y., Livanova, N. B.: J. molec. Biol. **62**, 537 (1971).
25. Morino, Y., Snell, E. E.: J. biol. Chem. **242**, 5591 (1967).
26. Adams, M. J., Haas, D. J., Jeffery, B. A., McPherson, A., Mermall, H. L., Rossman, M. G., Schevitz, R. W., Wonacott, A. J.: J. molec. Biol. **41**, 159 (1969).
27. Wiley, D. C., Lipscomb, W. N.: Nature (Lond.) **218**, 1119 (1968).
28. Josephs, R.: J. molec. Biol. **55**, 147 (1971).
29. Klapper, M., Klotz, I. M.: Biochemistry **7**, 223 (1968).
30. Boeker, E., Snell, E. E.: J. biol. Chem. **243**, 1687 (1968).
31. Valentine, R. C., Shapiro, B., Stadtman, E. R.: Biochemistry **7**, 2143 (1968).
32. Berns, D. S., Edwards, M.: Arch. Biochem. **110**, 511 (1965).
33. Fernandez-Moran, H., Marchalonis, J. J., Edelman, G. M.: J. molec. Biol. **32**, 467 (1968).
34. Bowers, W. F., Czubaroff, V. B., Haschemeyer, R. H.: Biochemistry **9**, 2620 (1970).
35. Harrison, P. M.: J. molec. Biol. **6**, 404 (1963).
36. Björk, I., Fish, W. W.: Biochemistry **10**, 2844 (1968).
37. Crichton, R. R.: Biochem. J. **126**, 761 (1972).
38. De Rosier, D. J., Oliver, R. M., Reed, L. J.: Proc. nat. Acad. Sci. (Wash.) **68**, 1135 (1971).
39. Reed, L. J., Cox, D. J.: In: The enzymes, 3rd ed., Vol. I, P. 213. (Boyer, P. D., Ed.).
40. Hanson, K. R.: J. molec. Biol. **38**, 133 (1968).

41. GUERRITORE, D., BONACU, M. L., BRUNORI, M., ANTONINI, F., WYMAN, J., ROSSI-FANELLI, A.: J. molec. Biol. **13**, 234 (1965).
42. FERNANDEZ-MORAN, H., VAN BRUGGEN, E. F. J., OHTSUKI, M.: J. molec. Biol. **16**, 191 (1966).
43. KELLENBERGER, E.: In: Principles of biomolecular organisation, Ciba Foundation Symposium, p. 192. (WOLSTENHOLME, G. E., Ed.). London: J. and A. Churchill 1966.
44. KLOTZ, I. M., LANGERMAN, N. R., DARNALL, D. W.: Ann. Rev. Biochem. **39**, 25 (1970).
45. HASCHEMEYER, R. H.: Advanc. Enzymol. **33**, 71 (1970).
46. VAN HOLDE, K. E.: In: Molecular architecture in cell physiology. Symposia of Society of Cell Physiologists, p. 89. New York: Prentice Hall 1966.
47. CASPAR, D. L. D.: Advanc. Protein Chem. **18**, 1 (1963).
48. MONOD, J.: In: Symmetry and function. Nobel Symposium, vol. II, p. 15. (ENGSTROM, A., STRANDBERG, B., Eds.). New York, London, Sydney: Wiley 1969.
49. VALENTINE, R. C., GREEN, N. M.: J. molec. Biol. **27**, 615 (1967).
50. GREEN, N. M., DOURMASHKIN, R. R., PARKHOUSE, R. M. E.: J. molec. Biol. **56**, 203 (1971).
51. YANAGIDA, M., AHMAD-ZADEH, C.: J. molec. Biol. **51**, 411 (1970).
52. ISHIKAWA, H., BISCHOFF, R., HOLZER, H.: J. Cell Biol. **43**, 312 (1969).
53. WOOD, H. G.: In: The Enzymes, 3rd ed., vol. 6, p. 83. (BOYER, P. D., Ed.) (1972).
54. JACOBSEN, B., GERWIN, B. I., AHMAD, F., WAEGELL, P., WOOD, H. G.: J. biol. Chem. **245**, 6471 (1970).
55. EDELMAN, G. M.: Biochemistry **9**, 3197 (1970).
56. REED, L. J., OLIVER, R. M.: Brookhaven Symposia in Biology **21**, 397 (1968).
57. VOGEL, O., HENNING, U.: Europ. J. Biochem. **18**, 103 (1971).
58. MANN, K. G., FISH, W. W., COX, A. C., TANFORD, C.: Biochemistry **9**, 1348 (1970).
59. YOURNO, J., KOHNO, T., ROTH, J. R.: Nature (Lond.) **228**, 820 (1970).
60. HATFIELD, G. W., BURNS, R. O.: Science **167**, 75 (1970).
61. TAKAGI, W., WESTHEIMER, F. H.: Biochemistry **7**, 895 (1968).
62. O'LEARY, M. H., WESTHEIMER, F. H.: Biochemistry **7**, 913 (1968).
63. NORTHROP, D. B.: J. biol. Chem. **244**, 5808 (1969).
64. MOSS, J., LANE, M. D.: Advanc. Enzymol. **35** (1972) (in press).
65. LOUDON, J., GOLDBERG, M. E.: J. biol. Chem. **247**, 1566 (1972).
66. KAGAMIYAMA, H., WADA, H., MATSUBARA, H., SNELL, E. E.: J. biol. Chem. **247**, 1571 (1972).
67. HAMMERSTEDT, R. H., MOHLER, H., DECKER, K. A., WOOD, W. A.: J. biol. Chem. **246**, 2069 (1971).
68. DURHAM, A. C. H., FINCH, J. T., KLUG, A.: Nature, New Biol. **229**, 37 (1971).

Probe Studies on the Role of Protein-Protein Interactions and Enzyme Conformation in Control

G. K. Radda and R. A. Dwek

Department of Biochemistry, University of Oxford, Oxford, Great Britain

With 25 Figures

These symposia at Mosbach are widely recognized because they attempt to summarize important developments in science in a specialized area in general terms. The organizers of this meeting have taken a great deal of trouble to point out to speakers the need for discussing their subject matter in rather more general terms than is customary at International Symposia. For this reason, one feels particularly honoured in being invited. Conscious of the need for giving a general background I would like to start at the beginning. In the year 100 B. C. the well known "biochemist" Lucretius in his treatise on "The Nature of the Universe" said. and I quote:

"The characteristics of the atoms of all substances are the extent to which they differ in shape and the rich multiplicity of their forms. Not that there are not many of the same shape but they are by no means all identical with one another. And no wonder. When the multitude of them, as I have shown, is such that it is without limit or count, it is not to be expected that they should all be identical in build and configuration".

It is this multiplicity of configuration of enzyme molecules that seems to be essential for the regulation of enzyme activity.

I. Biological Regulation

Before we examine the means by which we can study the role of conformation and protein-protein interaction in regulation let us briefly look at the requirements for efficient biological regulation.

There are essentially two immediate and short term ways of switching on (or off) an enzyme. The first utilizes "feedback" inhibition in that if we have a series of reactions such as that in Fig. 1 the product (D) can inhibit or activate the first enzyme

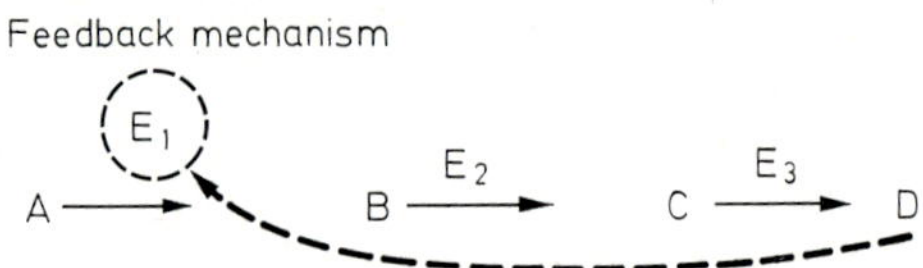

Fig. 1. Schematic representation of feedback inhibition

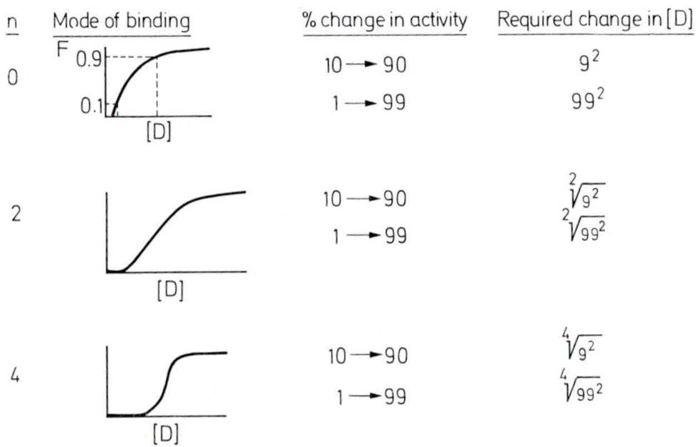

Fig. 2. Theoretical ligand binding curves and their relation to cooperative interactions

involved in the metabolic sequence. This then requires the binding of the ligand D to the enzyme E_1 thus altering its catalytic activity. Now, if this interaction follows the normal binding pattern, we can describe the interaction of E_1 with D by the usual hyperbolic binding curve. Fig. 2 lists the concentrations of D required to achieve a given change in activity. For example, for a change from

$10\% \rightarrow 90\%$ activity we need a sudden change in ligand concentration of 81 fold.

If, on the other hand, we alter the interaction in such a way that the binding is sigmoidal then we may achieve the same change over a much narrower ligand concentration range. We can express the "sigmoidicity" of a binding function in terms of the empirical Hill equation:

$$\text{Fraction Bound} = \frac{[D]^n}{K + [D]^n}$$

where n is a parameter that reflects this sigmoidicity, and this provides a further delicate control mechanism since the change in [D] for a given % change in activity now depends on n (Fig. 2).

Now it is quite obvious that when there is a sudden requirement in a biological system for inhibition or activation it would be unreasonable to expect a rapid, say 80 to 10,000-fold, change in the concentration of the regulatory ligand. However, a change of 3 to 10-fold is probably within a perfectly permissible limit. The first problem, therefore, is to understand the underlying physical and chemical reasons for the sigmoidal behaviour of regulatory enzymes.

Let me say at once that the secret lies in the fact that all these enzymes contain subunits and that binding of the regulator to one of these not only affects the activity of that subunit but the activity of the others, too. (The best-known example for this is of course haemoglobin where the binding of the first oxygen molecule increases the affinity of the other three subunits towards O_2) [1]. We can, therefore, reduce the question to how the information and energy derived from the binding of the first ligand is transmitted (presumably through the interface region of the subunits) to the other active sites. The Hill coefficient n, mentioned above in fact has a maximum value equal to the number of active centres.

However, even this method of regulation may not be sufficiently sensitive to relatively small signals that are available in a biological system. An alternative way is to amplify the initial input signal by some mechanism. One way of doing this is to have a cascade of enzymes of the kind that is present in the regulation of glycogen breakdown.

II. The Phosphorylase System

The signal is derived from a change in the concentration of cyclic AMP by the action of the hormone epinephrine [2] (Fig. 3). Here the activation of the enzyme kinase-kinase involves inter-

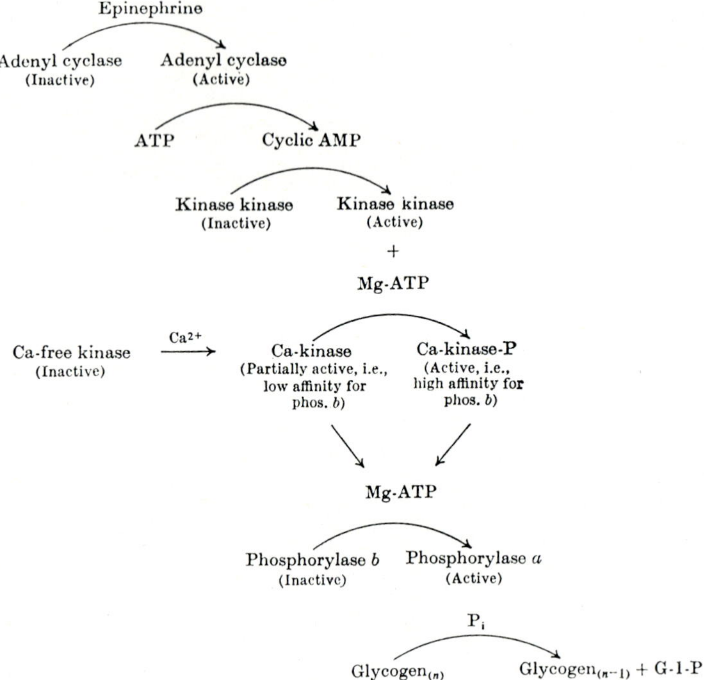

Fig. 3. Schematic representation of the activation of rabbit muscle phosphorylase. (Taken from Ref. [3] by permission of the Biochemical Society and the authors)

action with cyclic AMP and the active enzyme now catalyses the covalent modification of another enzyme (phosphorylase kinase) that leads to its activation. This enzyme in turn catalyzes the chemical modification of phosphorylase *b* (inactive) to the phosphorylated active form which then is responsible for the catalytic break-down of glycogen providing the necessary energy for muscle contraction.

The phosphorylase system thus provides us with a very remarkable example of a variety of protein-protein interactions which we have summarized in Fig. 4.

The questions we have to answer are as follows:

1. What is the nature of the sigmoidal "allosteric" activation by AMP?

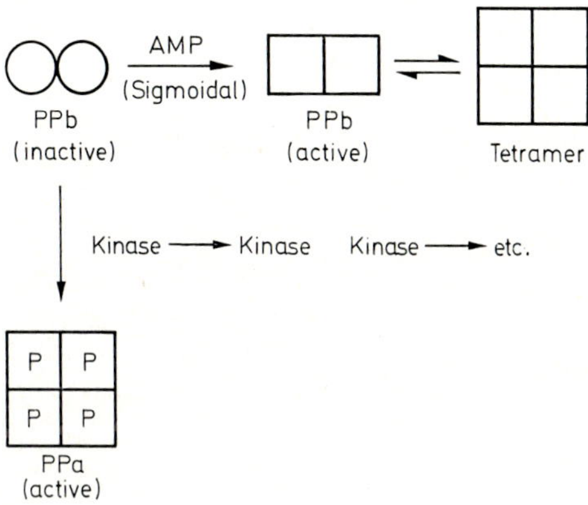

Fig. 4. Schematic representation of protein-protein interactions in the phosphorylase system

2. What is the mechanism of the phosphorylase-kinase interaction and how does phosphorylation of phosphorylase *b* lead to activation?

3. Is the increased tendency of both active forms to aggregate to the tetrameric structure the reason for, or a consequence of, the activation process?

III. Exploring Enzyme Conformation

It is now generally accepted (and Perutz's beautiful work on haemoglobin [3] amply justifies this) that an important feature in regulation is in the precise nature of the conformation of the protein particularly at the subunit interface region. We need, therefore, to detect and describe structural changes at "high resolution" that are associated with the activation process. We may say at once that a variety of physical and chemical methods are available for the detection of conformational changes but many of these are only capable of describing some overall features of the protein structure. Here we would like to discuss, with reference to the questions outlined above, the use of "probes".

"Probes" can be defined as small molecules or ions which can be attached to specific regions of the protein molecule and which through their chemical (e.g. reactivity) or physical (e.g. spectroscopic) properties are able to report some features of the structure at or close to their site of binding.

We shall discuss the use of four types of probe methods.

1. Chemical Reactivity

The reactivity of certain groups in an enzyme towards a specific reagent can be used as an empirical index of the enzyme's conformation around the functional group concerned. For example, the reagent (I) (7-chloro-4-nitrobenzo-2,3 oxa diazole: NBD-Cl) reacts specifically with

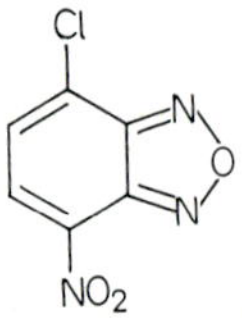

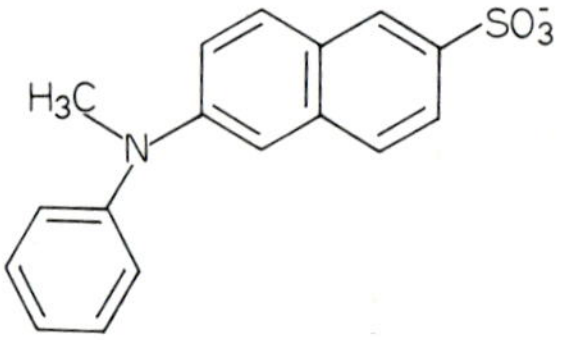

SH-groups and the reaction is accompanied by a large change in the absorption and fluorescence spectra of this chromophore [4]. Thus the rate and extent of the reaction can be monitored continuously.

2. Proton Relaxation Enhancement (PRE)

A metal ion such as Mn^{2+} (because of its paramagnetic property) will enhance the nuclear magnetic relaxation rates of protons of the water molecules in its hydration sphere. The extent of this enhancement (which is normally expressed as ε) depends on the nature of the other ligands around the metal ion. In particular, when the metal is bound to a macromolecule a significant increase in ε can generally be observed [5]. Without going into the details of the method and the theory behind it, we can simply state in a situation where free and bound Mn^{2+} are in equilibrium the observed enhancement (ε) will be a sum of two terms:

$$\varepsilon = \varepsilon_f\,[Mn^{2+}]_{free} + \varepsilon_b\,[Mn^{2+}]_{bound}$$

where ε_f and ε_b are characteristic spectroscopic properties of the free and bound Mn^{2+} ions respectively. By studying the dependence

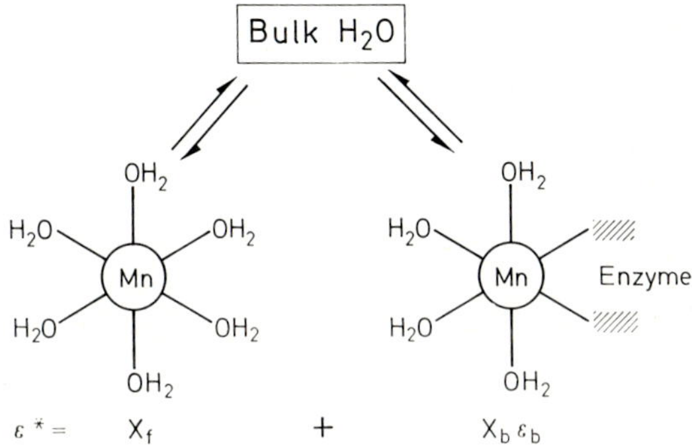

Fig. 5. Scheme for conditions of proton relaxation enhancement

of the observed enhancement on both metal and protein concentration these two characteristic parameters can be obtained and additionally the binding constant and numbers of binding sites for the metal can be determined. The method does not suffer from problems of sensitivity, as essentially one is looking at bulk water since the rate of proton exchange is so rapid that effectively every proton visits the metal ion, in the observation time of the experiment, and the relaxation is thus shared among them all. The measured relaxation rate thus corresponds to a weighted average of the protons in the different environments, Fig. 5.

3. Fluorescence Labels

There are two ways of attaching a fluorescent chromophore to a protein: covalently and non-covalently. In both instances (and particularly in the second) it is a considerable advantage if the quantum yield of the emission is higher for the bound than the unbound form. The two labels we shall use in our discussion are the NBD-group (I) and MNS (II). While for (I) the reagent (NBD-Cl) is not fluorescent, its derivative is significantly fluorescent. The quantum yield of MNS emission is very low in water but is enhanced in a less polar (hydrophobic) environment [6]. Since the emission of both these groups is environmentally sensitive, they may be used as probes reflecting the polarity (or changes in polarity) at the site of attachment. In addition, fluorescence can be used to study molecular motion (by combining polarization of fluorescence and life-time measurements).

4. Spin Labels

This method was first introduced by McConnell and his associates for the study of macromolecules [7]. It relies on the electron spin resonance (ESR) properties of the stable free organic nitroxide radicals of the type (III):

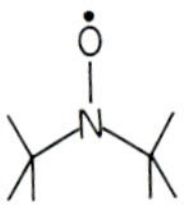

$I-CH_2-CO-NH-$ ... $N-O\cdot$

III 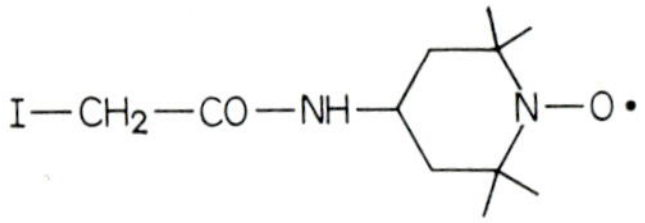 IV

The main feature of the ESR signal is that because of coupling between the free electron and the nitrogen nucleus the radical has three symmetrical lines when the molecule is tumbling rapidly ($\sim 10^{-11}$ sec), but when its motion is restricted the line-shape is considerably altered (Fig. 6). This can be used to derive information about the mobility of the labels. In addition the "splitting" of the lines (i. e. the distance in Hz between the three components) is a measure of the polarity of the environment of the radical. In favourable cases, too, the ESR-signal can be used to obtain information about orientation of the molecule with respect to a laboratory axis related to the magnetic field [8]. This type of label too can be linked to enzymes covalently and non-covalently. We shall describe some of the uses of covalent labels, e.g. (IV), that can be linked to proteins by standard modification procedures.

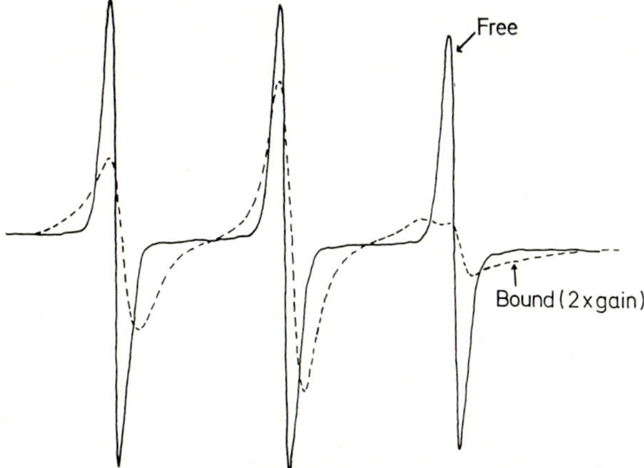

Fig. 6. Electron spin resonance spectra of free (continuous line) and bound (broken line) nitroxide radicals

IV. The Allosteric Properties of Muscle Phosphorylase b

Fig. 7 summarizes the ways we are going to study ligand-induced conformational changes in muscle phosphorylase b, which are essentially those described above. At this stage it is only necessary to point out that, while this enzyme has no metal requirement, it has one "accidental" Mn^{2+} binding site in each subunit [6]. The binding of this metal is surprisingly tight ($K_D = 150$ μM) and does not affect the activity of the enzymes or the binding of the ligands.

Taking chemical reactivity first, Fig. 8 represents the way NBD-Cl reacts under pseudo-first-order conditions with 3 SH groups/subunit in the native enzyme. Such a reaction curve (upper curve) can be easily analyzed in terms of sets of groups of different reactivities (dotted lines). In this particular case one SH reacts at a rate approximately 10 times faster than the two other groups. At the same time it can be shown that the activity of the enzyme is lost at a rate closely similar to that of the slowly reacting groups (lower curve). This immediately implies that the rapidly reacting SH is not "essential", while one or both of the slow groups (and it can be shown in different measurements that it is only one) are directly or indirectly involved in the activity of the enzyme.

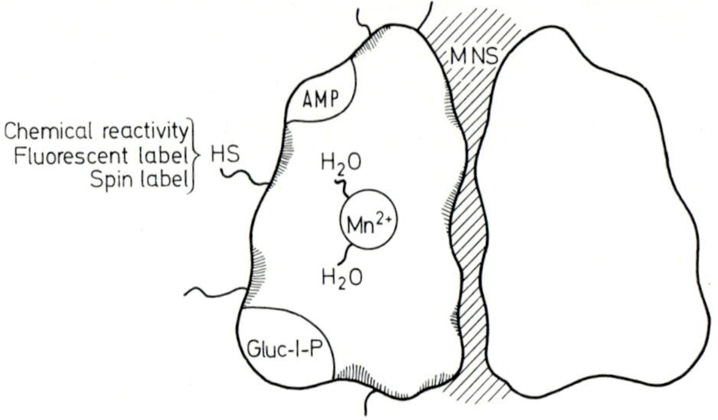

Fig. 7. Representation of ligand and probe sites on phosphorylase

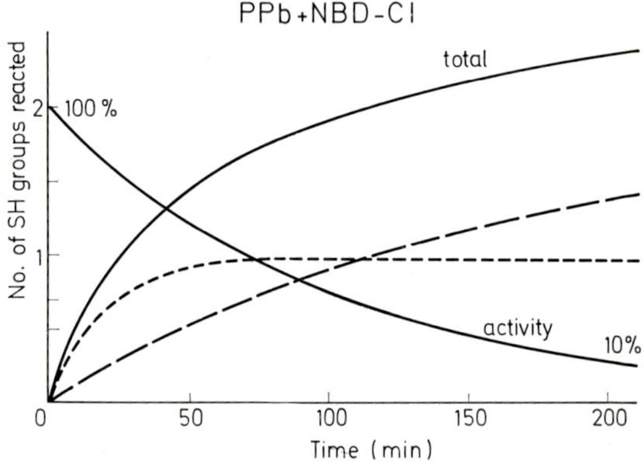

Fig. 8. Reaction of 7-chloro-4-nitrobenzo-2-oxa-1,3-diazole with phosphorylase *b*. Circles experimental points. The dotted lines are calculated for reaction of 1 fast and 2 slow SH-groups per protomer and the continuous line is the sum of these. Curve marked "activity" represents the time course of inactivation. Conditions: 1 mg/ml of phosphorylase *b*, 375 μM 7-chloro-4-nitrobenzo-2-oxa-1,3-diazole, 0.05 M triethanolamine -HCl buffer, pH 7.5, containing 0.2 M KCl and 1 mM EDTA at 15 °C. The reaction was followed by the change in absorbance at 420 nm

If we now add the activator ligand AMP to the enzyme, the combined reactivity of the SH-groups is decreased as shown by measuring the initial rates of the reaction as a function of AMP concentration. Using an analysis similar to that mentioned above it is easy to show that in the presence of excess AMP the modification rate of the rapid SH is unaffected but the slow-groups are "protected" with a concomitant protection against loss of activity (Table 1).

The same table shows that AMP increases the proton relaxation rate (by bound Mn^{++}) ions, thus providing evidence for some (probably small) rearrangements of the groups around the metal site.

Table 1. *Effects of ligands on phosphorylase b*

Ligand	ε	K for Mn^{2+} (mM)	Thiol-group reactivity (rate constants)		
			Rapidly reacting group ($l \cdot mole^{-1} \cdot min^{-1}$)	Slowly reacting groups $l \cdot mole^{-1} \cdot min^{-1}$)	
None	10.9	178	80	10	10
AMP	13.8	152	80	0	0
Glucose-1-phosphate	5.9	91	92	92	3.8
AMP + glucose 1-phosphate	7—10	90	70	0	0

In contrast to the changes in the probes resulting from addition of AMP, the effects of the substrate glucose-1-phosphate are the opposite. In particular the ε-value is decreased while the reactivity of the "slow" SH-groups is increased (that of the fast group remains constant). This increase in chemical reactivity is best explained by assuming that a conformational change has taken place around the essential SH as a result of substrate binding.

In the presence of AMP it is possible to label phosphorylase with the NBD-group on a single SH in each subunit without loss of enzymatic activity [6]. The fluorescence of this label is quenched when the enzyme is titrated with AMP and this quenching (expressed as % of the total change in Fig. 9) follows the same titration curve as can be obtained by (a change in the) proton relaxation enhancement. So here two different spectroscopic probes seem to detect the same ligand-induced structural change.

We were fortunate that we were able to label the same rapidly reacting SH with the spin label (IV) as reacts rapidly with NBD-Cl. This behaviour of the enzyme contrasts to the way it reacts with iodoacetamide when two fast SH-groups are modified [9] (This just emphasizes the importance of using a range of reagents in enzyme modification studies). The spin label on the enzyme has significant rotational mobility (the period of rotation is around

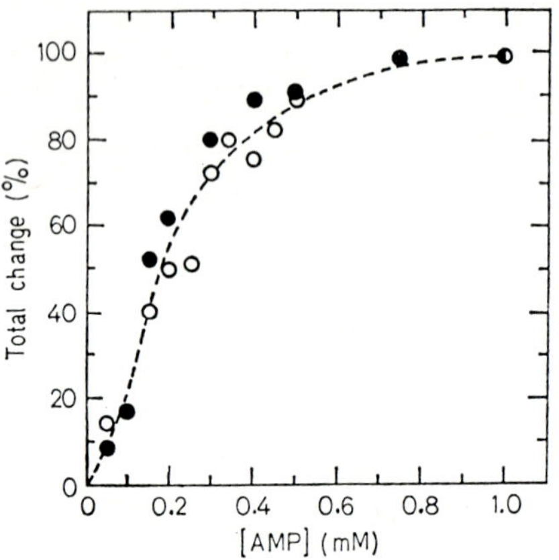

Fig. 9. Comparison of AMP binding to phosphorylase *b* as measured by nuclear magnetic resonance and by fluorescence. Open circles percentage maximal decrease in fluorescence of nitrobenzo-oxa- diazole-phosphorylase *b*. Excitation at 420 nm and emission at 520 nm. Closed circles percentage maximal increase in ε^* Conditions in nuclear magnetic resonance experiments 0.15 M Tris-HCl buffer pH 7.5, 0.1 M KCl and 1 mM EDTA at 25 °C

10^{-9} sec) which can be calculated from the ESR spectrum shown in Fig. 10. This figure also demonstrates that AMP decreases the rotational mobility of the label (by about 30%). The splitting of the ESR lines is consistent with the label experiencing an environmental polarity close to that of water. This, together with the facts that the spin-label rotates more rapidly than the overall tumbling time of the enzyme and that the SH reacts with the reagent (IV) at a rate similar to an unhindered SH-group in model compounds such as glutathione, implies that the particular protein group is on the enzyme surface.

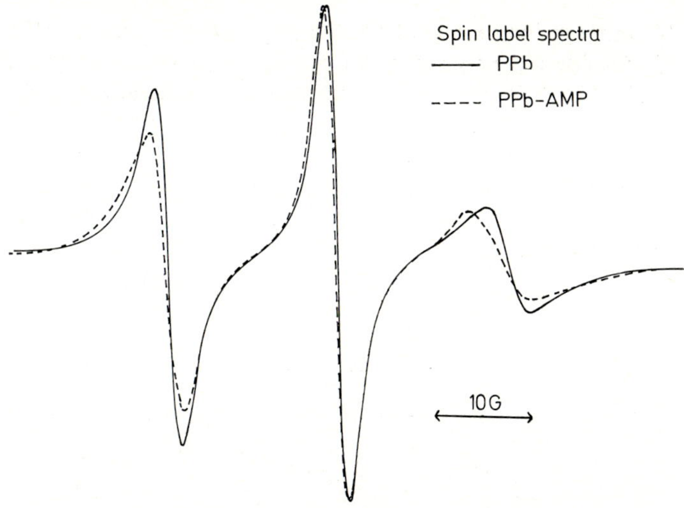

Fig. 10. Electron spin resonance spectrum of phosphorylase *b* labelled with spin label (IV) with and without AMP

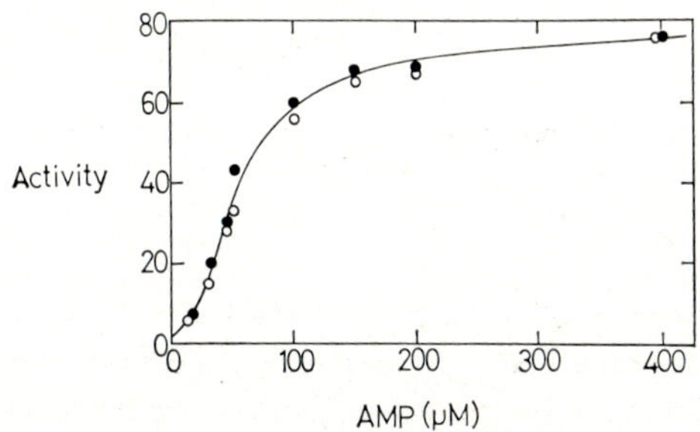

Fig. 11. Activity of native and spin-labelled phosphorylase *b* as a function of AMP concentration (○) native, (●) spin labelled

The modified enzyme is again just as active as the native enzyme (Fig. 11) and also responds to AMP the same way. Thus the spin label in this case is an ideal probe in that it responds to the ligand without measurably interfering with its action.

The change in the mobility of the spin label in response to AMP binding follows very closely the appearance in enzyme activity (Fig. 12). The Hill coefficient n for both these titrations is 1.34, implying significant subunit interaction between the two protomers of phosphorylase both with respect to activity and the spin-label detected conformational change. These experiments also etsablish an important relationship, namely that between the kinetic and physical properties of the enzyme.

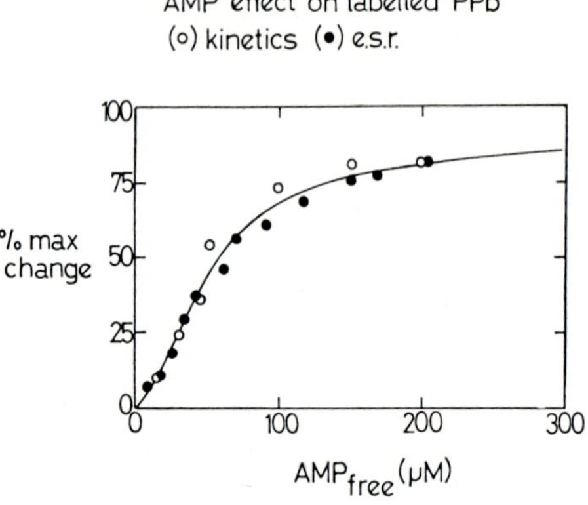

Fig. 12. Effect of AMP on spin-labelled phosphorylase, measured by activity and ESR

Using the ratio of intensities of the low field to middle line as a characteristic parameter of the ESR spectrum, the effect of the different ligands (at the limiting ligand concentrations) is summarized in Fig. 13. The implication of these observations is that we have to ascribe different conformations to the enzyme in the unliganded and liganded states and that G-1-P and AMP and the two together induce different transitions. These four conformational states are also detectable from the limiting proton relaxation enhancement values (Table 1). When these kinds of measurements are done over a range of temperatures, four additional conformational states (Fig. 14) must be postulated as a result of a temperature dependent transition around 13 °C.

Low field/centre ratios for the four conformations of PPb

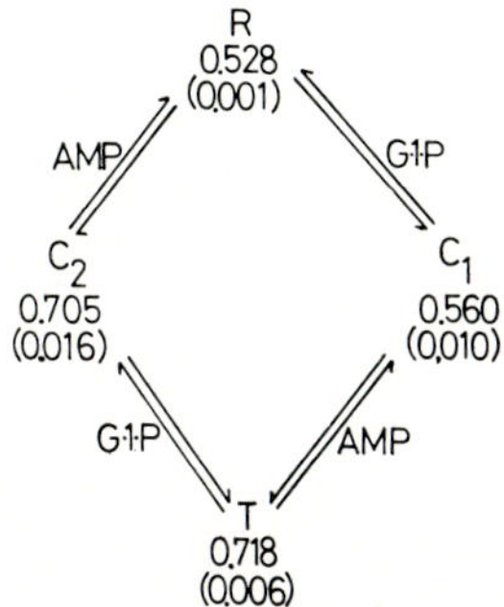

Fig. 13. A scheme for the four conformations of phosphorylase *b* as detected by the spin label

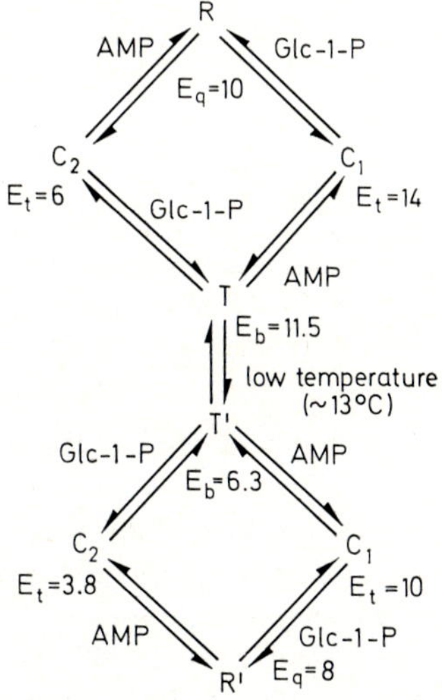

Fig. 14. Proposed scheme for allosteric transitions of phosphorylase *b* as defined by the proton relaxation enhancement data

What these experiments emphasize is that an enzyme like phosphorylase must be regarded as relatively "mobile" in solution though, of course, because of the "sensitivity" of the probe methods, the structural changes involved may indeed be very small. (We shall discuss this point more fully below).

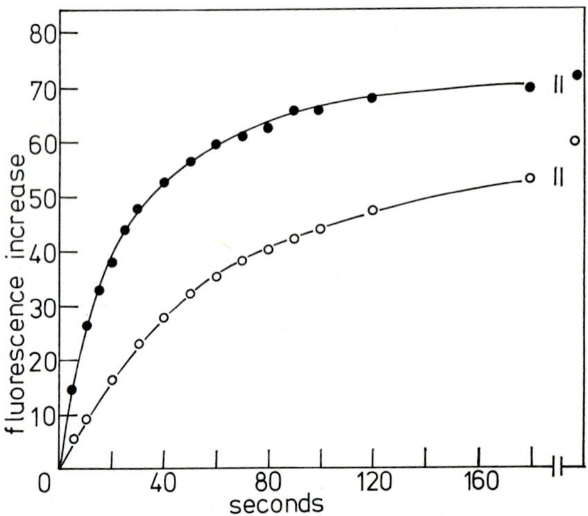

Fig. 15. The dimer-tetramer conversion of phosphorylase *b* as followed by the increase in fluorescence of MNS (20 μM). At zero time AMP and $MnCl_2$ were added at concentrations of 2 mM and 5 mM respectively. The concentrations of phosphorylase *b* were 2.6 mg/ml (●) and 0.7 mg/ml (○). Excitation was at 365 nm and emission at 450 nm. The buffer was 0.05 M triethanolamine containing 0.1 M KCl and 1 mM β-mercaptoethanol at pH 8.5 and 25°

So far we have said nothing about the role of tetramer formation in the activation process or, for that matter, whether the changes detected by the probes are a result of this aggregation. This can be relatively easily decided using the fluorescent probe MNS (II). This molecule binds to phosphorylase with an enhancement of its fluorescence. A further enhancement is observed when AMP is added to the system (Fig. 15), particularly in the presence of high concentrations of divalent metal ions. The rate of this enhancement is however, slow and second-order in enzyme concentration, showing (in agreement with sedimentation velocity studies) [6] that MNS detects the dimer tetramer equilibrium. The ligand-induced aggregation rate is considerably slower than the activation process. In addition, using this fluorescence method, we have been able to demonstrate that the changes observed by the other probes are not a result of the tetramer formation. Thus the ligand-induced conformational changes also precede the aggregation.

V. The Stereochemical Relation between the Different Sites

So far we have shown that four different probes can be used to detect ligand induced structural changes. We would now like to define these changes more precisely and also to find out something

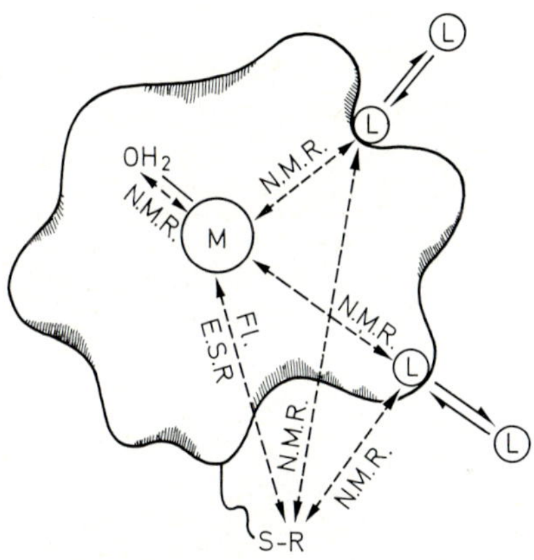

Fig. 16. Schematic representation of the methods used in defining probe sites and their relation

about the stereochemical relations between the different ligand sites. Fig. 16 shows schematically what we have so far: a metal-binding site, an SH-group that can be labelled either with a fluorescent label or a spin label, an allosteric site and a substrate site. To map out the distances between these various groups, we rely on those physico-chemical properties of the groups in question that can be perturbed by the other neighbouring groups. We shall not discuss the technical details of these measurements (they have been described elsewhere) but will give a few simple examples to illustrate the methods.

We can use three types of neighbour interactions to obtain distances:

1. The paramagnetic perturbation of the proton resonances of the ligands (AMP and G-1-P). The centre providing this perturbation may either be the Mn^{2+} ion or the covalently linked spin label.

2. The interaction of the spin label with the bound Mn^{2+} ion (which can be measured by the electron spin resonance properties of the spin label).

3. The quenching of NBD fluorescence by the paramagnetic Mn^{2+}.

We shall first illustrate these methods on some simpler model systems.

Model Systems

In the high-resolution proton magnetic resonance spectrum of AMP three lines are clearly resolvable. These are the aromatic resonances of the H_2 and H_8 hydrogens (Fig. 17) and the H_1' resonance (which is a doublet) of the ribose ring. Mn^{++} binds to AMP and because of its paramagnetic properties it broadens the proton resonances. The measured line width $\Delta\nu$ depends on the amount of Mn^{++} bound to AMP and the distance of the paramagnetic metal from the protons according to the equation:

$$\Delta\nu = C \frac{P_M}{r^6} \cdot f(\tau_c)$$

where C is a constant for the particular system, P_M represents the fraction of the ligand bound to Mn^{++}, and $f(\tau_c)$ is the time scale at which the interaction between the paramagnetic centre and the H nuclei is modulated. This latter quantity may be estimated from independent measurements (such as proton relaxation enhancement data) (see discussion in Ref. [10]). For the AMP-Mn^{++} complex these considerations allow us to derive the distances between the metal and the various hydrogens (Table 2).

Very similar considerations apply to deriving the distances between the paramagnetic centre provided by a spin label and various hydrogen nuclei.

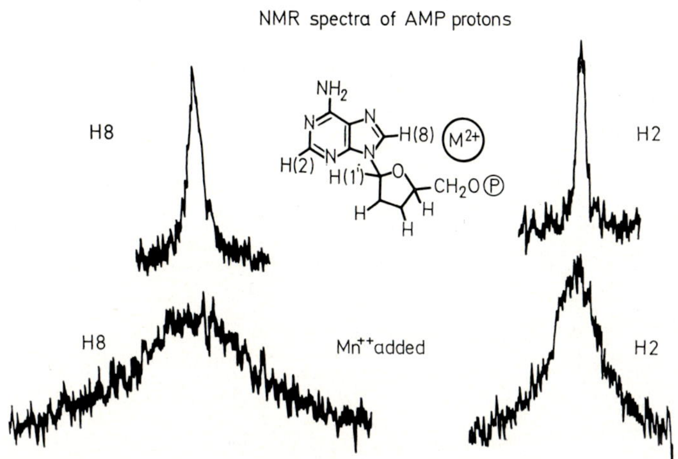

Fig. 17. Effect of Mn^{++} on the magnetic resonance spectra of the H_8 and H_2 protons of AMP. (AMP 20 mM), $MnCl_2$ (10 μM). Buffer: Tris-DCl (50 mM), KCl (100 mM) in D_2O; pH corresponding to same buffer in H_2O was 8.5. NMR experiments were done at 20 °C and at 90 MHz

Table 2. *Mn^{2+}-Proton Distances in the Mn^{2+}-AMP complex*

Nucleus	Distance in Å
H_8	3.4
H_2	4.9
H_1'	5.1

In favourable cases fluorescence can be quenched by paramagnetic ions but the mechanism is not yet fully understood. For example, if we attach an NBD-group to the SH-group of glutathione, then the binding of Mn^{++} to the molecule (probably to the α-amino-

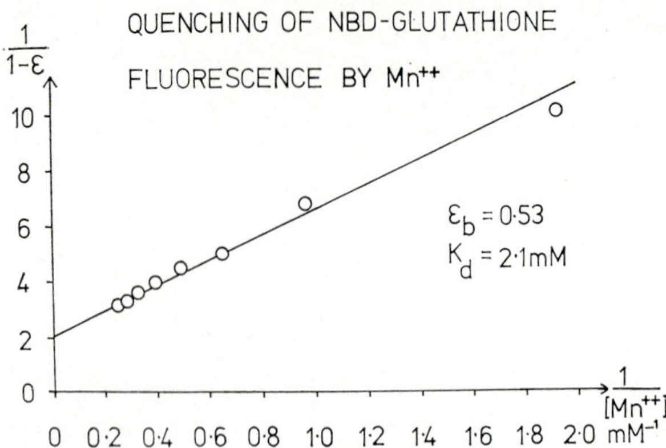

Fig. 18. Quenching of NBD-glutathione fluorescence by Mn^{++}

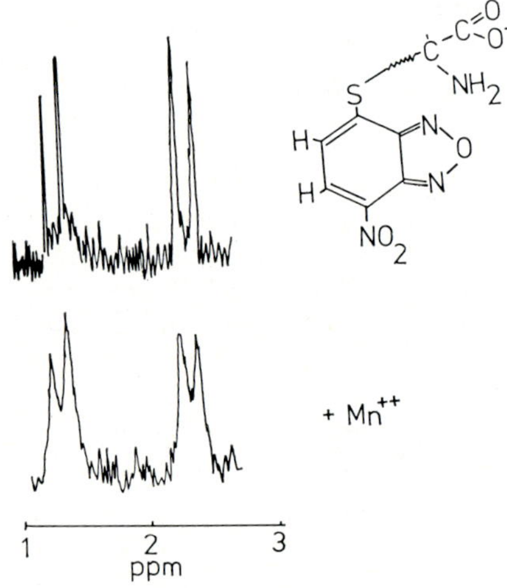

Fig. 19. NMR spectra of S-NBD-glutathione: aromatic protons at 60 MHz. NBD-glutathione was 38.5 mM in Tris-DCl (50 mM), KCl (100 mM) in D_2O; pH corresponding to same buffer in H_2O was 8.5; $MnCl_2$ added (lower curve) was 0.8 mM. Temperature was 35 °C

carboxyl end) quenches the NBD fluorescence, progressively with increasing Mn^{++} concentration, giving a limiting quenching of 50% (Fig. 18). Now we can use our first method to measure the distance between the NBD-group and the Mn^{++} binding site. Fig. 19 shows how the high resolution NMR spectrum of the NBD-group is affected by Mn^{++}. Calculations from these observations give us an NBD-Mn^{++} distance of about 7 Å. We have obtained some idea about the distance dependence of this kind of quenching using several systems described before [10] and some additional ones based on fluorescent labelled oligo-proline peptides. These (still rather preliminary data) point to a $1/(\text{distance})^3$ dependence of the paramagnetic quenching.

Mapping the Probe and Ligand Sites on Phosphorylase b

Just as in the first model system, the Mn^{++} ions bound to the enzyme affect the protons of the bound AMP. Essentially we have to know the binding constant of the metal and of the ligand to the enzyme and the Mn^{++}-AMP binding constant in solution. Then on the basis of four measurements (Fig. 20) we can calculate the line-broadening of the proton resonance for the bound ligand by the enzyme-Mn^{++} complex. From this we can estimate the Mn^{++}-AMP distances on the enzyme.

A simpler procedure can be applied to estimate the ligand-spin label distance. Here we only need two sets of measurements (Ref. [10], Fig. 7) involving the enzyme modified with the nitroxide spin label and the same enzyme with the spin label reduced with dithionite.

The distance between the spin label and the Mn^{++} ion on the enzyme [10] can be estimated [11] by measuring the reduction in intensities of the ESR lines of the nitroxide radical.

Finally, we can show that the fluorescence of the NBD-group attached to the enzyme (to the same SH that can be selectively modified by the spin label) is quenched when Mn^{++} is introduced into the modified enzyme. The limiting quenching in the absence of AMP is 29% while that in the presence of AMP is only 10% (Fig. 21). Using the calibrations we described earlier, the two quenching processes would correspond to an NBD-Mn^{++} distance on the enzyme of 8 and 9.5 Å. Thus the introduction of the allosteric ligand does indeed alter (but only by a small extent) this particular distance. (We shall show below that a similar increase in the NBD-Mn^{++} distance on the enzyme is brought about by phosphorylase $b \rightarrow a$ conversion).

Using these various observations and adding similar ones for the substrate glucose-1-phosphate, we can construct a model for the relevant parts of the enzyme (Fig. 22). In the triangulation of distances, and because we have three reasonably rigidly fixed points on AMP, the orientation of this ligand with respect to the Mn^{++}-spin label axis is fairly well defined. (In fact, since only two of the points,

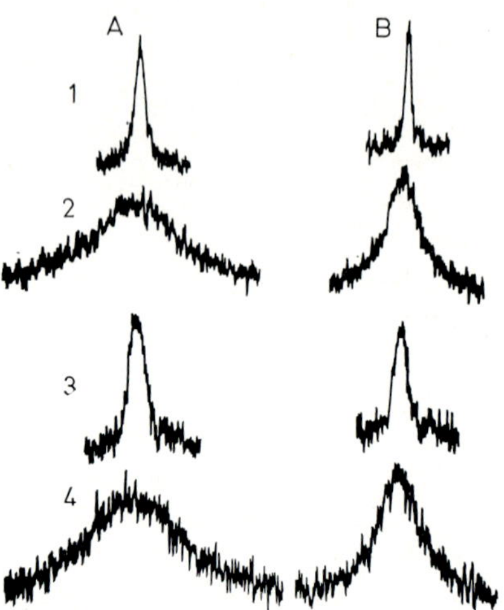

Fig. 20. Effect of Mn^{2+} on the magnetic resonance spectra of the H_8 (A) and H_2 (B) protons of AMP. Curves 1 AMP (20 mM) alone; curve 2 A AMP (20 mM) + $MnCl_2$ (9 μM), 2 B AMP (20 mM) + $MnCl_2$ (45 μM); curves 3 AMP (20 mM) + phosphorylase (245 μM); curve 4 A AMP (20 mM) + phosphorylase (245 μM) + $MnCl_2$ (9 μM), 4 B AMP (20 mM) + phosphorylase (245 μM) + $MnCl_2$ (45 μM). Solutions were in the buffer as in Fig. 1. NMR experiments were done at 25 °C and at 90 MHz

the H_2 and H_8 protons, are rigidly fixed while the third H_1', has one relatively restricted rotational freedom, there are a small number of other solutions). The position of glucose-1-phosphate is less well defined since we only have one H on the molecule located and this can be accommodated on a circle around the Mn^{++}-spin label axis. Nevertheless, the two ligand sites are in close proximity and it appears that the conformational changes that are associated with the regulatory function are also relatively small [10]. We wish to raise here a cautionary note by stressing that the information obtained from the techniques mentioned here make no allowance for non-equivalent binding sites for the ligands on the enzyme.

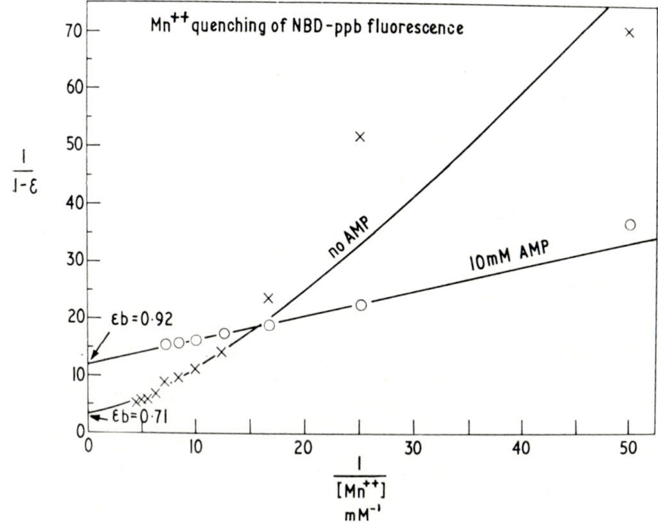

Fig. 21. Quenching of NBD-phosphorylase fluorescence on Mn^{2+} binding. NBD-phosphorylase in Tris-HCl (50 mM) buffer containing KCl (100 mM) at pH 8.5. Temperature was 25 °C. Excitation was at 420 nm and emission at 515 nm. ε on the ordinate is the fluorescence intensity relative to one at $Mn^{2+} = 0$

In such a situation more sophisticated theoretical treatments of the experimental results would be required if similar information is to be obtained. However, we feel justified at present in ignoring this complication since a reasonably self-consistent picture can be built up from the combination of all these techniques. If this model is correct one might expect this ligand-binding region of the enzyme to be close to the subunit interface. This would be required by the sigmoidicity in the probe responses to ligand binding which implies that the conformational changes must be transmitted at least to the interface and possibly further into the other subunit.

Perhaps it is appropriate to mention here that the co-operative oxygen binding in haemoglobin is also a result of fairly small changes in the protein structure [3], the most important ones being localised at the subunit contacts by way of specific salt bridges. It is also worthwhile to point out that an SH-group in haemoglobin, located close to this region of the protein, undergoes changes that can be observed by its reactivity towards specific reagents and by the mobility of a nitroxide spin label attached to it.

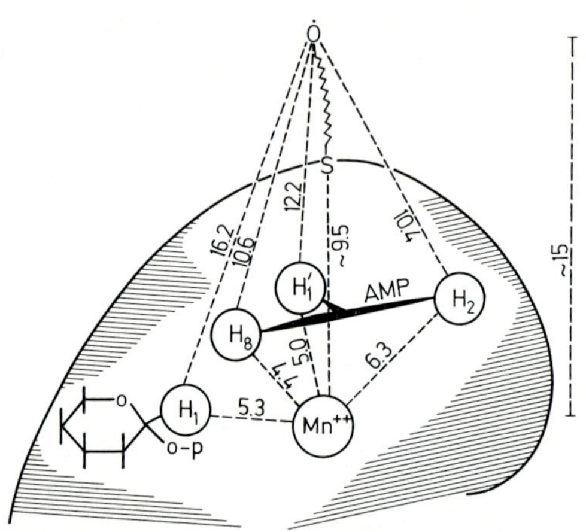

Fig. 22. The stereochemical relation of AMP, glucose-1-phosphate and probe sites on phosphorylase; numbers give distances in Å

VI. Enzyme-Enzyme Interaction in Regulation

We can now examine the second major mechanism of enzyme regulation, that involving an enzyme-catalyzed covalent modification. The activation of phosphorylase *b* by phosphorylase kinase and ATP can be followed by taking spin-labelled phosphorylase *b* and observing the ESR spectrum of the label during the conversion (Fig. 23). The label again becomes more immobilized (as in the activation by AMP). However the kinetics of the physicochemical change is complicated by the fact that the reaction product (ADP) and to a small extent the other components in the reaction (e.g. Mg^{++}) all affect the spin label both in the *b*- and *a*-forms of the enzyme to different extents. When all the ligands are removed after the conversion is completed the *a*-form shows very similar behaviour to that of the *b*-form with saturating amounts of AMP. Thus the ESR spectrum of spin-labelled *a* shows a mobility of the label close to that of the active *b*-form (*b* + AMP). Similarly, the limiting proton relaxation enhancement for the *a*-form (which still

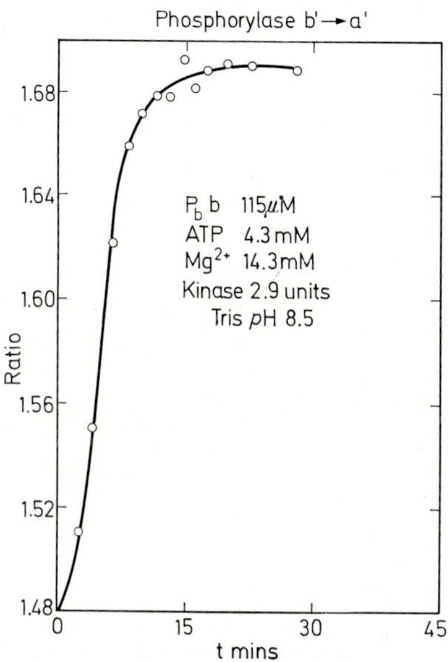

Fig. 23. Time course of phosphorylase *b* to *a* conversion as followed by the ESR spectrum of spin-labelled phosphorylase *b*. The vertical axis (ratio) represents the low field/centre line ratios of the ESR lines

binds one Mn^{++} ion per subunit) is 18 compared with 10 and 15 for the unliganded and liganded (APM) *b*-forms, respectively. Additionally, the Mn^{++} quenching of NBD-phosphorylase *a* fluorescence gives a limiting quenching of 9% (Fig. 24) corresponding to an NBD-Mn^{++} distance similar to that observed in phosphorylase *b* with AMP. These results indicate that phosphorylation of the enzyme locks the conformation in the same form as is obtained at saturating AMP concentrations.

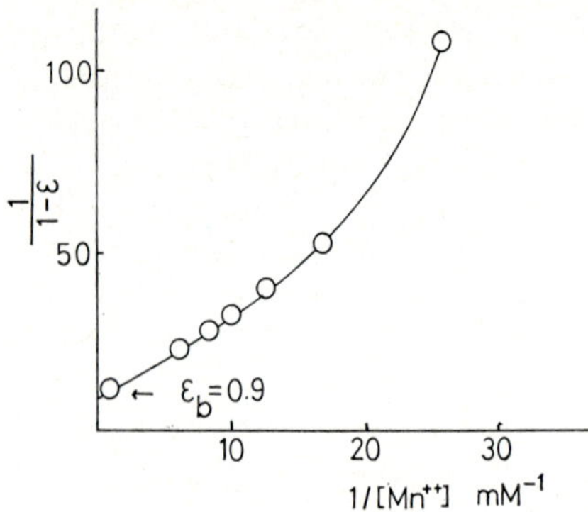

Fig. 24. Quenching of NBD-phosphorylase *a* fluorescence on Mn^{++} binding (c. f. Fig. 21)

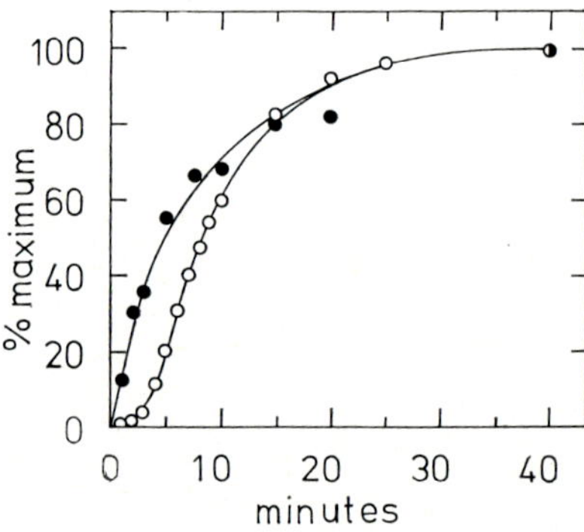

Fig. 25. The phosphorylase kinase catalyzed conversion of phosphorylase *b* to *a* as followed by MNS fluorescence (○), and by phosphorylase activity assayed in the absence of AMP (●). The reaction conditions were the same as in Fig. 2. MNS at a concentration of 20 μM has no effect on the phosphorylase *a* activity

The two active forms of phosphorylase have a higher tendency to aggregate than the unliganded *b*-form. Again we can easily show, using the fluorescence probe MNS, that the activation process in the *b* to *a* conversion is not a consequence of the tetramer formation. If we follow the rate of phosphorylase *b* to *a* conversion in activity measurements and by MNS (Fig. 25), it is seen that enzymic activity appears without a lag while the formation of the tetramer (as detected by the fluorescence enhancement) shows a lag, presumably because at low concentrations of the a-form the rate of tetramer formation is the rate-limiting step in the fluorescence enhancement process. This shows that the dimer is the active species, as has been demonstrated by others using different techniques [2].

VII. Considerations of the Energetics of Subunit Interactions

With the phosphorylase system we have a unique situation to consider the energetic requirements of the conformational perturbations by AMP, in that the a-form is already in the "active" conformation while the b-form is almost entirely in the inactive state. Using the spin-labelled forms of the enzyme we can easily measure the binding constants and mode of binding (i.e. cooperativity) of AMP to the two conformational states, as adding the ligand perturbs the ESR spectrum of the spin label. The binding of AMP to phosphorylase b is strongly sigmoidal with a Hill coefficient of 1.37 (Fig. 12), while binding to the a-form is closer to a hyperbolic function (Hill coefficient of ~ 1) [13]. The dissociation constants, too, are very different, 60 μM and 6 μM to phosphorylase b and a, respectively. If we assume exclusive binding (i.e. that AMP *only* binds to the active form), then the ratio of the "apparent" dissociation constants for the two forms will reflect the free energy difference between the two states. This suggests a ΔG-value of approximately 1.5 kcal/mole for the component of the subunit interaction that is responsible for the observed cooperativity.

It is interesting to compare this activation process with that observed for another nucleotide IMP. While IMP also activates phosphorylase b, this activation is different from the AMP-linked process in several ways: (i) IMP does not change the K_M of the substrate (glucose-1-phosphate) for the enzyme (in contrast with AMP) [12]. (ii) IMP does not affect the conformation of phosphorylase b as measured by proton relaxation enhancement and the spin label. (iii) The IMP activation displays very little co-operativity. (iv) IMP binds more tightly to phosphorylase b than to a (the dissociation constants are $\sim$ 600 μM and $\sim$5 mM respectively).

These observations imply that phosphorylase b is in a relatively more favourable conformation to bind IMP, but not AMP, while for phosphorylase *a* the reverse is true. It also follows that the cooperativity in regulation is linked to the ligand-induced conformational change and its transmission to the subunit contact region. At the same time activation (though less cooperative in nature) can be brought about by small changes in the individual subunits.

VIII. The Effect of Other Ligands and Medium

Regulatory enzymes seem to have a number of features in common: the multiplicity of conformations, subunit interactions,

Table 3. *The dissociation constants of various ligands from spin-labelled phosphorylase a and b and the % decrease in mobility of the spin label caused by the ligand*

Ligand	Phosphorylase *b* % decrease in mobility	Dissociation constant	Phosphorylase *a* % decrease in mobility	Dissociation constant
AMP	33	60 μM	18	6 μM
ADP	14	~200 μM	14	10 μM
ATP	4	2 mM	12	1.3 mM
Phosphate	21	7 mM	7	—
Glycerophosphate	16	4.7 mM	2	—
IMP	−1.5	~600 μM	−3	5 mM
UDPG	−2.0	5 mM	18	13 mM
Glucose-1-phosphate	4	2 mM	3	—
Glucose-6-phosphate	62	60 μM	18	8 mM
Glycogen	18	~0.5%	8	—
Glucose	1.0	~5 mM	—	—

polymerization and generally the marked sensitivity of the enzymic properties to the environment. Glutamine synthetase interacts and is regulated by a variety of ligands [14]. Phosphorylase also interacts with a whole range of ligands that must be present in the cell. Table 3 summarizes some of the effects (and dissociation constants) of different ligands on the spin-label spectrum in both forms of phosphorylase. Some of these ligands compete with each other while others produce positive heterotropic interactions. (We have studied competition between most of these ligands, but the details will not be presented here). Add to this the fact that many of these interactions will be highly dependent on pH, ionic strength, the nature of cations and anions in solution, and it is obvious that the biological situation is extremely complex. Indeed, one must also consider (as has been discussed in this symposium by HEILMEYER) [15] that phosphorylase is probably bound to glycogen and may interact with the other enzymic components in the glycogen particle. So the ultimate understanding of regulation in the cell may require ideas and techniques borrowed from "Colloid chemistry" and not from the chemistry of solutions. As HINSHELWOOD said (in: The Structure of Physical Chemistry, Oxford: Clarendon Press 1951, p. 467): "It is a long way from here to the shapes and structures of living nature. The direction of the path is somewhat as follows. When a crystal of a simple substance grows, the geometry of its solid state is essentially rectilinear. With macromolecular substances new possibilities arise. On the other hand, some of the structural properties of the large molecules themselves tend to impose symmetries of the same kind as the crystallographic symmetries. ... In systems with large size and flexibility, formed of sheets or fibres, or indeed of sheets enclosing actual fluid, the imposition of rounded forms by the surface becomes more and more marked. ... The drama which then unfolds is not for physical chemistry to record."

Acknowledgements. In summarizing our work here, we have drawn on experimental data obtained by a number of people in this laboratory and in particular Drs. A. BENNICK, D. J. BIRKETT, I. D. CAMPBELL, I. R. GRIFFITHS, N. C. PRICE and A. G. SALMON.

This work was supported by the Science Research Council and is a contribution from the Oxford Enzyme Group.

References

1. ANTONINI, E., BRUNORI, M.: In: Hemoglobin and myoglobin in their reactions with ligands. (NEUBERGER, A., TATUM, E. L., Eds.) Amsterdam-London: North Holland Publ. Co. 1971.
2. FISCHER, E. H., POCKER, A., SAARI, J. C.: In: Essays in biochemistry, **6**, 23. (CAMPBELL, P. N., DICKENS, F., Eds.) London: Acad. Press 1970.
3. PERUTZ, M. F.: Nature (Lond.) **228**, 726 (1970).
4. BIRKETT, D. J., PRICE, N. C., RADDA, G. K., SALMON, A. G.: FEBS Lett. **6**, 346 (1970).
5. MILDVAN, A. S., COHN, M.: Advanc. Enzymol. **33**, 1 (1970).

6. Birkett, D. J., Dwek, R. A., Radda, G. K., Richards, R. S., Salmon, A. G.: Europ. J. Biochem. **20**, 494 (1971).
7. Stone, T. J., Buckman, T., Nordio, P. L., McConnell, H. M.: Proc. nat. Acad. Sci. (Wash.) **54**, 1010 (1965).
8. McConnell, H. M., McFarland, B. G.: Quart. Rev. Biophys. **3**, 91 (1970).
9. Zarkadas, C. G., Smillie, L. B., Madsen, N. B.: Canad. J. Biochem. **48**, 763 (1970).
10. Bennick, A., Campbell, I. D., Dwek, R. A., Price, N. C., Radda, G. K., Salmon, A. G.: Nature (Lond.) **234**, 140 (1971).
11. Leigh, J. S., Jr.: J, chem. Phys. **52**, 2608 (1970).
12. Black, W. J., Wang, J. H.: J. biol. Chem. **243**, 5892 (1968).
13. Helmreich, E., Michaelides, M. C., Cori, C. F.: Biochemistry **6**, 3695 (1967).
14. Stadtman, E. R.: Advanc. Enzymol. **28**, 41 (1965).
15. Heilmeyer, L.: This Symposium.

Discussion

E. Helmreich (Würzburg): I have two questions: 1. Dr. K. Feldmann in Würzburg together with Dr. H. Winkler from Professor M. Eigen's laboratory in Göttingen have recently shown that the fluorescence enhancement of the pyridoxamine-P chromophore in reduced phosphorylase which is a consequence of a pH perturbation can be completely reversed by AMP. The nucleotide apparently stabilizes the catalytically active conformation which is the dimeric form of the enzyme with low fluorescence intensity. My question is therefore:

Do Mn^{++} ions also quench the fluorescence of phosphorylase reacted with your fluorescent probe?

2. We have recently presented evidence suggesting a possible participation of pyridoxal-phosphate in the reaction catalyzed by glycogen phosphorylases. If one of the protonatable groups of pyridoxal phosphate, perhaps the phosphate group, pK_2, should actually participate in catalysis, the distance between this group and the C_1 of the glycosyl residue in glucose-1-P or in the α-1,4-glycosidic linkage of glycogen should not be larger than a few Å. Therefore, would you care to comment on the distance between the glucose-1-P-site and the pyridoxal-P site in muscle phosphorylase.

G. K. Radda (Oxford): The answer to your first question is that Mn^{++} ions do quench the fluorescence of the pyridoxal in phosphorylase, but the extent of the quenching is very small and we have not yet been able to quantitate this in any meaningful way.

The answer to the second question is related to this. We have not yet any data that gives us any indication about the distances between the glucose-1-phosphate and pyridoxal phosphate sites in this enzyme, but I hope that we will have such information in the near future.

E. Helmreich (Würzburg): I would like to ask one more question in connection with earlier electronmicroscopic evidence which leads to the conclusion that the subunits in phosphorylase are very similar but not quite identical. Since at present the complete amino acid sequence and the terminal amino acids are not known, this question cannot be answered unequivocally on a chemical basis; on the other hand, there is also no chemical information to suggest that the subunits would not be identical. Would you care to comment on the identity of the subunits of phosphorylase?

N. M. Green (London): Dr. Eagles and Johnson at Oxford [J. Molec. Biol. **64**, 693 (1972)] have done optical diffraction measurements on electron micrographs of negatively stained phosphorylase crystals; with this method you can resolve the subunits. They observed the same rhombic lattice as Valentine and Chignell have described [Nature (Lond.) **218**, 950 (1968)]. However, the arrangement of subunits within each molecule appears rectangular rather than rhombic, which is consistent with the presence of four identical subunits arranged with 222 symmetry. There is therefore no suggestion that the subunits are not identical and the earlier interpretation of the rhombic lattice can be discounted.

Enzyme-Enzyme Interactions in Tryptophan Synthetase from E. coli

K. Kirschner and R. Wiskocil

Biozentrum der Universität Basel, Abt. für Biophysikalische Chemie, Basel, Switzerland

With 14 Figures

A. Multienzyme Complexes: General Concepts

1. Multifunctional Enzymes

Multienzyme complexes are multifunctional in the sense of being able to catalyze two or more different metabolic reactions [1, 2, 3]. In the majority of known examples (e.g. pyruvate dehydrogenase [4] and yeast fatty acid synthetase [5]) the different catalytic functions associated with the complex represent the members of a complete or partial metabolic chain. Physically speaking, multienzyme complexes are stable aggregates of different polypeptide chains (i.e. protein subunits assembled in well-defined proportions). To the extent that there are no covalent bonds between the constituent subunits, these statements describe the quaternary structure of multienzyme complexes. It is reasonable to conclude that the evolution of these organized enzymes utilized the inherent advantages offered by subunit — subunit (or quaternary) interactions between different proteins. What are these advantages?

2. Monofunctional Enzymes

It is helpful first to consider the effect of quaternary structure at the simpler level of monofunctional enzymes. In contrast to multifunctional enzymes, the latter class catalyzes only *one* metabolic reaction. It is a well-known fact [6] that most intracellular monofunctional enzymes occur in the form of aggregates (or "oligomers"; [7, 8]) of two or more identical subunits (or "protomers"). Lactate dehydrogenase [9] is a well-studied example of a tetrameric, monofunctional enzyme.

It is a tantalizing hypothesis that the quaternary structure is necessary because the active site(s) do not preexist in the monomers but are formed only in the contact region between two identical subunits [3]. However, the NAD+ binding sites of lactate dehydrogenase [9] are located almost entirely within the individual protomers. Moreover, they are well separated from each other in the

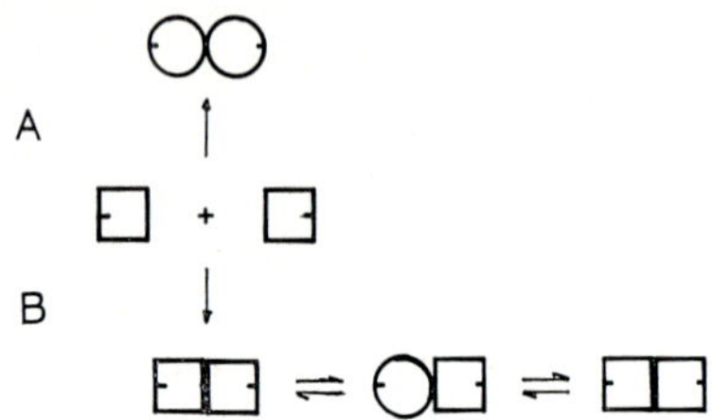

Fig. 1. Monofunctional Enzymes: The consequence of subunit association. The subunit is considered to exist in at least two different conformations (indicated schematically by squares and circles). Subunit association, when accompanied by a conformational change, can lead to stabilization and/or activation of individual subunits (A). Alternatively, the coupling of conformational changes *within* the oligomer can lead to cooperative binding of substrates, activators or inhibitors (i.e. allosteric regulation of enzyme activity, B). Note that the active sites (thick marks) are physically separated

tetrameric structure. While it is difficult to generalize from a few examples, consideration of the most probable pathways for the evolution of enzyme structure makes the generation of catalytic sites between subunits appear to be the exception rather than the rule. Thus the advantages apparently gained by monofunctional enzymes from quaternary structure are not immediately evident. The fact that some monofunctional enzymes are less active or inactive in the dissociated state (e.g. lactate dehydrogenase [10]) indicates that the individual active site might be stabilized via conformational changes accompanying the formation of quaternary structure. This is schematically indicated in Fig. 1 A.

On the other hand it is clear that, in the special case of regulatory (or allosteric [7, 8]) enzymes, the multiplicity of sites generated by association of identical subunits is a necessary prerequisite for indirect and cooperative interactions between identical but separated sites. There is evidence that these effects can be mediated via coupled conformational changes *within* the oligomeric structure as indicated in Fig. 1 B. Hemoglobin, [11]; aspartate transcarbamylase, [12]; and glyceraldehyde-3-phosphate dehydrogenase, [13], are a few well studied examples.

3. Multienzyme Complexes

It is clear that aggregates of different proteins, each contributing a specific catalytic site, must be able to profit from quaternary structure in the same general way as monofunctional oligomers. This is indicated schematically in Fig. 2. That is, the interaction between subunits can lead to a mutual stabilization of active conformations and/or to an indirect and coordinated control of enzyme activity. This type of regulation is illustrated by aspartokinase (I)-homoserine dehydrogenase (I) from E. coli [14], in which both associated catalytic activities are inhibited by threonine in a cooperative fashion.

Besides their heterologous composition, multienzyme complexes appear to differ from monofunctional enzymes in another important way: the different active sites of a number of complexes appear to be closely *juxtaposed* [1—3]. The inherent advantages of this spatial organization for promoting the efficiency of catalysis is clear in general terms. Juxtaposition of catalytic sites involved in the catalysis of sequential metabolic reactions would be expected to increase the overall catalytic efficiency by reducing the diffusion paths for products (=substrates of subsequent reactions). Moreover, metabolic intermediates could be prevented from diffusing into the surrounding medium ("channelling", cf. [15]). This would eliminate undesirable branching of the particular metabolic pathway and could also protect labile intermediates from spontaneous degradation [1—3]. Thus complex formation appears to be the most efficient means of concentrating and compartmentalizing different

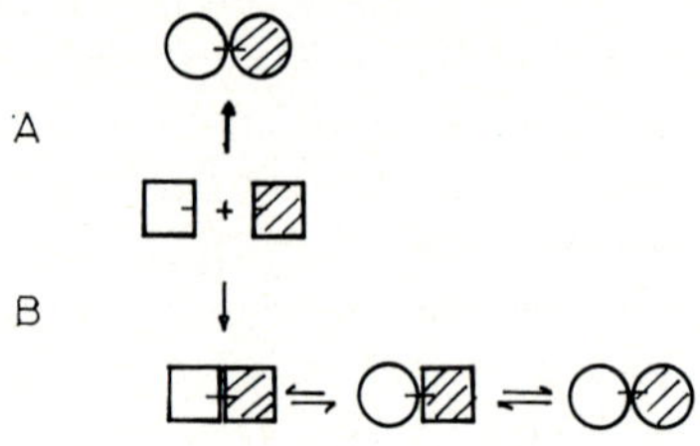

Fig. 2. Multifunctional Enzymes: The consequence of subunit association. Different proteins (enzymes) are characterized by open or shaded symbols. The shape of the symbols and the meaning of the interactions indicated by routes A and B are the same as in Fig. 1. Note, however, that the active sites are juxtaposed

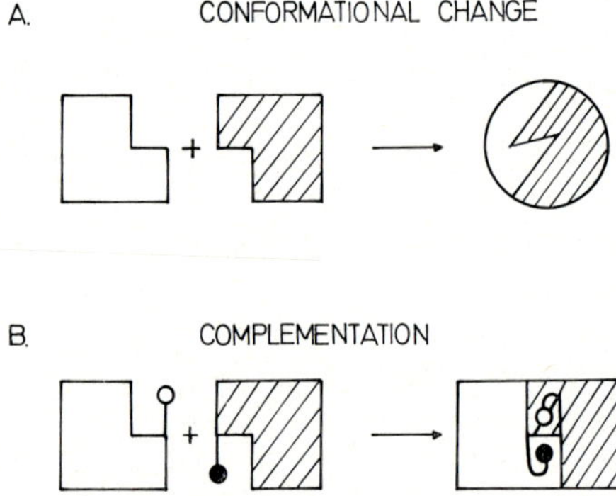

Fig. 3. Mutual activation of different enzymes by complex formation. Different shading indicates different proteins. Different shapes indicate different conformations. A. Mutual activation by mutually induced or stabilized changes of conformation ("conformational model"). B. Mutual activation by complementing side chains from the other protein respectively. The side chain is indicated by a small arm ending in a circle ("complementation model")

but metabolically related enzymes at the molecular level [16]. Moreover, the binding of the initial substrate could activate the subsequent catalytic sites by direct or indirect means (e.g. conformational changes) [17].

The concept of "composite active sites" [18] also allows for a mutual activation of different active sites by direct means (Fig. 3) In what is here termed "complementation", one subunit could donate a catalytically important amino-acid side chain to the active site of the contiguous subunit and vice versa.

B. Experimental Approaches

The brief introduction above to the interesting relationship between the structure and function of multienzyme complexes has emphasized the possible role of the structural flexibility of the constituent subunits. Thanks to the advance of various physicochemical techniques, it has recently become possible to ask meaningful questions (in the form of more or less direct experiments) about the involvement of discrete conformational changes in the function of enzymes (e.g. structure determination by X-ray diffraction [19, 20], nuclear magnetic resonance [21] or rapid reaction techniques [22, 23], to cite but a few). Hemoglobin is an excellent example of the deep insight gained into the mechanism of cooperative binding of oxygen and the structural changes of the protein involved by the application of most of the techniques cited above [11]. Certain questions (e.g. the structure of composite active sites and the nature and extent of conformational changes) can only be resolved by determining the molecular structure of multienzyme complexes by X-ray crystallography. However, since the more complex systems such as fatty acid synthetase and the α-ketoacid dehydrogenases are characterized by molecular weights of 2 to 5×10^6 daltons, the task is formidable. It is thus pertinent to look to the other methods for answering at least some of the important questions in the meantime.

In contrast to oligomers composed of identical subunits, multienzyme complexes afford the operational advantage of isolating the component (monomeric) enzymes after reversible dissociation of the complex. The basis for this approach is the chemical difference between heterologous enzymes. In the case of monofunctional oligomers, separate monomers can only be obtained at high dilution or after chemical modification. Hemoglobin is an exception because the α and β chains are chemically different although functionally equivalent to a first approximation [11]. Subsequently, various physical and chemical properties of the separated components can be compared with those observed in the complex *in situ*. The individuality of the different proteins (characterized by specific functions) provides specific "handles" by which the changes imposed upon each component during the assembly process can be studied separately. Aspartate transcarbamylase, for example, occupies a position intermediate between monofunctional oligomers such as lactate dehydrogenase on one hand and multienzyme complexes on the other. The separation of ATCase into catalytic and regulatory subunits (which have no known catalytic function) is easily achieved and the studies on the isolated components have led to a wealth of new information [12].

One can be reasonably optimistic that the results of such investigations on multienzyme complexes will throw new light on the possible role of conformational changes in stabilizing pre-existing and/or in generating new active or binding sites, on the changes of the catalytic mechanism induced by juxtaposition of several active sites and on the mechanism of coordinate regulation of the activity of multienzyme complexes. Moreover, the assembly of multienzyme complexes giving rise to well-defined closed structures can be regarded as a model for other assembly processes (for example, the morphogenesis of viruses [24]).

The ease with which multienzyme complexes can be reversibly dissociated differs from case to case. One of the simplest and best-studied examples is tryptophan synthetase from E. coli. (For recent reviews, see [1, 25]). This interesting type case of a multienzyme complex is currently under active investigation in a number of laboratories. The results obtained in using the general approach described above will now be discussed.

C. Tryptophan Synthetase

1. Partial Reactions and Properties of the Enzyme

Tryptophan synthetase catalyzes the last step in the biosynthesis of tryptophan from indole-glycerol-P and L-serine (Reaction 1) [1, 25].

$$\text{indole-glycerol-P} + \text{L-serine} \rightleftharpoons \text{L-tryptophan} + \text{glyceraldehyde-P} \qquad (1)$$

The enzyme from E. coli is a moderately stable complex of two quite different proteins, referred to as the α and β_2 subunits. The individual subunits can be separated easily by various methods. Reassociation of the enzyme from the subunits gives a preparation indistinguishable from the native $\alpha_2\beta_2$ complex. With no evidence to the contrary, tryptophan synthetase can be regarded as a dimer of functional $\alpha\beta$ dimers with no direct or indirect interactions between the $\alpha\beta$ dimers [18, 26]. It is also a comparatively small multienzyme complex (mol wts: α subunit $= 29000$, β_2 subunit $= 2 \times 45000$ daltons).

Neither the α nor the β_2 subunit is capable of catalyzing the physiological reaction (Reaction 1) by itself. However, both the α and β_2 subunits still possess distinct and measurable functions. The α subunit catalyzes the reversible aldolytic cleavage of indole-glycerol-P (Reaction 2). The β_2 subunit contains the two equivalents of pyridoxal-P found in the $\alpha_2\beta_2$ complex and catalyzes the practically irreversible condensation of indole with L-serine to form L-tryptophan (Reaction 3):

$$\text{indole-glycerol-P} \xrightleftharpoons{\alpha,\ \alpha_2\beta_2} \text{indole} + \text{glyceraldehyde-P} \qquad (2)$$

$$\text{indole} + \text{L-serine} \xrightarrow[\text{(pyridoxal-P)}]{\beta_2,\ \alpha_2\beta_2} \text{L-tryptophan} \qquad (3)$$

Reactions 2 and 3 can also be catalyzed by the $\alpha_2\beta_2$ complex. At first sight they appear to be the partial reactions, which, in sequence, result in the overall reaction (Reaction 1). Thus the allocation of the partial reactions to two different subunits serves to define tryptophan synthetase as a multienzyme complex.

Tryptophan synthetase from E. coli is characterized by further properties which make this enzyme particularly interesting as an experimental system and which might be representative for other multienzyme complexes. Formation of the $\alpha_2\beta_2$ complex leads to a 30 to 100 fold increase in the catalytic efficiencies (turnover numbers) of the partial reactions (Reactions 3 and 2) over those observed with the individual α and β_2 subunits respectively [25].

Interestingly, this phenomenon of activation by protein-protein interaction is also observed with many mutants of tryptophan synthetase in which either the α subunits or the β_2 subunits are enzymically inactive *per se* or in the complex [1, 25]. In other words, activation of one subunit by the other does not necessarily require the intactness of the active site of the latter.

2. Kinetic and Equilibrium Studies

Steady-state kinetics

CREIGHTON [18] has presented detailed studies of the rates of indole-glycerol-P cleavage to indole and glyceraldehyde-P (Reaction 2) and of the formation of tryptophan from indole-glycerol-P and serine (Reaction 1) as catalyzed by the $\alpha_2\,\beta_2$ complex of tryptophan synthetase. He also investigated the inhibition of both reactions by indole and tryptophan (the two products) respectively.

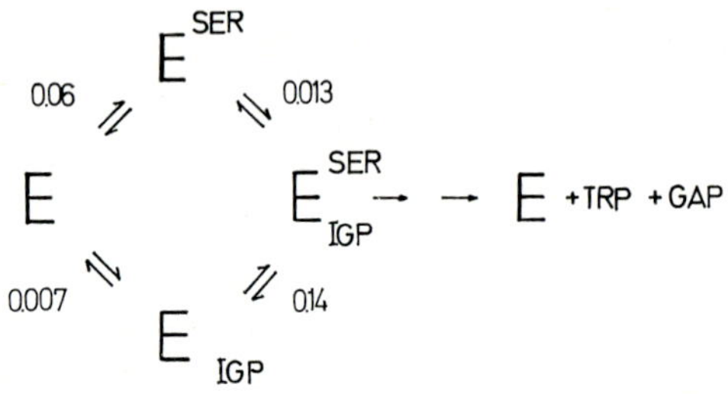

Fig. 4. Mechanism of catalysis for Reaction 1: indole-glycerol-P + serine → tryptophan + glyceraldehyde-P. E stands for the unoccupied active site of the $\alpha_2\,\beta_2$ complex. E · ser and E · igp indicate the binary complexes of the enzyme with serine and indole-glycerol-P respectively. E · ser · igp is the corresponding ternary complex of enzyme with both substrates. *Trp* is tryptophan, *gap* is glyceraldehyde-P. The numbers associated with the binding equilibria are the values of the corresponding apparent dissociation constants (mM) as obtained from the analysis of steady state kinetic data (from CREIGHTON [18])

The mechanism for the two-substrate reaction 1 is depicted in Fig. 4. Here the mechanism of substrate binding is compatible with a random, rapid-equilibrium process, leading to the ternary enzyme: indole-glycerol-P: serine complex. This is subsequently converted to the product, tryptophan, in the rate-determining catalytic step. It is not surprising that tryptophan is a competitive inhibitor with respect to serine. However, the inhibition of tryptophan is non-competitive with respect to indole-glycerol-P [18]. An interesting quantitative finding is that the apparent dissociation constant of tryptophan synthetase for indole-glycerol-P is increased when the enzyme is saturated with the other substrate, serine. The opposite is also true (compare the values on opposite sides of the quadratic scheme in Fig. 4).

The simplest mechanism compatible with the data [18] on Reaction 2 as catalyzed by the $\alpha_2\beta_2$ complex is summarized in the following

Scheme 1.

$$E \overset{0.04}{\leftrightharpoons} E.IGP \leftrightharpoons E.IND + GAP \leftrightharpoons E + IND + GAP$$

The notation is as in Fig. 4 with IGP = indole-glycerol-P and IND = indole. It consists of the ordered release of the products indole and glyceraldehyde-P after the rate-determining cleavage of the C-C bond.

Unfortunately, the mechanism of indole-glycerol-P cleavage as catalyzed by the subunit alone has not yet been studied in detail. The K_M for indole-glycerol-P is approximately 10 times larger than in the $\alpha_2\beta_2$-catalyzed reaction (K = 0.47 mM, [27], vs. K = 0.04 mM, [18] cf. Scheme 1). This indicates that the affinity of the α subunit for indol-glycerol-P increases upon incorporation into the $\alpha_2\beta_2$ complex. More definitive evidence on this point has been obtained by binding and inhibition studies with indole-propanol-P (Fig. 5). This compound is a substrate analogue of indole-glycerol-P, but cannot undergo aldolytic cleavage [28].

Indole-propanol-P is a strictly competitive inhibitor with respect to indole-glycerol-P in Reaction 2 catalyzed by the $\alpha_2\beta_2$ complex (Fig. 6). The inhibition constant (K_I = 0,005 mM) is one-half to one-third as large as the thermodynamic dissociation constants measured by equilibrium dialysis and spectrophotometric titration (K_D^{IPP} = 0.011 to 0.015 mM, [28]). It is identical to the apparent dissociation constant (K_{ic} = 0.007 mM, [18], cf. Fig. 4) of the enzyme for indole-glycerol-P, indicating that the hydroxyl groups of the latter are not important for binding.

In contrast, the apparent thermodynamic dissociation constants of indole-propanol-P and indole-glycerol-P to the α subunit

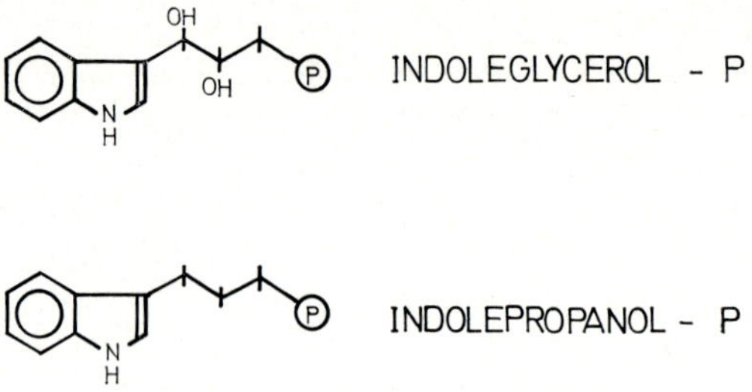

INDOLEGLYCEROL - P

INDOLEPROPANOL - P

Fig. 5. Indole-propanol-P is an analogue of indole-glycerol-P

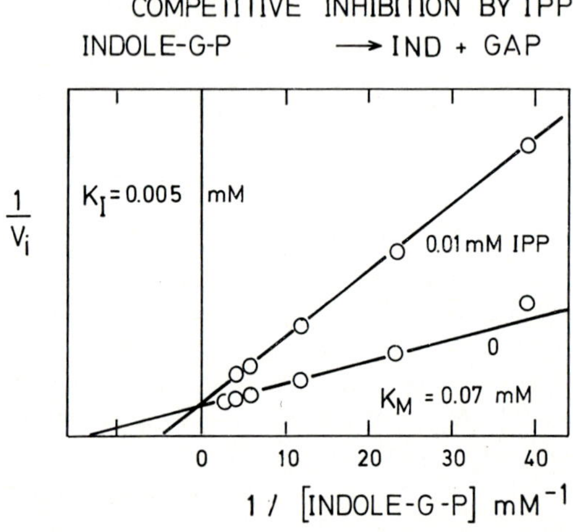

Fig. 6. Competitive inhibition, by indole-propanol-P, of the cleavage of indole-glycerol-P to indole (Reaction 2) as catalyzed by the $\alpha_2\,\beta_2$ complex. The reaction was followed as described by Creighton [18]. The concentrations of substrate and inhibitor were varied as indicated in the Figure. K_M is the Michaelis constant for indole-glycerol-P and K_I the inhibition constant for indole-propanol-P. The intercepts of the reciprocal plots on the ordinate are consistent with the concept of competitive inhibition by indole-propanol-P

alone ($K^{IPP} = 0.1$ mM, $K^{IGP} = 0.11$ mM, measured again by equilibrium dialysis and spectrophotometric titration [28]) are approximately 10 times larger. This shows that the 100-fold increase in the catalytic activity of the α subunit upon complex formation with the β_2 subunit is accompanied by an increase in affinity for the substrate and its analogue, albeit to a lesser degree.

Complementary studies with the β_2 subunit and the $\alpha_2\beta_2$ complex of tryptophan synthetase as catalysts for the synthesis of tryptophan from indole and serine (Reaction 3) have recently been

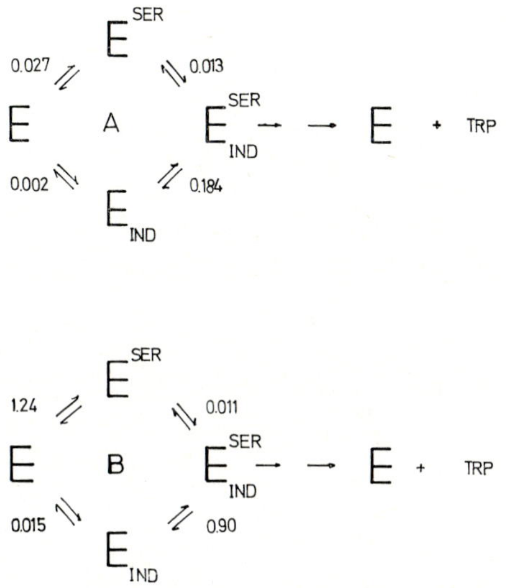

Fig. 7. Mechanism of catalysis for Reaction 3: indole + serine → tryptophan. The numbers and symbols have the same general meaning as in Fig. 4, except that E stands for the active site of the $\alpha_2\beta_2$ complex in A and for that of the β_2 protein alone in B. Data from FAEDER and HAMMES [29, 30]

published by FAEDER and HAMMES [29, 30]. Although the evidence from steady-state kinetics is not conclusive, it is compatible with the rapid-equilibrium, random addition mechanism shown in Fig. 7 A and B. It is interesting to note that the mechanism apparently does not change when the β_2 subunit is incorporated into the $\alpha_2\beta_2$ complex. Moreover, the apparent dissociation constants of indole and serine to form the respective binary complexes decrease 10 to 50 fold when the β_2 subunit is incorporated in the $\alpha_2\beta_2$ complex.

In the case of serine binding, the increase of affinity has been directly confirmed by spectrophotometric titration experiments [29, 30].

Finally, indole-propanol-P is a *non*-competitive inhibitor with respect to indole in Reaction 3 as catalyzed by the $\alpha_2 \beta_2$ complex.

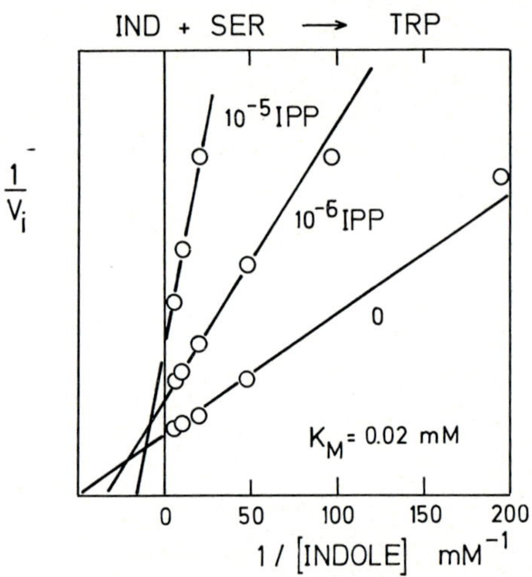

Fig. 8. Non-competitive inhibition, by indole-propanol-P, of the synthesis of tryptophan from indole and serine (Reaction 3) as catalyzed by the $\alpha_2 \beta_2$ complex. The reaction velocity was measured as described by FAEDER and HAMMES [29]. The concentrations of substrate and inhibitor were varied as indicated in the Figure. The L-serine concentration was 30 mM. K_M is the Michaelis constant for indole. The intersection of the reciprocal plots to the left of the ordinate indicates non-competitive inhibition

(Fig. 8 [28]). This observation is completely analogous to the non-competitive inhibition of the overall Reaction 1 [18] by tryptophan with respect to indole-glycerol-P.

It supports the previous conclusion that indole-glycerol-P and tryptophan can be bound simultaneously to the $\alpha_2 \beta_2$ complex of tryptophan synthetase.

In spite of several open questions, the information from steady-state kinetics justifies the following tentative conclusions:

1. The mechanism of the partial Reaction 3 [29, 30] and perhaps also that of Reaction 2 remains *unchanged* when the α and β_2 subunits form the $\alpha_2\beta_2$ complex.

2. The non-competitive inhibition by tryptophan towards indole-glycerol-P [18] and of indole-propanol-P towards indole [28] is evidence for the concept that the indole subsites which must exist in the separate α and β_2 subunits are *retained* in the $\alpha_2\beta_2$ complex (cf. also [31]).

3. The increase in catalytic efficiencies ("mutual activation") of the α and β_2 subunits observed upon complex formation are *accompanied by an increase in affinity* (i.e. in specificity) for the substrates indole-glycerol-P, indole and serine [30].

4. The failure in detecting indole as an obligatory intermediate in the synthesis of tryptophan from indole-glycerol-P (Reaction 1) and other evidence has led CREIGHTON [18] to propose a new, concerted mechanism for this reaction. It is distinct from the sum of the partial reactions in that the indole moiety of indole-glycerol-P is transferred to serine in one step (push-pull mechanism). However, it is not clear whether and how the transfer of the indole ring from the α subsite to the β subsite (cf. conclusion 2) occurs. Alternatively, Reaction 1 could still represent the sequence of the partial reactions with the difference that the indole liberated in Reaction 2 is rapidly and completely transferred (i.e. "channelled") to the high-affinity indole subsite on the β_2 subunit This model has been developed by DEMOSS [31] for the tryptophan synthetase from *Neurospora crassa*. In the presence of serine the indole moiety would react to form tryptophan before it has had a chance to escape into the solvent.

5. In the light of the evidence discussed above, it is almost inescapable to conclude that the preexisting sites of the α and β_2 subunits are brought into *direct contact* within the $\alpha_2\beta_2$ complex of tryptophan synthetase (formation of a "composite catalytic center" [18, 31] Fig. 9).

Given the existence of a composite active site (Fig. 9), at least two models can explain the phenomenon of mutual activation. In the first model (here called the "complementation model"), the active sites for each partial reaction catalyzed by the $\alpha_2\beta_2$ complex are augmented by amino-acid side chains donated by the other subunits [3], (cf. Fig 3. B). The artificial formation of active dimers of ribonuclease [32] can be taken as an analogy for this model. In the extreme case, the overall structure of the component subunits could remain essentially rigid during formation of the complex.

The second model (here called the "conformational model") invokes the popular concept of the flexibility of protein structure [1, 8, 30] (cf. Fig. 3 A). The relatively low catalytic efficiencies of the active sites of the α and β_2 subunits in the monomeric state

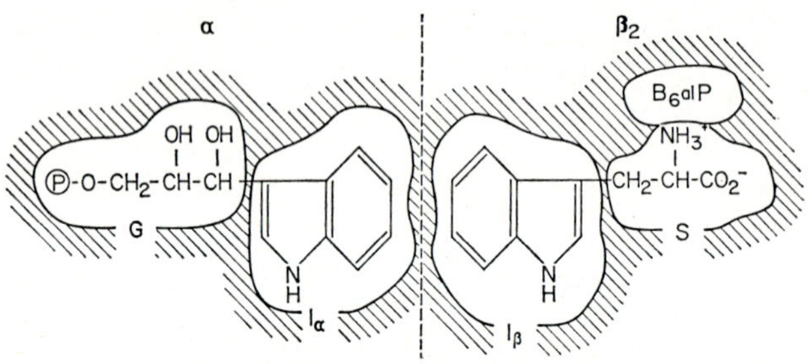

Fig. 9. The composite catalytic center of the tryptophan synthetase $\alpha_2\beta_2$ complex from E. coli. The binding sites for indole-glycerol-P to the α subunit and of L-tryptophan to the β subunit are schematically subdivided into subsites. For example, Iα and Iβ are the indole subsites in the α and β subunits. The dotted line represents the area of contact between the α and β subunits. After Creighton [18]

and their high efficiency in the complex are characterized by *different* conformations of the respective subunits. In forming the complex from the monomers, the previously nonexisting, efficient conformations would be induced, or the preexisting but nonprevalent conformations would be stabilized (cf. Fig. 2). An intermediate model can be construed in which both complementation *and* conformational changes are needed to explain the mutual activation phenomenon.

The suggestive evidence from steady-state kinetics, voluminous as it is, suffers from the general limitations of this approach. For example, the apparent dissociation constants for substrates have different physical meaning, depending on the assumed mechanisms.

At the present moment it is not possible to determine the mechanism in an unequivocal manner. Moreover, reaction schemes of the kind shown in Figs. 4 and 7 must be considered minimal mechanisms since they are not resolved into the elementary steps. Finally, no definitive conclusions can be drawn either about the physical structure of the postulated composite active site, or about the origin of the observed increase in catalytic efficiency and affinity for substrates upon formation of the complex from the monomers.

The elucidation of the structure of tryptophan synthetase by X-ray crystallography is under way [33] and can be expected ultimately to clarify these questions. However, the possible involvement of conformational changes in the formation of the $\alpha_2 \beta_2$ complex can be studied in absence of structural information, for example, by rapid reaction techniques.

Since the increase in catalytic efficiency is coupled to an increase in affinity for substrates, the experimental system can be simplified to the study of the kinetics of binding of substrates (e.g. serine) and of substrate analogs (e.g. indole-propanol-phosphate) at high enzyme concentrations. The transients (observed in rapid-mixing experiments) or relaxation processes (observed in chemical-relaxation experiments) can be expected to detect existing conformational equilibria in the form of discrete isomerization steps. Moreover, the analysis of the kinetics of binding steps can lead, in principle, to the direct determination of thermodynamic dissociation constants of individual binding steps.

3. Fast Reaction Studies

a) The catalytic mechanism

The presence of pyridoxal-P at the active site of the β_2 subunit provides a sensitive, chromophoric indicator for the formation of intermediates participating in the various reactions of tryptophan synthetase. Four such intermediates have been detected via their characteristic spectra [34—36]. Stopped-flow kinetic studies of the formation and disappearance of three of these species [36] indicate that all are intermediates in pyridoxal-P-dependent Reactions 1 and 3 catalyzed by the $\alpha_2 \beta_2$ complex. These intermediates have been tentatively identified with certain chemical species [25]. The major effect of association of the β_2 subunit with the α subunit is to enhance the rate of Reaction 3 by accelerating the formation of the Schiff base formed between serine and enzyme-bound pyridoxal phosphate [36].

b) Binding of substrates to the β_2 subunit and the $\alpha_2\,\beta_2$ complex

FAEDER and HAMMES [29, 30] have used stopped-flow and temperature-jump techniques in studying the rates of binding of substrates to the isolated β_2 subunit and to the $\alpha_2\,\beta_2$ complex of tryptophan synthetase. Measurements with the β_2 subunit alone revealed the existence of an isomerization equilibrium which disappears when α subunit is added. This finding is consistent with a preequilibrium of the β_2 subunit between two states which is shifted towards one side upon forming the $\alpha_2\,\beta_2$ complex. The observation that addition of NH_4^+ ions can mimic the effect of α subunit in stimulating the β_2 subunit to approximately 65% of the level attained with excess α subunit [37] supports the interpretation that the β_2 subunit is a flexible protein.

The kinetics of serine (or tryptophan) binding to the β_2 subunit or to the $\alpha_2\,\beta_2$ complex is not yet completely understood. The data [29, 30] can be tentatively summarized as follows: The binding of ligands is accompanied by at least two isomerization steps. The faster one is related to the first-order process observed with the β_2 subunit alone. The slower one, can be interpreted as an isomerization of the primary enzyme ligand complex. This is illustrated in Scheme 2.

$$\mathrm{L} + \mathrm{E} \rightleftharpoons \mathrm{L} + \mathrm{E}' \rightleftharpoons \mathrm{L}\cdot\mathrm{E}' \rightleftharpoons \mathrm{L}\cdot\mathrm{E}''$$

where L is the ligand (serine or tryptophan) and E, E′ and E″ are different (conformational) isomers of the enzyme. The smaller overall dissociation constant of the $\alpha_2\,\beta_2$ complex for serine with respect to the β_2 subunit appears to be due to a relative shift of the final isomerization equilibrium ($\mathrm{L}\cdot\mathrm{E}' \rightleftharpoons \mathrm{L}\cdot\mathrm{E}''$) towards the right.

c) Binding of indole-propanol-P to the α subunit

The availability of the substrate analogue indole-propanol-P [28] allows similar measurements to be carried out with the α subunit. As mentioned previously, the binding of indole-propanol-P has been studied by equilibrium dialysis (using ^{32}P-labelled material [28].)

Fig. 10 shows the results of equilibrium dialysis studies with the α subunit, the β_2 subunit and the $\alpha_2\beta_2$ complex. It is clear that only one mole of indole-propanol-P is bound per equivalent of α subunit and that the β_2 subunit binds only negligible amounts in the relevant concentration range. Moreover, the affinity for indole-propanol-P of the $\alpha_2\beta_2$ complex ($K_D = 0.01$ mM) is increased

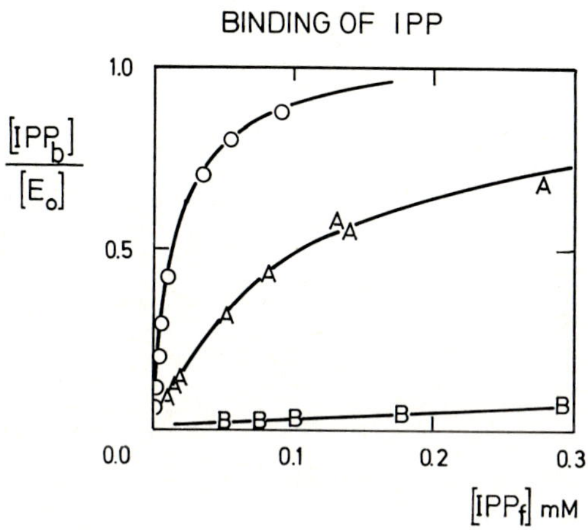

Fig. 10. Binding of ^{32}P-indole-propanol-P to tryptophan synthetase. [IPP b] = concentration of bound indole-propanol-P; [IPP f] = concentration of free indole-propanol-P; [Eo] = concentration of total enzyme. The binding was measured by equilibrium dialysis in 0.1 M phosphate buffer pH 7.6 at 20 °C. The concentration of total enzyme ranged from 0.05 to 0.2 mM. *A*: α subunit, *B*: β_2 subunit, ○: $\alpha_2\beta_2$ complex

by about a factor of 7 over that of the separate α subunit ($K_D = 0.07$ mM).

Fig. 11 shows the ultraviolet difference spectrum of indole-propanol-P bound to the α subunit. This property can be used to measure the binding by spectrophotometry and to follow the kinetics of binding in stopped-flow and temperature-jump experiments (e.g. at 295 nm).

Equilibrium mixtures of α subunit and indole-propanol-P in 0.1 M phosphate buffer pH 7.6 at 20° respond to a rapid temperature jump with a single relaxation process [38, 39]. The dependence of the reciprocal relaxation time $1/\tau$ on concentration (Fig. 12) clearly shows that the process is associated with a binding reaction Scheme 3.

$$L + E \underset{k_D}{\overset{k_R}{\rightleftharpoons}} L \cdot E\,.$$

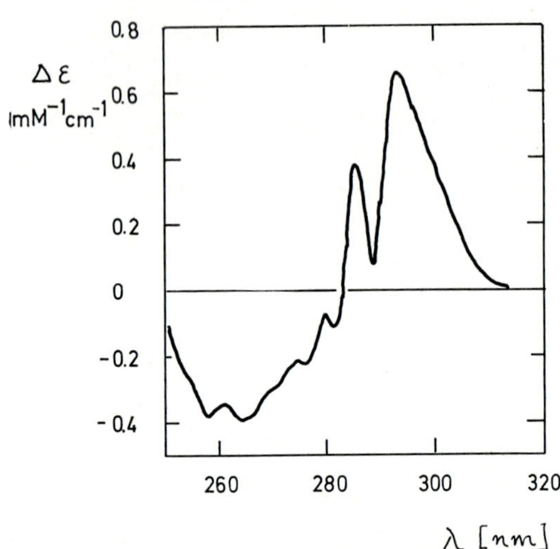

Fig. 11. Ultraviolet difference spectrum of indole-propanol-P bound to the α subunit of tryptophan synthetase. The difference spectrum was measured in the same buffer as described in Fig. 10, using tandem cells. The difference extinction coefficients were calculated from the known concentrations of enzyme and ligand and from the known dissociation constant (cf. Fig. 10)

Eq. 1 describes the quantitative relationship between the reciprocal relaxation time, the rate constants of recombination (k_R) and of dissociation (k_D) and the concentration of free binding sites of the α-subunit ($\overline{E}$) and of free ligand ($\overline{L}$, here identical with indole-propanol-P).

$$1/\tau_1 = k_D + k_R\,(\overline{E} + \overline{L}) \tag{1}$$

Fig. 12 shows that the thermodynamic dissociation constant calculated from the ratio of rate constants

$$K_{kinet} = k_D/k_R = 0.02 \text{ mM}$$

is *smaller* than the equilibrium constant K measured by equilibrium dialysis and spectrophotometric titration (cf. Fig. 10).

Tentatively, this observation can be accounted for if the α subunit in the absence of ligands exists as a preequilibrium between

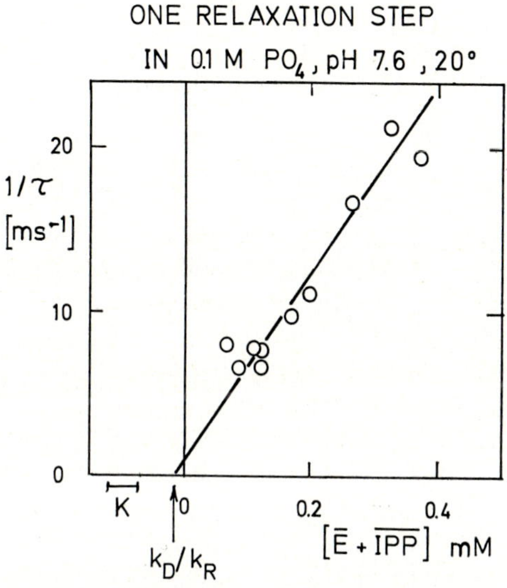

Fig. 12. The kinetics of binding of indole-propanol-P to the α subunit of tryptophan synthetase in phosphate buffer. The reciprocal relaxation time $1/\tau$, obtained from oscilloscope traces showing only a single relaxation process, is plotted versus the concentration parameter (cf. Eq. 1). A linear increase indicates a second-order (i.e. a binding) reaction. The bar indicates the range of values found for the overall thermodynamic dissociation constant K in binding studies at equilibrium (cf. Fig. 10)

two states, with the least prevalent form binding indole-propanol-P preferentially

Scheme 4.

$$L + E \underset{k_1'}{\overset{k_1}{\rightleftharpoons}} L + E' \underset{k_D}{\overset{k_R}{\rightleftharpoons}} E' \cdot L.$$

Here E and E′ are the two isomeric (conformational) states of the α-subunit and L is indole-propanol-P. It can be shown that the equilibrium constant for the isomerization preequilibrium must be of the order of

$$K_{\text{isom}} = \frac{k_1'}{k_1} = \frac{\overline{E}}{\overline{E}'} \sim 10.$$

Furthermore, the data depicted in Fig. 12 can only be interpreted in a consistent manner if the isomerization equilibrium relaxes at a slower rate than the binding step. Eq. 2 is the analytical expression for the relaxation time of the postulated second relaxation process.

$$1/\tau_2 = k_1 + k_1' \left(\frac{K + \overline{E}'}{K + \overline{E}' + \overline{L}} \right) \tag{2}$$

where $K = k_D/k_R$ is the intrinsic thermodynamic dissociation constant for the rapid binding step. Moreover, Eq. 2 predicts that the specific relaxation rate ($1/\tau_2$) should decrease wit inh creasing concentration of the ligand.

Unfortunately, the postulated slow relaxation process was not detectable, neither in the absence of indole-propanol-P nor in its presence. However, a change in buffer ion from phosphate to tris-acetate under otherwise identical conditions of temperature, pH and ionic strength immediately revealed a process with the predicted concentration dependence (Fig. 13 B). The equilibrium dissociation constant for indole-propanol-P is the same for the two buffer systems [28].

The simplest explanation for the different relaxation spectra is the fact that the enthalpies of ionization of phosphate and tris as a primary amine are drastically different. While the pK values of the second and third ionizations constants of phosphoric acid are practically independent of temperature, the pK value of tris drops by approximately 0.3 units for every 10° increase in temperature (Sigma technical bulletin). Thus a temperature change in tris buffer will perturb equilibria in a different manner to that obtained with phosphate buffer.

Under these conditions the faster relaxation process (Fig. 13 A) does not show the linear dependence of rate on ligand concentration observed in phosphate buffer (cf. Fig. 12). Furthermore, in the absence of ligand, the α subunit alone is characterized by two relaxation steps (not shown here). It is unlikely that these processes are associated with an aggregation phenomenon since the α subunit is known to be monomeric within the range of concentrations employed in the experiments reported here [25, 26].

While a more thorough study (under varied conditions of buffer ion, pH and temperature) is required to clarify the dynamic inter-

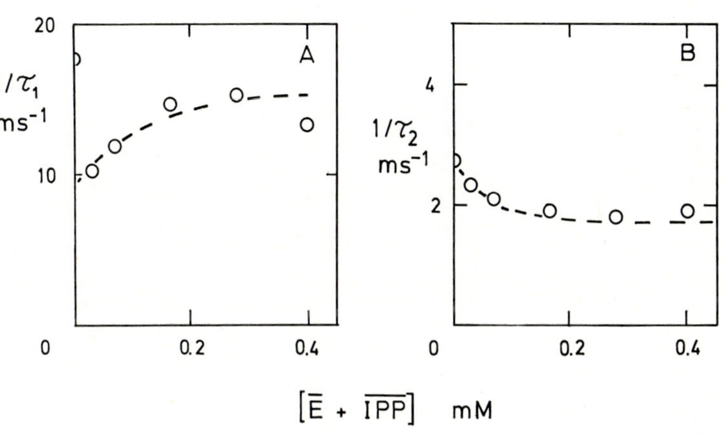

Fig. 13. The kinetics of binding of indole-propanol-P to the α subunit of tryptophan synthetase in tris-acetate buffer. Two relaxation processes were observed. The reciprocal relaxation times $1/\tau_1$ and $1/\tau_2$ are plotted versus the concentration parameter of Eq. 1. Since $L > \overline{E}, \overline{E}'$, the error introduced is small, particularly at high concentrations of L. The decrease of $1/\tau_2$ agrees qualitatively with the prediction of Eq. 2

action of the α-subunit with indole-propanol-P, the minimal mechanism for accommodating the hitherto fragmentary information is shown in

Scheme 5.

$$\begin{array}{ccc} S + E & \rightleftharpoons & ES \\ \upharpoonleft\downharpoonright & & \upharpoonleft\downharpoonright \\ S + E' & \rightleftharpoons & E'S \\ \upharpoonleft\downharpoonright & & \upharpoonleft\downharpoonright \\ S + E'' & \rightleftharpoons & E''S \end{array}$$

The α subunit must exist in at least three different states in order to account for the observation of two relaxation processes in absence of ligand. For symmetry reasons each state is presumed to bind indole-propanol-P. The ligand probably binds preferentially to one or two of these states, thereby leading a shift in the conformational equilibria.

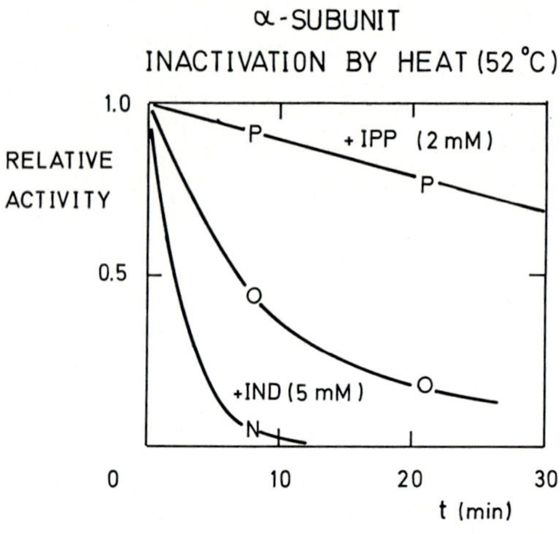

Fig. 14. The α subunit of tryptophan synthetase is protected against thermal denaturation by indole-propanol-P. The enzyme was heated to 52 °C in 0.1 M phosphate buffer 7.6 (containing 1 mM dithiothreitol and 5 mM EDTA) in the absence and presence of ligands as indicated in the figure. Enzyme activity [25] was normalized with respect to the unheated control sample

Indirect support for this hypothesis is the fact that indole-glycerol-P protects the sulfhydryl groups of the α subunit against reaction with N-ethyl maleimide [27]. Since the protective effect of substrate might be merely due to local hindrance of the access of the sulfhydryl reagent, the effect of indole-propanol-P and indole on the thermal denaturation [40] of the α subunit was studied. As can be seen from Fig. 14, indole-propanol-P strongly stabilizes the α subunit. Interestingly, indole has the opposite effect.

Indole and indole-glycerol-P also show opposite effects on the inactivation of the enzyme by sulfhydryl reagents. Bound indole renders the three cysteine sulfhydryl groups of the α subunit equally susceptible towards reaction with 5,5′-dithiobis (2-nitrobenzoic acid) (ELLMAN's reagent), whereas indole-glycerol-P protects almost completely [27]. Although this kind of evidence is indirect, there is ample precedent for the protection of enzymes by substrates against chemical modification and heat denaturation. Particularly in the latter case, it is difficult to envisage protection without the stabilization of the native structure by the substrate (i.e. by the involvement of conformational changes).

4. Conclusion

The information on the mechanism of binding of specific ligands such as indole-propanol-P to tryptophan synthetase is still rather scanty. However, taking these data together with the work of FAEDER and HAMMES [29, 30], one can tentatively conclude that the component α and β_2 subunits of tryptophan synthetase are flexible proteins. Moreover, the assembly to the $\alpha_2 \beta_2$ complex appears to shift the pre-existing conformational equilibria in the way depicted in Fig. 2 A. It remains to be determined whether one of the conformations accessible to the isolated subunits is identical to the one stabilized in the active $\alpha_2 \beta_2$ complex.

Acknowledgement. One of the Authors (K. K.) is greatly indebted to the Department of Biochemistry, Stanford University, Stanford, California, USA for the hospitality and material support received during the tenure of a two-year special fellowship.

References

1. GINSBURG, A., STADTMAN, E. R.: Ann. Rev. Biochem, **39**, 429 (1970).
2. REED, L. J., COX, D. J.: The enzymes, 3. ed., Vol. 1, p. 213 (BOYER, P. D., Ed.). New York: Academic Press 1970.
3. HENNING, U.: Angew. Chem. **78**, 865 (1966); cf. also: Angew. Chem. Internat. Ed. Engl. **5**, 785 (1966).
4. — Mosbach Colloquium **23**, 343. Berlin-Heidelberg-New York: Springer 1972.
5. SUMPER, M., LYNEN, F.: Mosbach Colloquium **23**, 365. Berlin-Heidelberg-New York: Springer 1972.
6. KLOTZ, I. M., LANGERMANN, N. R., DARNALL, D. N.: Ann. Rev. Biochem. **39**, 25 (1970).
7. WHITEHEAD, E.: Progr. Biophys. molec. Biol. **21**, 323 (1970).
8. KOSHLAND, D. E., Jr.: The enzymes, 3. ed., Vol. 1, p. 342 (BOYER, P. D., Ed.). New York: Academic Press 1970.
9. ROSSMANN, M. G., ADAMS, M. J., BUEHNER, M., FORD, G. C., HACKERT, M. L., LENTZ, P. J., Jr., MCPHERSON, A., Jr., SCHEVITZ, R. W., SMILEY, I. E.: Cold. Spr. Harb. Symp. quant. Biol. **36**, 179 (1971); cf. Nature (Lond.) **227**, 1098 (1970).

10. JAENICKE, R., KOBERSTEIN, R., TEUSCHER, B.: Europ. J. Biochem. **23**, 150 (1970).
11. PERUTZ, M. F.: Nature (Lond.) **228**, 726 (1970).
12. GERHART, J. C.: Curr. Top. Cell. Regul. **2**, 275 (1970).
13. KIRSCHNER, K., GALLEGO, E., SCHUSTER, I., GOODALL, M.: J. molec. Biol. **58**, 51 (1971).
14. COHEN, G. N.: Curr. Top. Cell. Regul. **1**, 183 (1969).
15. DAVIS, H. R.: In: Organizational biosynthesis, p. 303 (VOGEL, H. J., LAMPEN, V. O., BRYSON, V., Eds.). New York: Academic Press 1967.
16. LYNEN, F.: New Perspectives in Biology **4**, 132 (1970).
17. GAERTNER, F. H., ERICSON, M. C., DEMOSS, J. D.: J. biol. Chem. **245**, 595 (1970).
18. CREIGHTON, T. E.: Europ. J. Biochem. **13**, 1 (1970).
19. HESS, G. P., RUPLEY, J. A.: Ann. Rev. Biochem. **40**, 1014 (1971).
20. BLOW, D. M., STEITZ, T. A.: Ann. Rev. Biochem. **39**, 63 (1970).
21. JARDETZKY, O., WADE-JARDETZKY, N. G.: Ann. Rev. Biochem. **40**, 605 (1971).
22. KIRSCHNER, K.: Curr. Top. Cell. Regul. **4**, 167 (1971).
23. GUTFREUND, H.: Ann. Rev. Biochem. **40**, 315 (1971).
24. EISERLING, F. A., DICKSON, R. C.: Ann. Rev. Biochem. (in press).
25. YANOFSKY, C., CRAWFORD, I. P.: The enzymes, 3rd. ed. (BOYER, P. D., Ed.). New York: Academic Press (in press).
26. GOLDBERG, M. E., CREIGHTON, T. E., BALDWIN, R. L., YANOFSKY, C.: J. molec. Biol. **21**, 71 (1966).
27. HARDMAN, J. K., YANOFSKY, C.: J. biol. Chem. **240**, 725 (1965).
28. KIRSCHNER, K., WISKOCIL, R.: Unpublished experiments.
29. FAEDER, E. J., HAMMES, G. G.: Biochemistry **9**, 4043 (1970).
30. — — Biochemistry **10**, 1041 (1971).
31. DEMOSS, J. A.: Biochim. biophys. Acta (Amst.) **62**, 279 (1962).
32. CRESTFIELD, A. M., FRUCHTER, R. G.: J. biol. Chem. **242**, 3279 (1967).
33. SCHULTZ, G. E., CREIGHTON, T. E.: Europ. J. Biochem. **10**, 195 (1969).
34. MILES, E. W., HATANAKA, M., CRAWFORD, I. P.: Biochemistry **7**, 2742 (1968).
35. GOLDBERG, M. E., YORK, S. S., STRYER, L.: Biochemistry **7**, 3662 (1968).
36. YORK, S. S.: Biochemistry **11**, 2733 (1972).
37. HATANAKA, M., WHITE, E. A., HORIBATA, K., CRAWFORD, I. P.: Arch. Biochem. **97**, 596 (1962).
38. FRENCH, T. C., HAMMES, G. G.: Meth. Enzymol. **15**, 3 (1969).
39. EIGEN, M., DEMAEYER, L.: Techniques of organic chemistry, Vol. VIII, Part. II, pp. 895—1054 (FRIESS, S. L., LEWIS, E. S., WEISSBERGER, A., Eds.). New York: Wiley (Interscience) 1963.
40. MALING, B., YANOFSKY, C.: Proc. nat. Acad. Sci. (Wash.) **47**, 555 (1961).

Discussion

H. Holzer (Freiburg): As far as I know from the literature the tryptophan synthetase of yeast and Neurospora catalyzes the same three reactions as does the synthetase of E. Coli. However, it does not exist in the two-subunit structure; instead it seems to consist of one polypeptide chain which cannot be dissociated. Is there good evidence that this is really true. Have you any idea what are the consequences of this fact for the reaction mechanism.

K. Kirschner (Basel): I cannot answer this question in detail. However, Carsiotis et al. [Biochem. biophys. Res. Commun. **18**, 877 (1965)] have presented evidence in favour of the Neurospora crassa tryptophan synthetase consisting of four polypeptide chains. In general, there appears to be an analogy between the tryptophan synthetases and the fatty acid synthetases of E. coli and fungi. That is to say, the homologous enzymes form rather loose complexes in E. Coli and very tight complexes in yeast. As far as the kinetic analysis goes, the results obtained by De Moss (Ref. [31]) with the Neurospora crassa enzyme are qualitatively similar to the observations made by Creighton (Ref. [18]) and Faeder and Hammes (Refs. [29, 30]) on the E. Coli tryptophan synthetase. It is almost inescapable to conclude that in both enzymes two separate indole binding sites exist, i.e. that we have an active site composed of two juxtaposed α- and β-type active sites.

Heterologous Protein Interactions

Heterologous Enzyme-Enzyme Interactions

B. Hess and A. Boiteux

Max-Planck-Institut für Ernährungsphysiologie, Dortmund, Germany

With 17 Figures

I. Introduction

Heterologous enzyme-enzyme interactions are defined as interactions between enzymes having different structures and catalytic functions. Heterologous protein-protein interactions have been detected in all functions of the cellular organization. There are qualitative and quantitative aspects associated with such functions. The high functional capacity of protein suggests that proteins are not only suitable for controlled catalysis but also sufficiently complex to regulate cellular space as well as time. Heterologous interactions might be essential for the chemical operation of bioenergetic or biosynthetic pathways: it might be necessary to pass on the products of an enzymic reaction to the next enzyme in a reaction pathway in order to prevent their diffusion into the cellular space or secondary uncontrolled, non-enzymic reactions of labile intermediates, such as hydrolysis, hydration, oxidation, anomerization. They might also perform "measurement" of chainlength, as in the case for fatty acid synthesis.

One quantitative aspect is concerned with the timing and control of a reaction sequence. For instance, it might also be necessary to "inform" a nearest- or distant-neighbor enzyme in an enzyme reaction chain that it has to be active or inactive. Such information could easily be transmitted by enzyme-enzyme interactions, resulting in either steric hindrance or fitting of an active center, which would ensure the coordination of such an information transfer with the fluxes and transient time of pathways.

Taking the opposite viewpoint, we should also ask how uncontrolled protein-protein interactions are avoided; how an active or regulatory site of an enzyme is protected against a perturbation which is essentially unspecific; how an enzyme is isolated within the narrow cellular space in which the collision frequency between macromolecules computed for glycolysis in yeast is about 30 nsec (see below).

Interactions between proteins could be quite rapid and efficient and under the correct conditions, should occur with second-order rate constants of the order of 10^5—$10\ M^{-1} \times sec^{-1}$ (GUTFREUND [1]), thus allowing fluxes of $10^{-4}\ M \times sec^{-1}$, which are comparable with fluxes of bioenergetic pathways. Furthermore, it should be kept in mind that, with such second-order rate constants, possible dissociation constants of 10^{-3} M, and first-order rate constants of $10^3\ sec^{-1}$, complexes between heterologous enzymes might be quite weak and would readily dissociate on dilution of cellular systems. Attempts to isolate such species, e.g. the isolation of enzyme-substrate complexes have shown that it is not an easy task. Yet such complexes obviously must exist. Against the background of these general considerations, we would like to discuss the interactions of enzymes occurring in the cytosol as well as the membrane space of mitochondria.

II. Glycolysis

Glycolytic reactions in general take place in a clearly defined cellular space, the cytosol. The high concentration of macromolecular proteins with their respective water and ion shells means that the glycolytic system is densely packed and favors the formation of protein-protein complexes. A number of authors have questioned whether and in what form such complexes may exist (for summary, see DE DUVE [2]). It might be fitting to discuss the kinetic consequences of high density. On the basis of antibody titrations as well as kinetic measurements, we have found that the glycolytic enzymes in *S. carlsbergensis* represent roughly 65% of the total soluble cellular protein [3—5]. Within the cytosol the concentration of catalytic sites of most glycolytic enzymes is in the range between 10^{-5} and 10^{-4} M, which is equimolar with many glycolytic intermediates. The glycolytic system is a highly concentrated, viscous protein solution. Assuming that all enzymes exist in monomeric forms of average size, and adding up the different species, one obtains an approximately 1.3 mM solution of monomeric units.

Assuming, furthermore, a homogeneous distribution of monomeric globular species, a rough calculation of the average molecular separation of such a solution of proteins gives a distance of about 40 to 50 Å between the individual units [6].

Here, the problem of transit time — the time of transfer of a product of one enzyme to the substrate binding site of the next enzyme in a metabolic pathway as defined by DIXON [7] — is of great interest. It might be useful to estimate the mean transit time required for molecules like pyruvate to travel from pyruvate kinase to pyruvate decarboxylase. Using EINSTEIN's law with the

Table 1. *Formation of Complexes Investigated by Ultracentrifugation*

System	Condition	Complex Formation	Author
Phosphorylase particle (muscle)	—	+	HEILMEYER et al. [8]
Rat liver supernatant (GAPDH, PGK, ALD, TIM, LDH)	11—140 mg protein/ml	–	DE DUVE et al. [2]
Isolated enzymes GAPDH, PGK (yeast)	10^{-5}—10^{-4} M	–	BISCHOFBERGER, HESS, SCHACKNIES [9]

diffusion coefficient of 10^{-5} for pyruvate, and bearing in mind that only a fraction of the total surface of an enzymic sphere can be hit by a point-like substrate for the initiation of a reaction, we compute a transit time in the microsecond range between collisions on the active site [6]. Thus the mean transit time, at least for pyruvate, is short enough to be neglected in a system composed of enzymes with turnover rates in the range of 100 per sec per binding site. Thus, the glycolytic enzymes in the yeast cell are sufficiently densely packed to exclude any transport problems by diffusion. And indeed, if glycolytic enzymes were to be organized in the form of complexes, this would only be important if it conferred additional properties for controlling the overall pathway.

The occurrence of protein-protein interactions in glycolysis has been tested by ultracentrifugation as well as by kinetic experiments in cellular extracts, as summarized in Table 1. The analysis of a number of enzymes of the lower glycolytic pathway of rat liver supernatant fractions, as well as of isolated glyceraldehyde-3-phosphate dehydrogenase and phosphoglycerate kinase from yeast, indicated complete molecular separation of the species under the conditions of high enzyme concentrations used; yet complex formation was discovered in ultracentrifugation studies of the glycogen phosphorylase particles of muscle (see HEILMEYER et al. [8]).

Any structural organization of enzymes, such as aggregation or intermolecular induction of conformation changes, ought also to be reflected in the concentration dependence of fluxes and flux-dependent parameters of glycogenolytic and glycolytic enzyme sequences. In the phosphorylase system [8], heterologous protein-protein interactions are indicated; interaction of hexokinase with mitochondria was also observed [10].

The problem of the control of the aldolase and triose phosphate isomerase reaction is of great interest. In an attempt to understand the steady-state behavior of the reactants of both enzymes in ascites tumor cells, the question of possible interactions between the two enzymes was raised [11]. A computer model was developed describing the steady-state properties of this part of glycolysis; however, the reality of the model has not been proven. In an analysis of the steady-state concentrations of the triose phosphates and fructose diphosphate in rat-liver supernatant DE DUVE et al. [2] did not find any evidence of "abnormalities" which would point directly to the occurrence of interactions between the two enzymes. Thus there is no evidence that glycolysis is controlled by protein-protein interactions at this reaction step, neither in ascites tumor cells nor in liver. Therefore, it was of great interest when PETTE and coworkers [12] reported the interaction of aldolase and F-actin. The increase of $K_{1/2}$ for fructose diphosphate by about one order of magnitude in the actin-bound enzyme suggests that the aldolase activity might be controlled by binding to F-actin.

In an attempt to define the activity states of enzymes under near-physiological conditions, we have analyzed the kinetic properties of coupled glycolytic enzymes isolated from yeast or muscle as well as in extracts of *S. carlsbergensis* [4, 13—17]. Such studies are of great help in solving, not only problems of the interaction of enzymes under in-vivo conditions, but also the problem of the stability of oligomeric structures over a wide concentration range. They are likely to shed some light on the largely unknown field of enzymic properties under in-vivo conditions. However, it should be kept in mind that with these techniques a physiological concentration range — at least for the glycolytic systems — can be approached only within roughly one order of magnitude, so that the physiological situation has to be estimated by extrapolation.

Table 2. *Equations describing the overall transient time (ΔT) of a three-enzyme system in terms of the transient times (τ_2, τ_3) of the second and third enzyme in a sequence on the basis of the* MICHAELIS-MENTEN *theory. The specific case investigated is: E_1 (zero-order conditions) = hexokinase, E_2 (first-order conditions) pyruvate kinase, E_3 (first-order conditions) lactate dehydrogenase (from* [15]*)*

$$A \overset{E_1}{\rightarrow} B \underset{\tau_2}{\overset{E_2}{\rightarrow}} C \underset{\tau_3}{\overset{E_3}{\rightarrow}} D$$

$$\mathrm{NADH}\,(t) = k_1 \left[-A + t + B \cdot e^{-\frac{t}{\tau_2}} - C \cdot e^{-\frac{t}{\tau_3}} \right]$$

$$A = \frac{k_2 + k_3}{k_2 \cdot k_3} = \tau_2 + \tau_3 = \Delta T$$

$$\tau = \frac{K_m}{V_{max}}$$

In order to detect interactions between the enzymes and proteins present in a given system, the sensitivity of V_{max} and K_m on variation of the enzyme concentration was tested on the basis of the Michaelis theory. This can be illustrated with the three-enzyme system hexokinase – pyruvate kinase – lactate dehydrogenase, as defined in Table 2.

The concentration of NADH at any time is given in the equation. Within a transient period (or lag time = ΔT) the exponential terms of the equation vanish, and the rate is described by the pseudo-zero-order constant related to the activity of the first enzyme of the sequence. The transient time (τ) is the time required for each subsequent enzyme to approach to within 1/e of a new steady-state. The total lag of the system is the sum of the transient times of the second and third enzymes of the chain. Each transient time is equated to the ratio of K_m/V_{max} of the respective enzymes.

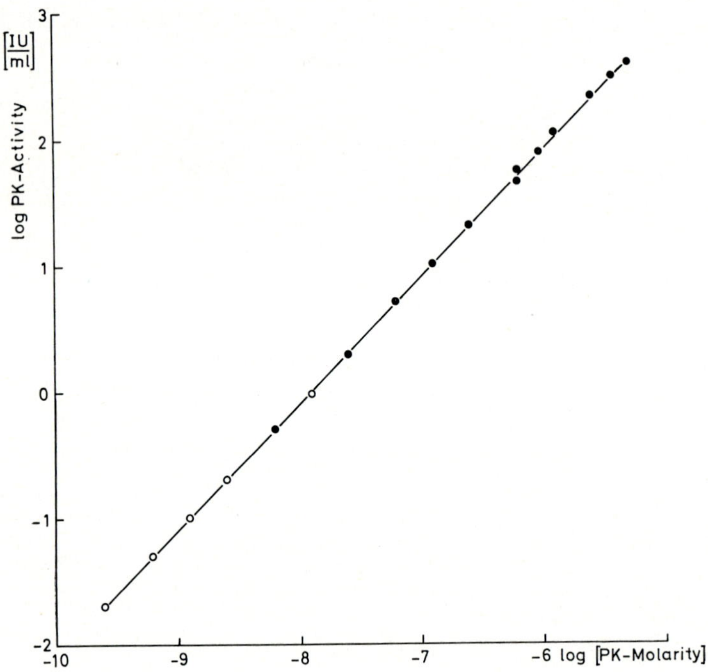

Fig. 1. Relationship between the logarithm of the activity of highly purified pyruvate kinase of yeast and the logarithm of its molarity measured with lactate dehydrogenase in a coupled assay (from [17])

If one transient time is known from independent experiments, the other can be computed. If interactions affecting the activity parameters occur, changes in the transient time or the flux rate are observed to depend on the concentrations of the enzymes.

With this technique, the concentration dependence of the following enzymes was tested: hexokinase (purified from yeast) [15], glucose-6-phosphate dehydrogenase (purified from yeast) [16], glyceraldehyde-3-phosphate dehydrogenase (purified from muscle and yeast, in raw extract of yeast), aldolase (purified from muscle, in raw extract of yeast) [18], pyruvate kinase (purified from muscle [15], purified from yeast [17], in raw extract of yeast), pyruvate

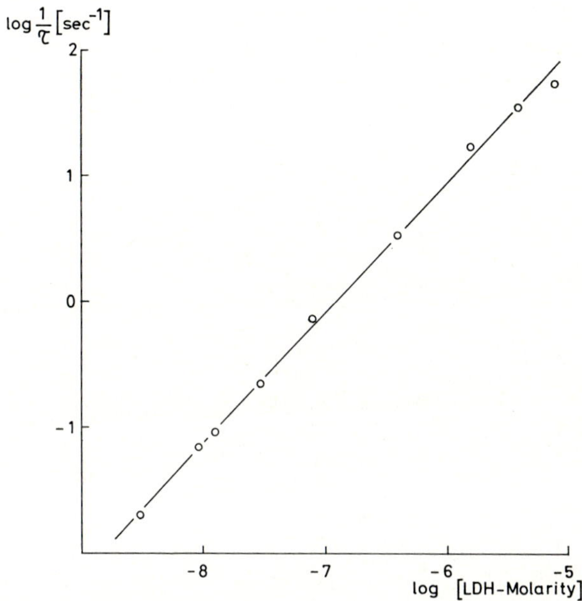

Fig. 2. Relationship between the logarithm of the reciprocal τ-value and the logarithm of the molarity of lactate dehydrogenase in a coupled assay with pyruvate kinase (Muscle enzymes) (from [14])

decarboxylase (purified from yeast, in raw extract of yeast), alcohol dehydrogenase (purified from yeast, in raw extract of yeast) [18].

As an example, the concentration dependence of the allosteric pyruvate kinase of yeast coupled to lactate dehydrogenase is given in Fig. 1, where activity versus molarity is plotted over the concentration range 2.5×10^{-10} to 5×10^{-6} M and a strict proportionality is observed. Also, the reciprocal τ-value for lactate dehydrogenase from the system pyruvate kinase (muscle) — lactate dehydrogenase (muscle) is plotted against the molarity over the concentration range 10^{-8} to 10^{-5} M (see Fig. 2). Again, a strict proportionality was obtained. These experiments show that highly purified enzymes mixed in solution operate in sequence, completely

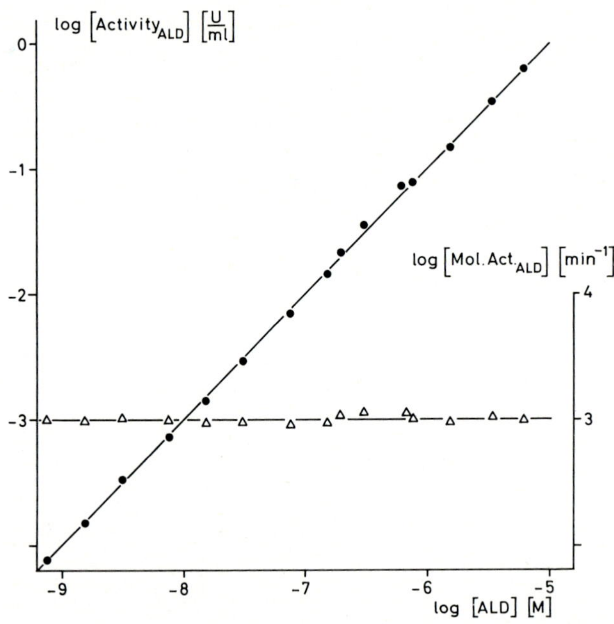

Fig. 3. Relationship between the activity of purified muscle aldolase and its molarity in a system where the enzyme was coupled to muscle glyceraldehyde-3-phosphate dehydrogenase in a molar ratio of 1.5 (GAPDH/ALD) (●●●). The test was carried out in the presence of 40 mM $HAsO_4^{--}$. The right ordinate represents the turnover within the concentration range tested (△△△) [18]

independent of each other, and strictly according to Michaelis-Menten kinetics.

In addition, the activity of aldolase from muscle coupled to glyceraldehyde-3-phosphate dehydrogenase from muscle, as well as the reciprocal τ-value for the latter enzyme, were found to be strictly linear to the concentration of both enzymes over a wide range, as shown in Figs. 3 and 4.

We wondered whether there is any difference in the overall kinetics of glycolytic enzymes tested in a highly purified form as compared to their function in a natural or near-natural environment, e.g. in a highly concentrated yeast extract prepared from *S. carlsbergensis* according to the method given in [19]. An analysis

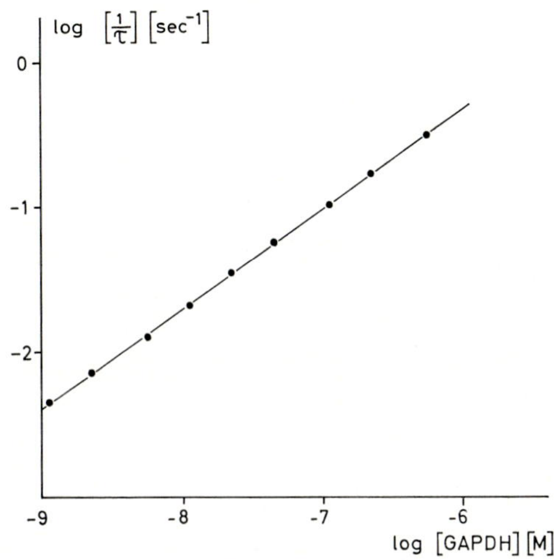

Fig. 4. Relationship between the reciprocal transient time of purified glyceraldehyde-3-phosphate-dehydrogenase and its molarity. For conditions, see legend of Fig. 3 [18]

of the activity of aldolase coupled endogeneously to glyceraldehyde-3-phosphate dehydrogenase in a yeast extract was carried out under conditions similar to those used in the experiment shown in Figs. 3 and 4; the activity of aldolase was proportional to the total enzyme concentration in the range between less than 0.01 and 10 g protein/l (see Fig. 5). Maximum aldolase activity was tested with concentrations of 1.1×10^{-6} M aldolase and 1.1×10^{-5} M glyceraldehyde-3-phosphate dehydrogenase, for technical reasons not covering quite the same range as the experiments of Figs. 3 and 4. Also, a slightly higher turnover rate was found. This experiment, too, indicates that the two enzymes act independently of each other. However, it should be borne in mind that our experimental conditions represent only about 10% of the physiological protein concentration of the living cellular system.

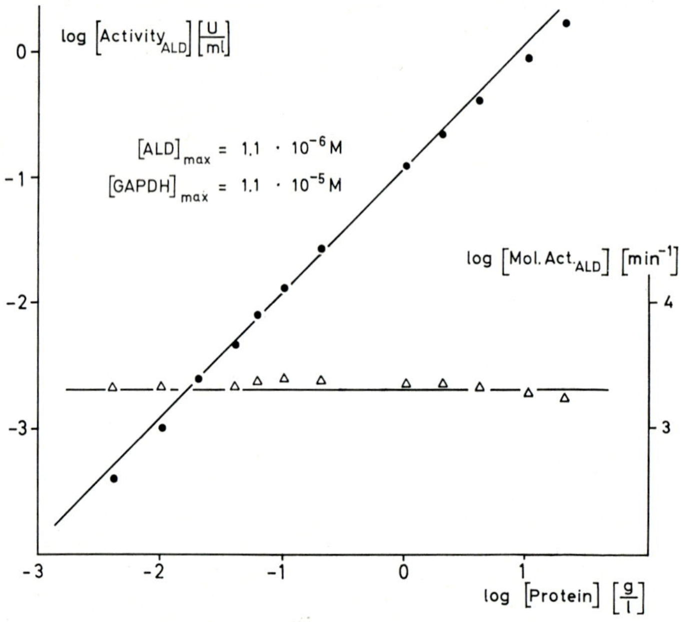

Fig. 5. Relationship between the activity of endogeneous aldolase coupled to endogeneous glyceraldehyde-3-phosphate dehydrogenase in a yeast extract (●●●). The right ordinate represents the turnover within the concentration range tested (△△△). The test was carried out in the presence of 40 mM $HAsO_4^{--}$ [18]

A similar analysis of the enzyme system pyruvate decarboxylase — alcohol dehydrogenase in a yeast extract gave the result plotted in Fig. 6. We observed inhibition of the overall activity of pyruvate decarboxylase and a corresponding decrease in the turnover rate starting from concentrations somewhat below 1 g protein/l up to higher concentrated systems. How strong the inhibition might become at a physiological protein concentration of approximately 150 g/l is a matter for surmise. When the activity of alcohol dehydrogenase alone was tested, however, a far smaller decrease in

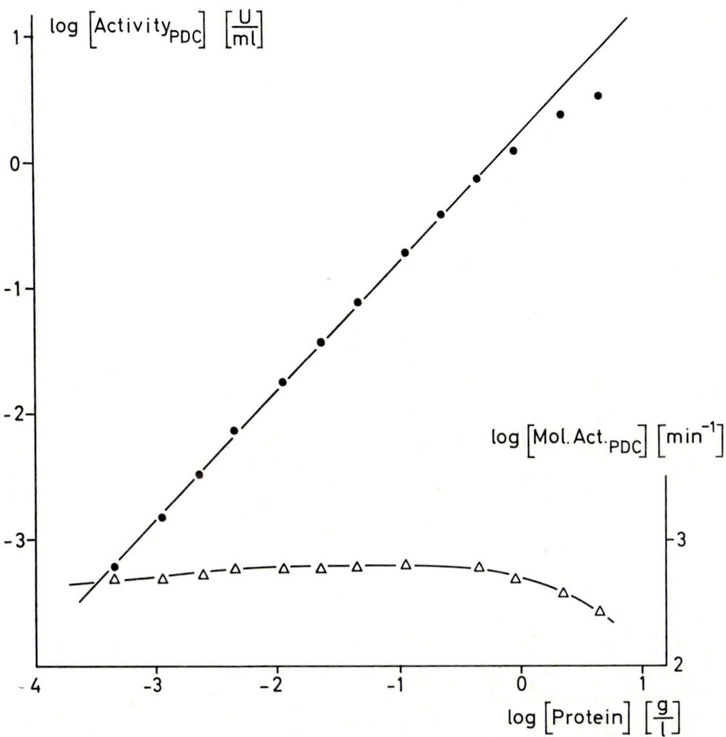

Fig. 6 Relationship between the activity of pyruvate decarboxylase assayed in a test where it was coupled to endogeneous alcohol dehydrogenase in a yeast extract in the presence of 100 mM HPO_4^{--}, and the total protein concentration in the extract (●●●). The right ordinate represents the computed turnover within the concentration range (△△△) [18]

turnover rate was observed. Since the same phenomenon was also recorded on testing the purified pyruvate decarboxylase and alcohol dehydrogenase from yeast under comparable conditions, as shown in Fig. 7, there may well be some interaction between the two enzymes, especially under physiological conditions. The significance of this interaction will be studied in further experiments [18].

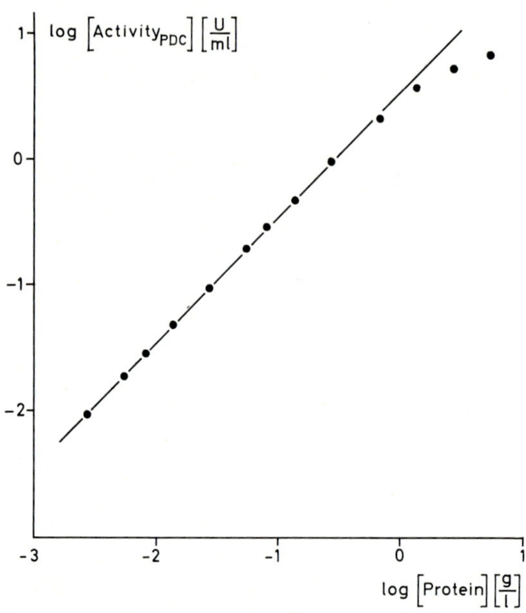

Fig. 7. Relationship between the activity of purified pyruvate decarboxylase and protein concentration. Measurement as in Fig. 6. Assays were carried out with purified yeast enzymes, activity ratio 10 (ADH/PDC) [18]

III. Pyruvate Dehydrogenase

Recently, we have become interested in the kinetic interactions of the various components of the pyruvate dehydrogenase complex. In an attempt to localize the results of the ATP inhibition of pyruvate dehydrogenase, we have analyzed the states of NAD/NADH as well as of bound FAD/FADH in a preparation of pyruvate dehydrogenase complex from heart [20] and observed an interesting interaction between pyruvate dehydrogenase and lipoamide dehydrogenase. Both components were continuously analyzed within

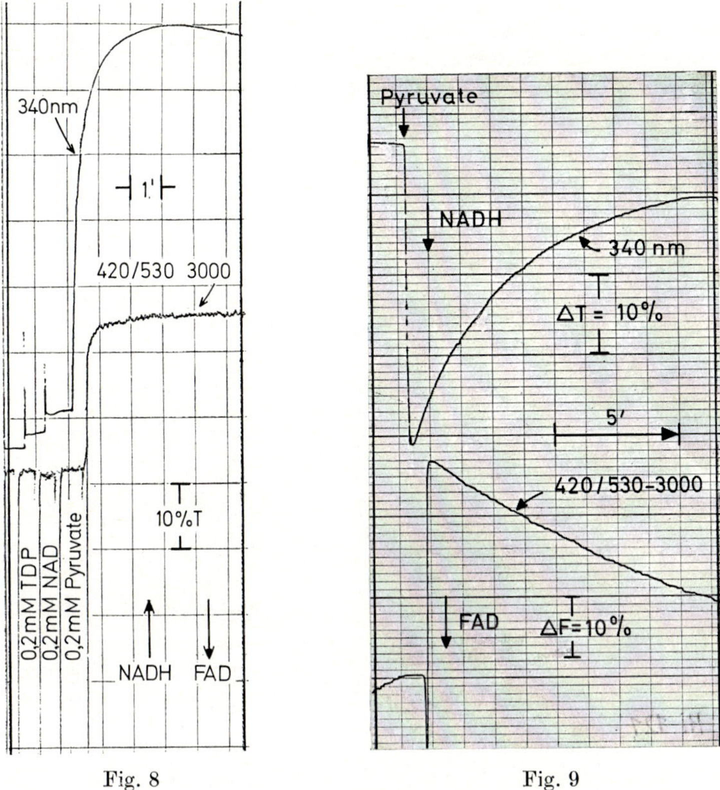

Fig. 8 Fig. 9

Fig. 8. Record of the reduction of NAD and FAD in a pyruvate dehydrogenase complex of heart muscle (2.1 mg/ml). Time proceeds from left to right. Recorder traces of the two-pen-recorder are slightly displaced in this and subsequent records. d = 1 cm (Exp. 146/IV/72) [20]

Fig. 9. Record of a reduction oxidation cycle of NAD and FAD in a pyruvate dehydrogenase complex of heart muscle (1.1 mg protein/1.5 ml). Fluorescence unit (Δ F): 10% of total fluorescence (titrated with dithionite, also in Fig. 8, 10, 11). d = 1 cm, 22 °C [20]

the complex of pyruvate dehydrogenase by a double-beam technique, as shown in Fig. 8, where the record of an experiment is represented. The transmission change of NADH is recorded at 340 nm, the fluorescence emission of FAD is simultaneously measured with an excitation beam selected with a monochromator set at 420 nm and a fluorescence emission filtered at 530 to 3000 nm. The rapid rise in the traces on addition of pyruvate to the suspension indicates a reduction in FAD as well as NAD. From the record in Fig. 8 it can be seen that the steady state of bound FAD is rapidly reached whereas the production of NADH continues until the added pyruvate is exhausted.

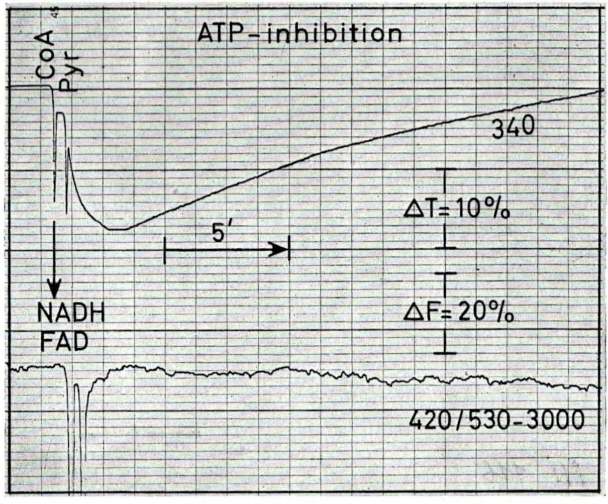

Fig. 10. Conditions as stated in Fig. 8, ATP 0.33 mM [20]

When a purified preparation of pyruvate dehydrogenase was incubated in the presence of small amounts of lactate dehydrogenase in order to recycle the NADH formed during pyruvate dehydrogenation, the addition of pyruvate to the system produced cyclic kinetics. After rapid initial formation of NADH and FADH, subsequent oxidation of both components is observed until the added pyruvate is completely converted (see Fig. 9).

The addition of ATP to the dehydrogenase inhibits the system, so that the subsequent addition of pyruvate no larger influences the steady state of FAD and only a slow cycle of NADH formation and consumption is observed (see Fig. 10). This is expected on the basis of the experimental results of REED et al. [21] and WIELAND et al. [22] who observed the inhibition of pyruvate dehydrogenase complex by ATP. Since it has been demonstrated that the inhibition

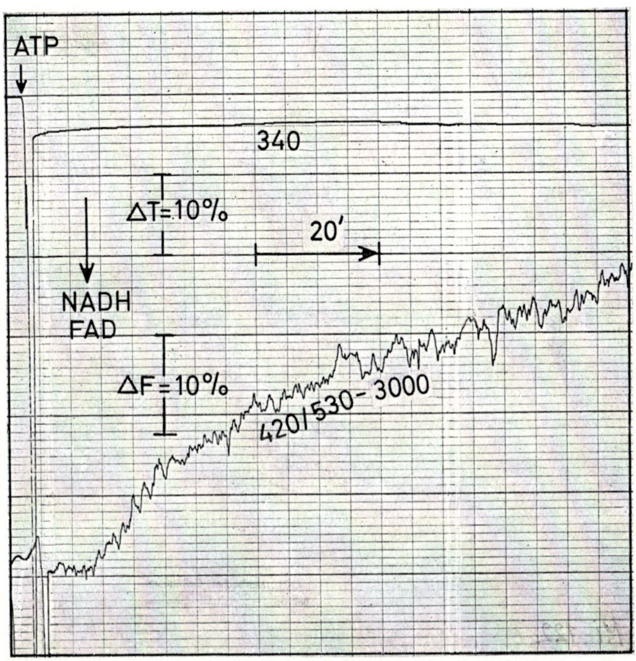

Fig. 11. Record of reduction of NAD and FAD in a pyruvate dehydrogenase complex of heart muscle (1.2 mg protein/1.5 ml) [20]

occurs by phosphorylation of the dehydrogenase moiety of the complex [21], we must now ask whether the phosphorylation of this moiety affects the other components of the enzyme complex. A comparison of the kinetics in Figs. 9 and 10 shows that the initial level of FAD fluorescence is appreciably lower with the inhibited enzyme than with the uninhibited preparation. Therefore, we analyzed in more detail the reaction of the pyruvate dehydrogenase complex upon addition of ATP. A clear response of FAD fluorescence to the addition of ATP was observed, as shown in Fig. 11. The addition of ATP in the absence of pyruvate to a sample of pyruvate dehydrogenase complex of heart is followed by a slight but rapid change in NADH absorbancy. After a lag time, the FAD fluorescence slowly disappears. This coincides with the

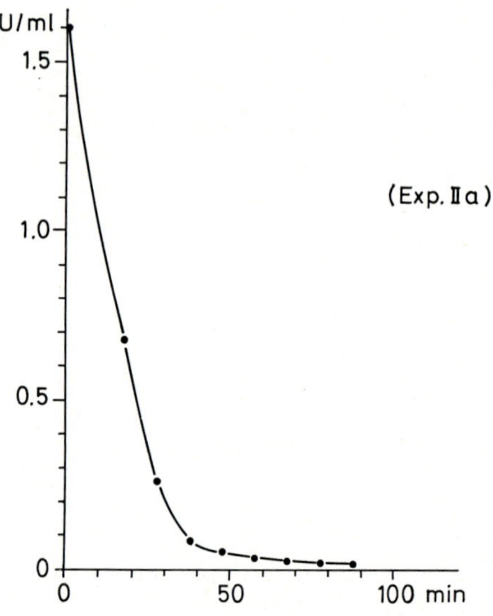

Fig. 12. Time course of the inhibition of pyruvate dehydrogenase complex in the presence of 0.5 mM ATP. For explanation, see text (III) [20]

inhibition of the overall activity of the pyruvate dehydrogenase complex which was followed during the time course of the reaction in samples taken from the reaction mixture and analyzed according to WIELAND et al. [22]. The result is given in Fig. 12.

On the basis of the chemical analysis carried out by the groups of WIELAND [22] and REED [21], it is suggested that the fluorescence quenching of FAD on addition of ATP to the complex is due to a change of conformation at the binding site for FAD, following the phosphorylation of the pyruvate dehydrogenase moiety. A direct interaction between the pyruvate dehydrogenase and lipoamide dehydrogenase moiety of the complex is demonstrated.

IV. Cytochrome Interactions

Cytochrome interactions are very fast and imply an extremely precise and rapid fitting of protein-protein interactions, resulting in electron transfers between the heme-bound irons of various species. One of the simplest electron-transfer systems is illustrated in the reaction between cytochrome c and yeast cytochrome c peroxidase, which has been studied by YONETANI and CHANCE [23]. This reaction occurs in solution. Two moles of ferro-cytochrome c react with one mole of monomeric peroxidase with a second-order reaction velocity constant in excess of $5 \times 10^8\ M^{-1} \times sec^{-1}$, closely approaching a diffusion-controlled step, the boundary being given by collisions between the two proteins. The first-order velocity constant of the reaction is $10^4\ sec^{-1}$ and the half-time for cytochrome c oxidation by the peroxidase complex is 70 μsec. CHANCE has pointed out that the first-order velocity constant indicates a maximum value for the residence time of cytochrome c being bound to peroxidase and must be considered as the overall time of the whole process. He also stressed that the value of the second-order reaction velocity constant is larger for the two reacting proteins compared to the reaction of cytochrome oxidase or peroxidase, respectively, with a small ligand like oxygen or hydrogen peroxide ($10^7\ M^{-1} \times sec^{-1}$). This observation suggests protein-facilitated electron transport between acceptor and donor sites of two partner molecules [23].

Even faster electron transfer between protein-bound hemes has been observed in the photosynthetic bacterium *Chromatium*. Here, activation of chlorophyll by laser flash results in the donation from cytochrome c to chlorophyll of an electron of high redox potential ($c_{423,5}$) with a half-time of two microseconds and a constant of $2 \times 10^6\ sec^{-1}$ at room temperature. This experiment is of special interest because of its temperature sensitivity; it has an activation energy of 3.3 kcal, indicating that thermal energy is required for the interaction between the two components [24]. In any case, such a rapid transfer of electrons implies a perfect fitting of the two active sites of the electron-donating and -accepting components; this, again, is only possible on the basis of controlled interactions between the two enzymes.

Complexes between cytochromes have been isolated in highly purified forms by a number of authors. However, the types of interaction between the various components of the complexes are poorly understood. LABEYRIE et al. [25, 26] recently described the isolation of a stable complex of cytochrome c/cytochrome b_2 (yeast L(+) lactate-cytochrome-c-oxidoreductase) in a heme/heme ratio of 1:4. Cytochrome b_2 has a molecular weight of 230,000 and contains four heme groups and four FMN groups, the latter not being required for the binding of cytochrome c. The complex has been obtained in a crystallized state. Formed under low ionic strength conditions, it is of special interest for its unexpected stoichiometry: one molecule cytochrome c per tetraheme cytochrome b_2 unit. Thus, a set of 12 electrons of cytochrome b_2 is channeled to cytochrome c when a flux is activated by the addition of lactate.

One of the forces fixing the cytochromes to each other is electrostatic in nature. It is known that the cationic properties of cytochrome c may be important for binding. LABEYRIE et al. [25, 26] found strong pH-dependence as well as dependence on the ionic strength of the binding process between cytochromes b_2 and c. They stress that electrostatic interactions between complementary charges of the proteins might be partly responsible for their association. Under conditions where cytochrome c is cationic, cytochrome b_2 is anionic. This illustrates the types of interaction which result in the positional orientation of cytochromes to ensure electron transfer.

The kinetic constants which describe the binding process of the two heme enzymes as well as the electron transfer are of great interest. LABEYRIE and coworkers [25] reported that one molecule of cytochrome c binds with a second-order velocity constant of 10^8 $M^{-1} \cdot sec^{-1}$ to one of four presumably identical sites of the tetraheme cytochrome b_2. The dissociation of this complex has a K_D value of 10^{-8} M. One out of twelve electrons of the reduced cytochrome b_2 molecule is transferred to the cytochrome c molecule with a first-order rate constant higher than 1000 sec^{-1}. LABEYRIE et al. also found that after formation of the complex between one molecule of cytochrome c and one molecule of tetraheme cytochrome b_2, there is a structural transition of the whole species which yields a very inefficient binding of more cytochrome c molecules to the three remaining binding sites of cytochrome b_2.

Such structural changes associated with the occupation of one binding site of a tetrameric heme enzyme and electron transfer might be important for the interpretation of interactions between cytochromes in the respiratory pathway of mitochondria. It was observed long ago that in mitochondria the half-time of electron transfer from cytochrome c to cytochrome a and cytochrome c_1, respectively, is of the order of 2 msec which is long compared to pure electron-transfer systems in solution, as mentioned before [24]. There is thus a strong inhibition of electron transfer in the mitochondrial pathway. The mechanism which controls electron transfer in such systems not only governs the linkage between electron transfer and energy transfer, it also regulates the interaction between the proteins carrying the redox components. Indeed, it has been suggested by SLATER [27] and CHANCE [28] that "the controlled electron flow is due to the impeded reactions of an inhibited form of their respiratory carriers with its adjacent members" (CHANCE).

It has long been known that electron flow through the cytochrome pathway of mitochondria is dependent on the energy state of the mitochondria. The flow is slow (state 4) if excess of ATP or "energy" is maintained; the flow is fast if ADP is in excess and energy being regenerated. Thus, cytochromes show a different reactivity to each other, depending on the energy state. A typical case is observed in yeast cells where mitochondria are located in their natural environment. In a preparation of baker's yeast (10%), the addition of ethanol leads to a rapid reduction of cytochrome c_1, cytochrome b_T and cytochrome b_K, as shown in Figs. 13 to 15. Whereas cytochrome b_T and c_1 remain reduced throughout the observation period, cytochrome b_K is slowly oxidized to its original level. But why do cytochrome b_T and c_1 not equilibrate rapidly with cytochrome b_K, as might be expected from the oxidation half-time of cytochrome b in isolated mitochondria in the presence of uncoupling agents [24]? Clearly, the interactions between cytochrome c_1, b_T and cytochrome b_K are inhibited according to the metabolic state of the system.

The differential reactivity of cytochromes towards each other can also be demonstrated by tracing the reaction kinetics of cytochromes in isolated mitochondria. The situation is illustrated in an experiment recently carried out by CHANCE (to whom we owe special thanks for supplying us with Figs. 16 and 17). Under uncoupled conditions, the cytochrome b is oxidized within about 170 msec half-time, followed by the oxidation of cytochrome c_1 in 5 msec. The cytochrome b kinetics, however, are observed to be completely different under coupled aerobic [30] and anerobic [31, 32] conditions. When the system is in a coupled state, cyto-

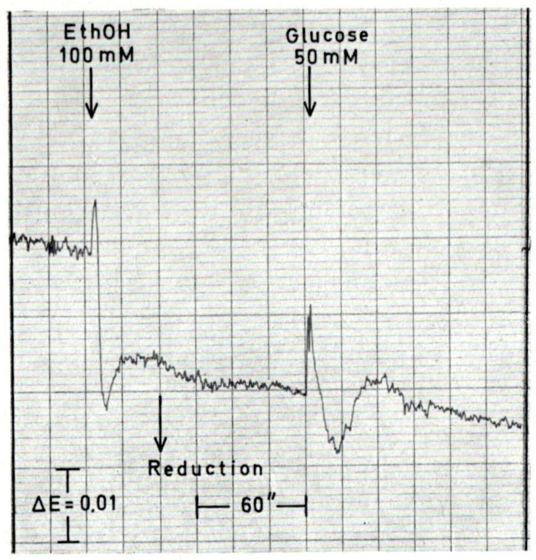

Fig. 13. Spectrophotometric record of the extinction change at 552—540 nm (±1 nm) in a 10% suspension of *S. cerevisiae*, d = 1 cm. Time proceeds from left to right. For explanation, see text [29]

chrome b is not oxidized but reduced after addition of oxygen. It is obvious that, for some structural reasons, electron transfer cannot occur between cytochromes b and c_1.

Recently CHANCE has shown that even cytochrome b_T reduction depends on the state of the respiratory chain and that this phenomenon can be observed under many conditions. A typical experiment (Fig. 16) shows that when an oxygen pulse is added to pigeon-heart mitochondria, after a rapid pre-steady state, cytochrome c_1 is oxidized and cytochrome b_T slowly reduced. A similar result is obtained when the system is energized by addition of ATP under anaerobic conditions. Thus, electrons are donated either to cytochrome c_1 or b_K, depending on the energy state of the system. If the system is uncoupled, cytochrome b_T immediately becomes oxidized.

Why do cytochromes in a reaction chain sometimes react rapidly with each other and sometimes not? Let us assume that cyto-

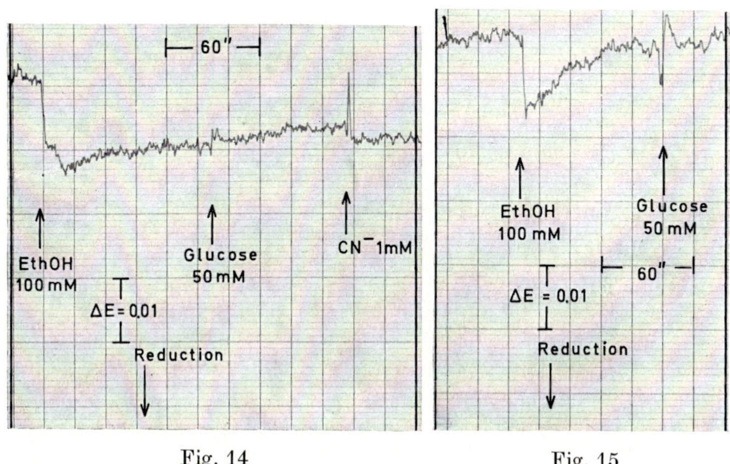

Fig. 14 Fig. 15

Fig. 14. Spectrophotometric record of the extinction change at 563—575 (±1 nm) in a 10% suspension of *S. cerevisiae*. d = 1 cm. Time proceeds from left to right. For explanation, see text [29]

Fig. 15. Spectrophotometric record of the extinction change at 558—575 (±1 nm) in a 10% suspension of *S. cerevisiae*. d = 1 cm. Time proceeds from left to right. For explanation, see text [29]

chromes react with each other by collision and that they are physically bound to the membrane with a spacing of the same order as their molecular diameter, producing a large effective local concentration, as suggested 20 years ago by Chance [33]. It is also reasonable to suppose that the cytochromes are organized in sets roughly 100 Å apart, as computed by Klingenberg [34]. The diffusion of cytochromes with a molecular weight of 200,000 in the polylipid network of the membrane might be quite restricted as

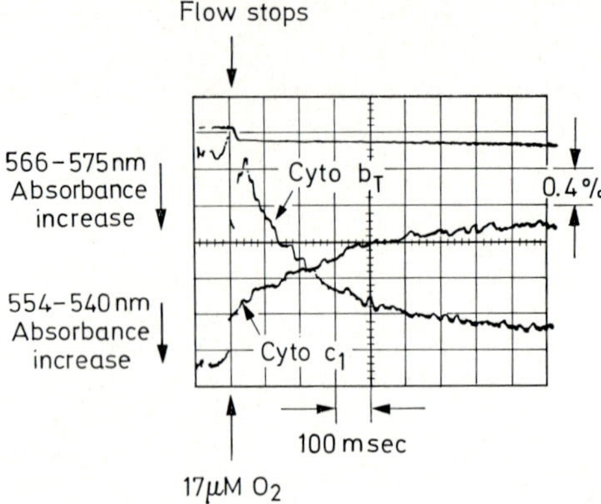

Fig. 16. Spectrophotometric records of the kinetics of cytochrome c_1 oxidation (absorbancy decrease at 554—540 nm) and cytochrome b_T reduction (absorbancy increase at 566—575 nm) in oxygen pulse experiments with pigeon heart mitochondria (2.8 mg protein/ml). The mitochondria were washed in KCl in order to deplete cytochrome c. 6 mM succinate, glutamate and malonate, 0.15 µg antimycin A/mg protein, 5 µM rotenon. Ordinate: transmission, abscissa: time in m sec, proceeding from left to right. For explanation, see text (Courtesy of CHANCE, 1972)

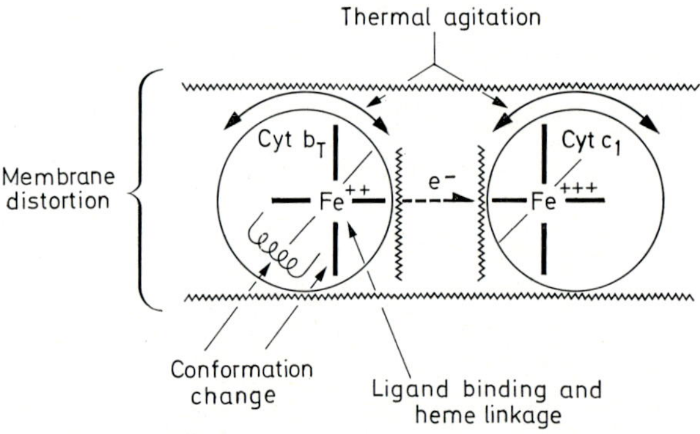

Fig. 17. Schematic representation of parameters which could effect electron transfer between cytochrome b_T and cytochrome c_1. For explanation, see text (Courtesy of CHANCE, 1972)

compared with the diffusibility of such small-electron transporting systems as ubiquinon, or even cytochrome c. It should be mentioned here that, according to LIEB and STEIN, model experiments in polymethylacrylate proved that doubling the mass of a molecule results in a 14-fold decrease of their specific diffusion rates [35]. Thus, in a mitochondrial system, some of the cytochromes might be separate from each other, and others at a permanent spacing of a few Å.

What types of interactions could be expected between nearest-neighbor cytochromes? As demonstrated in Fig. 17, a number of mechanisms might be involved. It has long been assumed that the rotational relaxation times of proteins in solution also apply to membrane-bound proteins [33]. Such rotational motion might vary the relative positions of the cytochromes with respect to one another, achieving a face-to-face or half back-to-face position (distal and proximal configurations). There are quite a number of data showing how fast the rotational relaxation time of the membrane-bound protein component can be. For instance, rhodopsin, a lipoprotein of molecular weight 40,000, is capable of rotational diffusion in the range of 20 μsec in the retinal membrane at 20 °C [36]. The mobility of cytochrome c by rotational and translational diffusion has been established in experiments showing that cytochrome c is simultaneously accessible to membrane-bound cytochrome and to membrane-bound peroxidases [37, 38].

Finally, how is the reactivity of the cytochromes themselves controlled? As indicated in Fig. 17, a conformational change is to be expected and should be indicated by a change in the redox state or even the redox potential of an enzyme. WILSON and DUTTON have shown that the redox potential of cytochrome b_T varies between -35 and $+245$ mV depending on the presence of ATP and oxygen [39]. A change in redox potential may only reflect a change in the heme environment, most probably a conformational change in the tertiary structure of the heme protein, since the six-coordinated ligand field of iron is recognized by its immediate neighbours. This seems very likely, not only because of the analogy with hemoglobin, but also because DICKERSON and MARGOLIASH [40, 41] have recently shown that oxidized cytochrome c has a different conformation from reduced cytochrome c. In oxidized cytochrome c an open channel leads to the heme on the histidine 80 site, whereas in reduced cytochrome c this channel is closed. Therefore, a simple picture emerges: in the case of cytochrome b_T, oxygen might generate a conformational change in cytochrome b_T, shifting its redox potential towards its high-potential form and at the same time sterically hindering the interaction of cytochrome b_T with cytochrome b_K.

The diagram in Fig. 17 also suggests that the state of the membrane itself may change in response to a change in the redox state of cytochrome species. Indeed, in a study of the interaction of reduced and oxidized ubiquinon in the mitochondrial membrane with a fluorescent probe incorporated into the hydrocarbon region of the membrane, it was found that oxidized ubiquinon strongly ($k \sim 2 \cdot 10^8\ sec^{-1}$) interacts with the fluorescent probe (the anthroyl group of 12-[9-anthroylstearic acid]), whereas reduced ubiquinon does not. From this result it was concluded that the membrane structure changes its state depending on the redox state of its ubiquinon component [42]. In model studies employing X-ray low-angle diffraction [43] the redox state of ubiquinon has been observed to have a direct influence on the structure of the lipid bilayers. Thus, a detailed understanding of protein-protein interactions occurring in lipid membranes might not be possible unless the state of the membrane units which carry the reaction protein structures is taken into consideration.

Summary

Heterologous protein-protein interactions can be observed in many functions of the cellular organisation. The significance of such interactions is discussed with regard to glycolysis, to pyruvate dehydrogenase as well as in cytochrome interactions. Protein-protein interactions may function to control the activity of enzymic catalysis. Furthermore, protein-protein interactions may be an essential part of the catalytic mechanism itself, as in the case of interactions between cytochromes where electron transfer does not occur if the cytochromes are not approaching each other efficiently enough to allow oxidation and reduction.

Abbreviations

ADH = alcohol dehydrogenase
ALD = aldolase
EthOH = ethanol

FMN = flavinmononucleotide
GAPDH = glyceraldehyde-3-phosphate dehydrogenase
LDH = lactate dehydrogenase
PDC = pyruvate decarboxylase
PGK = phosphoglycerate kinase
PK = pyruvate kinase
Pyr = pyruvate
TDP = thiamine diphosphate
TIM = triose phosphate isomerase

Acknowledgments. The skillful assistance of Miss M. Böhm, Mrs. R. Müller and Miss B. Käufer in the hitherto unpublished experiments is gratefully acknowledged. The authors are also heavily indebted to Prof. Dr. Britton Chance for supplying unpublished manuscripts, experimental data and Fig. 16 and 17, and to Dr. D. Kuschmitz for his helpful criticism of the manuscript.

References

1. Gutfreund, H.: Ann. Rev. Biochem. **40**, 315 (1971).
2. de Duve, C.: Wennergren Symposium on structure and function of oxidation reduction enzymes. Wennergreen Center, Stockholm Aug. 1970.
3. Hess, B., Boiteux, A., Krüger, J.: Advanc. Enzyme Regulat. **7**, 149 (1969).
4. — Allgemeine Prinzipien der Regulation der Glykose, II. Gemeinschaftstagung der Gesellschaften in der Deutschen Gesellschaft für experimentelle Medizin, Leipzig. Berlin: Akademie Verlag 1970 (im Druck).
5. — Boiteux, A., Krüger, J.: Fed. Proc. Abstr. 1556, **28**, Teil 1, p. 539 (1969).
6. — Organisation of Glycolysis. (Locker, A., Ed). Berlin-Heidelberg-New York: Springer (im Druck).
7. Dixon, M., Webb, E. C.: Enzymes, 2nd Ed. London: Longmans, Green and Co., Ldt. 1965.
8. Heilmeyer, L., Haschke, L. H.: Mosbach Colloquium **23**, 299. Berlin-Heidelberg-New York: Springer 1972.
9. Bischofberger, H., Hess, B., Schaknies, U.: Unpublished experiments.
10. Kosow. D. P., Rose, J. A.: J. biol. Chem. **243**, 3623 (1968).
11. Garfinkel, D., Hess, B.: J. biol. Chem. **239**, 971 (1964).
12. Arnold, H., Henning, R., Pette, D.: Europ. J. Biochem. **22**, 121 (1971).
13. Wurster, B., Hess, B.: Hoppe-Seylers Z. physiol. Chem. **351**, 869 (1970).
14. Hess, B., Wurster, B.: FEBS-Letters **9**, 293 (1970).
15. Barwell, C.-J., Hess, B.: Hoppe-Seylers Z. physiol. Chem. **351**, 1531 (1970).
16. Wurster, B., Hess, B.: Hoppe-Seylers Z. physiol. Chem. **351**, 1537 (1970).
17. Hess, B., Johannes, K.-W., Kutzbach, C., Bischofberger, H. P., Barwell, C. J., Röschlau, P.: Structure and function of pyruvate kinase. Joint Biochemical Symposium USSR-DDR. Berlin: Reinhardsbrunn Academie-Verlag 1971 (im Druck).

18. Boiteux, A., Hess, B.: Unpublished experiments.
19. Hess, B., Boiteux, A.: Hoppe-Seylers Z. physiol. Chem. **349**,1567 (1968).
20. — Siess, E., Wieland, O.: Unpublished experiments.
21. Linn, T. C., Pettit, F. H., Reed, L. J.: Proc. nat. Acad. Scie. (Wash.) **62**, 236 (1969).
22. Wieland, O., Siess, E.: Proc. nat. Acad. Sci. (Wash.) **65**, 947 (1970).
23. Chance, B.: Biochem. J. **103**, 1 (1967).
24. — De Vault, D., Legallais, V., Mela, L., Yonetani, T.: Kinetics of electron transfer reactions in biological systems. In: Fast reactions and primary processes in chemical kinetics, p. 437 (Claesson, S., Ed.). New York: Interscience Publishers 1967.
25. Baudras, A., Krupa, M., Labeyrie, F.: Europ. J. Biochem. **20**,58 (1971).
26. — Capeillere-Blandin, C., Iwatsubo, M., Labeyrie, F.: In: Oxidation reduction enzymes (Akeson and Ehrenberg, Eds.). Oxford, England: Pergamon Press 1972 (in press).
27. Slater, E. E., Colpa-Boonstra, J. F.: In: Haematin enzymes, p. 575, (Falk, J. E., Lemberg, R., Morton, R. K., Eds.). Oxford: Pergamon Press 1961.
28. Chance, B.: In: Haematin enzymes, p. 597 (Falk, J. E., Lemberg, R., Morton, R. K., Eds.). Oxford: Pergamon Press 1961.
29. Hess, B.: Unpublished experiments.
30. Erecinska, M., Chance, B.: Unpublished experiments.
31. Chance, B.: FEBS Letters **23,** (1972) 3.
32. Kovac, L., Smigan, P., Hrusovska, E., Hess, B.: ABB. **139**, 370 (1970).
33. Chance, B.: Nature (Lond.) **169**, 215 (1952).
34. Kröger, A., Klingenberg, M.: Vitam. and Horm. **28**, 533 (1970).
35. Lieb, W. R., Stein, W. D.: Nature (Lond.) New Biology **234**, 220 (1971).
36. Cone, R. A.: Nature (Lond.) New Biology **236**, 39 (1972).
37. Chance, B., Erecinska, M., Wilson, D. F., Dutton, P. L., Lee, C. P.: The function of cytochrome in mitochondrial membrane (in press).
38. Chance, B.: Personal communication.
39. Wilson, B. F., Dutton, P. L.: Biochem. biophys. Res. Commun. **1970**, 3959.
40. Dickerson, E. R., Takano, T., Eisenberg, D., Kallai, O. B., Samson, L., Cooper, A., Margoliash, E.: J. biol. Chem. **246**, 1511 (1971).
41. Margoliash, E.: Gordon research conference on energy coupling mechanism. Aug. 1971.
42. Radda, G., Erecinska, M., Chance, B.: Personal communication.
43. Cain, J., Santillan, G., Blasie, J. K.: In: Membrane research (Proc. 1972 ICN-UCLA Symp. Molecular Biology (Fox, C. Fred, Ed.). New York: Academic Press (in press).

Discussion

Th. Bücher (München): Did you find direct evidence for an interaction between triose phosphate isomerase and fructose diphosphate-aldolase?

B. Hess (Dortmund): Up to now no direct evidence for an interaction has been detected.

G. K. Radda (Oxford): You propose that the apparently anomalous behaviour of cytochrome b in becoming reduced with an oxygen pulse may be explained by the presence of some "constraint" in the coupled membrane. In the last year B. Chance and I have studied the mobility of coenzyme Q in mitochondria using a fluorescent probe method and have shown that, depending on the redox state of the mitochondrial membrane, the diffusion (translational motion) of coenzyme Q can be drastically altered.

Would you like to say if what you propose is a "thermodynamic constraint" (of the kind Wildon and Dutton put forward) or is it in fact a "kinetic constraint" that reduced the motion (translational or rotational motion of cytochrome c)?

B. Hess (Dortmund): I do not think that at the present time one can distinguish between the two alternatives since exact knowledge of the flux conditions is not available. I could visualize a situation in which the flux through the cytochrome system is very small and the thermodynamic constraint is really the explanation for the phenomenon being observed. However, if the flux passes a critical threshold value, the situation might be completely different since we pass into a domain far from equilibrium.

H. Holzer (Freiburg): In the case of yeast pyruvate decarboxylase, we have shown that extrapolation from the rate measurements in vitro (optical test of the purified enzyme under conditions simulating the conditions in intact cells) to the rate in vivo (CO_2 formation during anaerobic fermentation of glucose) is allowed (Holzer: IV. Mosbacher Kolloquium. Berlin-Göttingen-Heidelberg: Springer-Verlag 1953).

B. Hess (Dortmund): Our question was whether your results of 1953 were not purely accidental, especially since we have shown that the overall kinetic constants of the purified enzyme are quite dependent on a number of conditions, e.g. the phosphate concentration [Boiteux, A., Hess, B.: FEBS-Letters **9**, 293—295 (1970)]. Furthermore, since your in vitro experiments have been carried out with a highly dilute enzyme solution, the enzyme ought to be reinvestigated under near-physiological conditions. As I have discussed before, the result is not unequivocal.

Influences of Heterologous Protein Interactions on the Control of Glycogen Phosphorylase

L. HEILMEYER Jr. and R. H. HASCHKE

Physiologisch-Chemisches Institut der Universität Würzburg, Germany, and Department of Biochemistry, University of Washington, Seattle, Wash., USA

With 10 Figures

Even though extensive information concerning glycogen-phosphorylase has been accumulated since its discovery in 1938, the regulatory processes in the muscle leading to glycogen breakdown in response to hormonal or nervous impulses are not yet fully understood. From the fundamental work of CORI and his group [1] phosphorylase was the first enzyme shown to exist in an active *a* and an inactive *b* form. Interconversion of the two forms was demonstrated by FISCHER and KREBS and consists of phosphorylation (catalysed by phosphorylase kinase) and dephosphorylation (catalysed by phosphorylase phosphatase) [2]. Phosphorylase *b* was also the first enzyme shown to have an absolute requirement for an effector (AMP) which is not structurally related to substrates or products and which does not participate in the catalytic events [1]. Some basic information, relevant for the purified enzyme, is summarized in Fig. 1. Intracellularly, phosphorylase catalyzes the breakdown of glycogen to G-1-P which enters the glycolytic pathway and which is used to produce energy for muscle contraction. The enzyme is composed of two subunits with the dimer having a molecular weight of 200,000 [3]. The *b* form exists in an inactive conformation (squares) which can be shifted to an active state by AMP (circles). A second mode of activation through *b* to *a* conversion is provided by phosphorylation of one specific serine residue of each subunit (circles marked P). This information has been obtained by working with highly purified enzyme whereby the maximum enzymatic activity and the regulatory potential have been determined.

If these known parameters of the purified enzyme are taken to calculate phosphorylase activity in its natural environment, major discrepancies are revealed. For example, in resting muscle it is known that phosphorylase exists in its *b* form and is converted to the *a* form during muscle contraction [4]. However, it can be calculated from the rate of glycogen breakdown under anaerobic tetanic conditions that only 5% of the maximum potential enzymatic activity is needed [5]. During rest, phosphorylase is reconverted to the *b* form and shows only 1/10,000 of its maximum

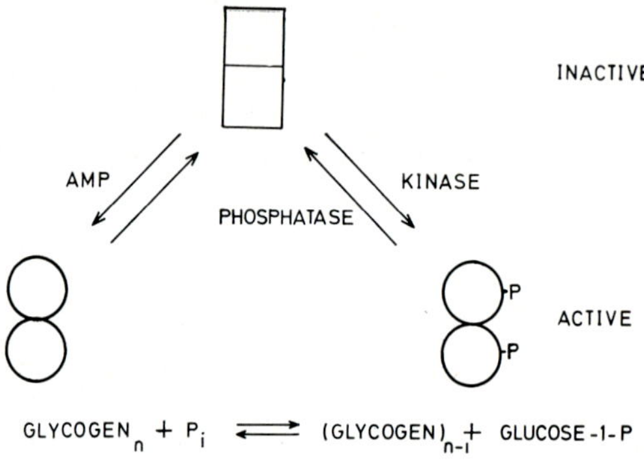

Fig. 1. Scheme of phosphorylase activation

potential activity in spite of the presence of 0.5 mM AMP, which would fully activate the enzyme under *in vitro* test conditions. Measurements of phosphorylase *b* activity under conditions simulating intracellular concentrations of the activator AMP and the inhibitors ATP and G-6-P show that it still exhibits activities about 100-fold higher than those observed in the intact cell [6]. A comparison of the results obtained from work on intact tissue and purified enzymes with respect to the activity and regulation of phosphorylase kinase and phosphatase also reveal major discrepancies [5]. It may be possible that these difficulties are due to the removal of the enzymes from their natural environment. Therefore, 2 years ago in the laboratory of Dr. FISCHER in Seattle a search was undertaken for an integrated enzyme system in between the two levels of organization of the intact cell and the purified enzymes that would retain some of the enzymatic properties of the intact muscle.

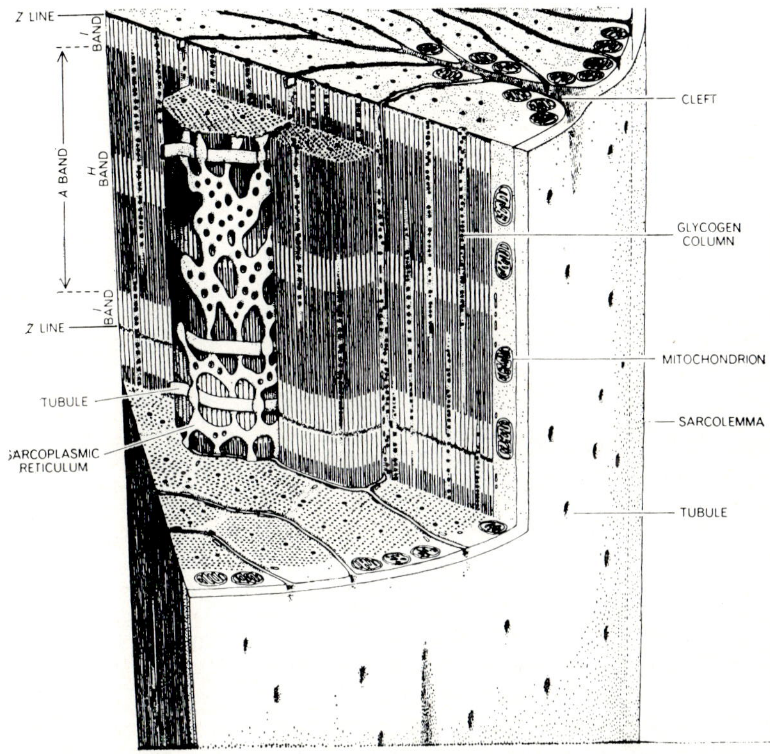

Fig. 2. Scheme of a muscle cell, HOYLE, G.; Sci. Amer. p. 222 85, (1970)

A schematic representation of a muscle cell deducted from electron micrographs is shown in Fig. 2 and taken from an article of Hoyle [7]. Glycogen particles are located in the interspace between the myofibrils and the sarcoplasmic reticulum. Many indications have been obtained that phosphorylase and other enzymes of glycogen metabolism are firmly associated with these glycogen particles [8]. A glycogen particle of the same size and staining behavior as seen in the electron micrographs of muscle slices can be obtained from crude muscle extracts by precipitation with either acetone or acid, or by differential centrifugation. All three preparations have essentially identical compositions of roughly 70% protein and 30% carbohydrate. One half of the protein is attribut-

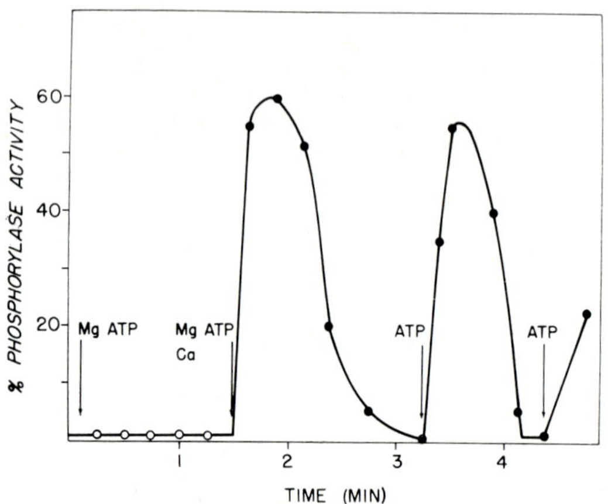

Fig. 3. Flash activation of phosphorylase. At the times indicated, the various effectors were added to a suspension of the complex (20 mg/ml phosphorylase) in glycerophosphate 50 mM, EDTA 1 mM, pH 7.0. Final concentrations were 3 mM Mg, 1 mM ATP, 0.1 mM free Ca. Phosphorylase assays were performed after 100-fold dilution at different time points according to [9]

able to fragments of the sarcoplasmic reticulum and the other is associated to glycogen. Phosphorylase accounts for 50% of the polysaccharide bound enzymes the remainder being glycogen synthetase and the respective interconverting enzymes.

If ATP/Mg alone is added to a highly concentrated suspension of this protein-glycogen complex, no response of the phosphorylase system can be seen (Fig. 3). If, however, free Ca ions are present together with ATP/Mg, an instantaneous conversion of phosphorylase *b* to *a* is observed, followed by an immediate reconversion to the *b* form. Phosphorylase activity was tested after high dilution in the absence of AMP in order to see only the activity of the *a* form. One hundred percent is defined as the activity measured in presence of AMP.

The very rapid increase in phosphorylase activity is therefore due to an activation of phosphorylase kinase, which in turn phosphorylates phosphorylase. Due to the highly active ATPase of the vesicles of the sarcoplasmic reticulum, the 1 mM ATP is hydrolyzed

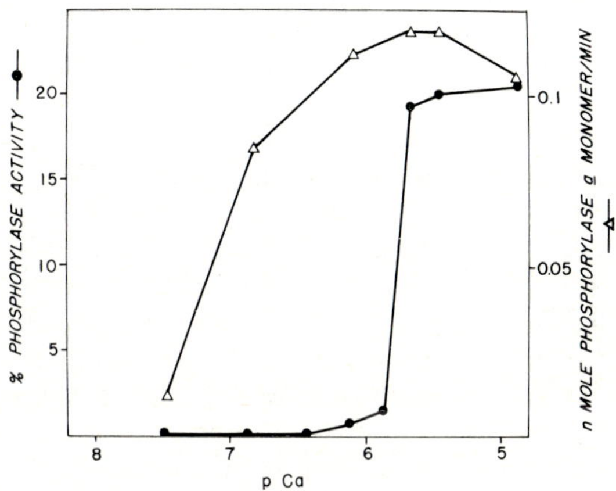

Fig. 4. Dependence of kinase activity on free Ca concentration in the complex suspension (●) and purified kinase solution (△). The complex suspension contained 5 mM EGTA and purified kinase (0.4 μg/ml) 2.5 mM EGTA. Ca was added to produce free Ca concentrations as indicated in form of the negative logarithm in the *abscissa*. Phosphorylase was tested 30 sec after start of the flash activation as described in Fig. 1 and kinase according to [9]

in about 30 sec. After exhaustion of ATP, phosphorylase kinase becomes inactive and the formed phosphorylase *a* is reconverted to the *b* form through the action of phosphorylase phosphatase. Readdition of ATP produces a second burst of phosphorylase activity and the whole process, called "flash activation" [10] of phosphorylase, can be repeated many times in succession with little change in the overall pattern.

The activation of kinase by the Ca ions is completely reversible since addition of excess EDTA over the Ca nearly completely abolishes the flash activation of phosphorylase. Using Ca/EGTA buffers, which stabilize a certain free Ca concentration, it can be shown (Fig. 4) that approx. 10^{-6} M free Ca ions are needed to trigger this flash activation. For comparison, the behavior of purified phosphorylase kinase is shown under the same conditions. Half maximal activation of the purified enzyme occurs at a free Ca concentration which is about 10 times lower.

In addition to the Ca activation of kinase, a well-known hormonal mediated phosphorylation of the enzyme occurs [10]. Since neither epinephrine nor cyclic AMP is present in the system described above, this type of activation mechanism is not operative under the experimental conditions employed.

Experiments in intact muscle have shown that maximal contraction occurs upon injection of Ca/EGTA buffers containing approx. 10^{-6} M free Ca [12]. Therefore, the free Ca concentration needed to trigger the flash activation is identical to that needed to initiate muscle contraction. The Ca-dependent activation-inactivation reaction of phosphorylase induced in the protein glycogen complex is probably very similar to the rapid changes in activity of this enzyme observed in the muscle during a contraction relaxation cycle.

The activation of phosphorylase kinase by Ca is due to an enhancement of the affinity of the enzyme for phosphorylase *b* (Fig. 5). The change in K_M caused by 10^{-5} M Ca was approximately 13-fold (decreasing from 1 to 0.08 mM). However, in the suspension of the protein glycogen complex which contains approximately 20 mg/ml phosphorylase (about 0.2 mM), no phosphorylase kinase activity can be shown in the range between 10^{-8} to 10^{-6} M of free Ca. Therefore, in addition to the inhibition of kinase by lack of Ca, the activity of this enzyme is further decreased, possibly because phosphorylase *b* is less available. The experiments indicate that the affinity of phosphorylase kinase for both Ca and phosphorylase *b* is modulated by other components of the protein glycogen complex. Therefore, in intact cells, the same concentration of free Ca released from the sarcoplasmic reticulum will stimulate both muscle contraction and glycogen breakdown.

Of other enzymes, phosphorylase phosphatase is directly involved in the flash activation of phosphorylase and is tightly bound to this complex. If both kinase and phosphatase were active at the same time, the two enzymes would be acting in opposition to each other, thus interfering with phosphorylase *b* to *a* conversion and producing a wasteful ATPase. Extensive work on purified phosphorylase phosphatase revealed few clues as to its possible involvement in the regulatory process [13]. A new type of phosphatase regulation can be observed if the enzyme is integrated in the protein-glycogen complex (Fig. 6). Phosphorylase phosphatase

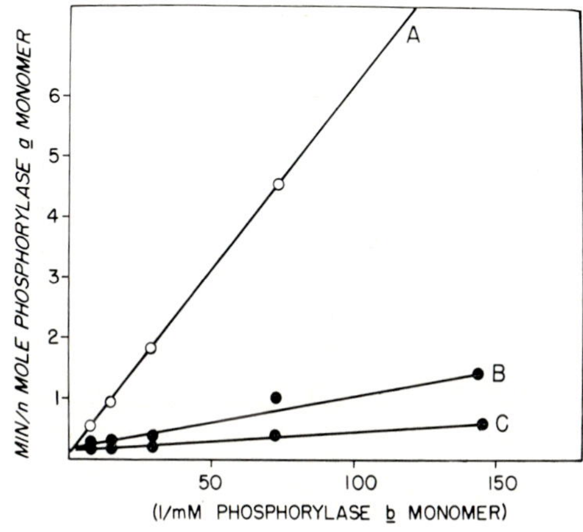

Fig. 5. Determination of the K_M of phosphorylase kinase for phosphorylase *b* as a function of free Ca concentration. Phosphorylase kinase was 0.7 μg/ml. Ca concentrations were A 10^{-8} M; B 10^{-7} M; and C 10^{-5} M. Conditions were similar to that described in legend of Fig. 4

activity was determined during the flash activation by adding ^{32}P labelled phosphorylase *a* and measuring the release of radioactive inorganic phosphate. As shown here in synchronization with flash activation, the activity of the phosphatase is inhibited by about 80%. This inhibition, which can only be seen in the intact protein-glycogen complex, is reversible and after a few minutes the full activity is regained. In absence of free Ca no flash activation occurs and only a minor inhibition of the phosphatase is observed. Disruption of the complex simply by high dilution causes dissociation of the proteins and completely abolishes this reversible phosphatase inhibition. Since in this system ATP is used up very quickly, a possible explanation would be an intermediate accumulation of AMP. AMP inhibits purified phosphatase activity by binding to the substrate phosphorylase *a*. It can be excluded that this type of AMP action is responsible for the reversible phosphatase inhibition (Table 1). Addition of 1 mM AMP or IMP (formed from AMP by

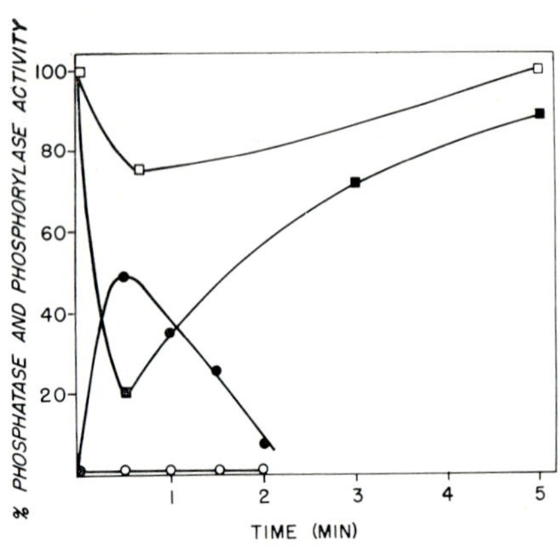

Fig. 6. Phosphorylase phosphatase activity during flash activation of phosphorylase in a suspension of the complex. Conditions and assays for the flash activation were as described in legend for Fig. 3. For assays for the phosphatase activity aliquots were removed at the indicated time points and ^{32}P labelled phosphorylase *a* was added. The rate of release of radioactive inorganic phosphate was determined according to [13]. ● and ○ phosphorylase, □ and ■ phosphorylase phosphatase activity; ● and ■ 0.3 mM free Ca and ○ and □ excess EDTA

AMP deaminase) produces only a minor inhibition of the complexed phosphatase. However, in the disrupted complex these nucleotides are powerful inhibitors. This characteristic of complex-bound phosphatase is in agreement with *in vivo* observations where a 100-fold higher concentration of AMP over the K_i (about 10^{-6} M) of the phosphatase is not inhibitory. If 20-fold diluted phosphatase was assayed using a phosphopeptide derived from radioactively labelled phosphorylase *a* as substrate, the reaction is not inhibited by IMP.

However, if phosphatase activity was assayed with phosphopeptide during the flash activation, an inhibition is observed, as with the measurement with phosphorylase *a* as substrate. This inhibition is due to interactions of phosphatase with some other component of the complex, as shown in Fig. 7. The affinity of the phosphatase for phosphorylase *a* increases about 15-fold upon liberation of the enzyme from the protein glycogen particle by 50-fold dilution.

It appears that AMP cannot act on complex-bound phosphorylase *a*. Therefore, one would expect that phosphorylase *b* is not

Table 1. *Effect of Nucleotides and Ca on phosphorylase phosphatase activity in a diluted and undiluted protein-glycogen complex. Phosphatase was tested 30 sec after addition of the effectors as described in the legend to Fig. 6 with either phosphorylase a or a peptide derived from this enzyme by tryptic digestion*

Effectors	% Inhibition			
	undiluted suspension		20-fold diluted suspension	
	phosphorylase a	phosphopeptide	phosphorylase a	phosphopeptide
NONE	0	0	0	0
ATP Mg Ca	73	51	85	11
ATP Mg	12	19	74	11
AMP Mg Ca	5	—	64	—
IMP Mg Ca	7	0	62	0

activated allosterically by this nucleotide. In order to measure activation of this enzyme in the complex, AMP-1-N-oxide was substituted for AMP to avoid degradation of the effector by AMP deaminase. This derivative of AMP is not acted on by the deaminase. The activation of phosphorylase *b* in the protein-glycogen complex was compared to the behavior of the purified enzyme.

The data shown in Fig. 8 demonstrate that upon 20-fold dilution the V_{max} increases ca. 3-fold, whereas the K_A changes only slightly. Phosphorylase *b* in the diluted material shows approximately the same specific activity as the purified enzyme. The residual phosphorylase *b* activity in the protein glycogen complex calculated as 33% of the activity found upon dilution could be explained by a fraction of fully active phosphorylase which is dissociated from the complex. The enzyme remaining in the complex could be inactive because it does not bind the nucleotide. Another explanation would be that all the phosphorylase binds the nucleotide as strongly as the purified enzyme but the resulting conformational change, which induces activity, is restricted by other components of the complex.

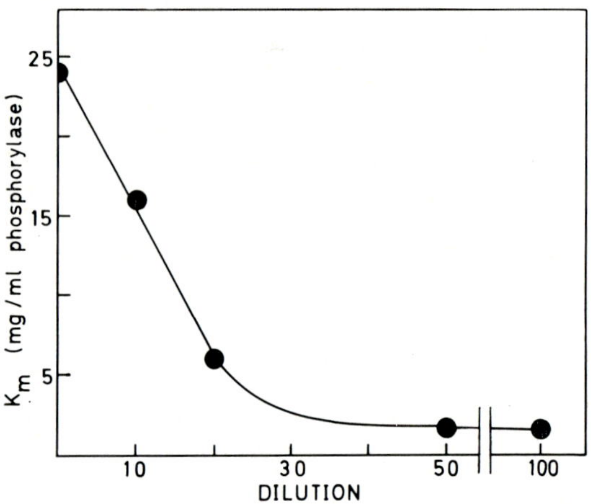

Fig. 7. Effect of dilution of the complex suspension on the K_M of phosphorylase phosphatase for phosphorylase *a*. At the indicated dilutions, phosphorylase phosphatase activity was determined as a function of phosphorylase *a* concentration and the K_m value estimated in a plot of reciprocal velocity against reciprocal substrate concentration. The assays were performed according to [13]

A means to differentiate between these two cases was provided by the reaction of phosphorylase *b* with a thioether analog of AMP (6-deamino-6-(3′-carboxy-4′-nitro phenylthio) adenosine-5′-phosphate). As shown by FASOLD et al., upon binding of this AMP, analog to phosphorylase, a substitution reaction occurs with a cysteine residue and the AMP moiety is covalently linked to the enzyme. Concurrent with the covalent binding, a stably activated form of phosphorylase *b* is produced [14].

The amount of derivative specifically bound to the AMP binding site was measured by assaying the enzyme after 10-fold dilution in

AMP-free substrate. Under these conditions, no allosteric activation by this analog of the *b* form occurs. Therefore, only covalently modified enzyme where the AMP is linked to its specific binding site shows activity. The results in Fig. 9 show that the increase in phosphorylase activity is linear with time and occurs at the same

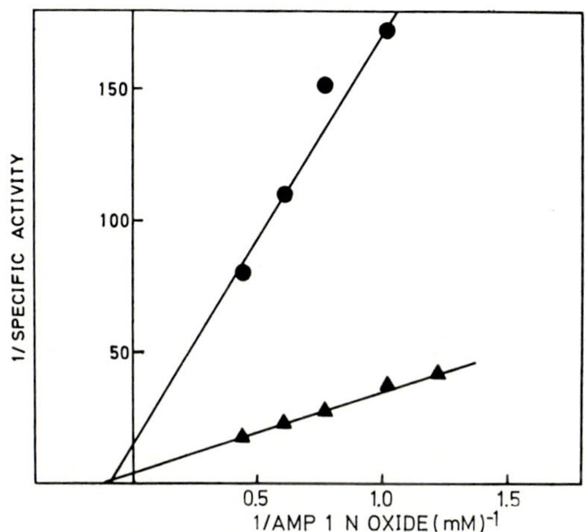

Fig. 8. AMP-1-N oxide activation of phosphorylase *b* in an undiluted complex suspension (●) and a 20-fold diluted sample (▲). The reaction mixture contained 50 mM sodium arsenate, 20 mg/ml glycogen, 50 mM glycerophosphate, 1 mM EDTA pH 7.0 and AMP-1-N oxide in the concentrations indicated in the *abscissa*. At desired times, aliquots were removed and the reaction stopped by mixing with an equal volume perchloric acid. After 10 min another volume of 1 M K_2CO_3 was added and the supernatants were collected. The amount of glucose produced was determined enzymatically

rate with both complexed and purified phosphorylase. Since the AMP derivative which attaches specifically at the AMP binding site reacts with both states of phosphorylase at the same rate, it can be concluded that all of the complexed enzyme is available to bind the nucleotide.

Neither the vesicles of the sarcoplasmic reticulum nor glycogen can be responsible for the suppression of the action of AMP-1-N oxide on phosphorylase. Similar effects can be seen in a fraction which was obtained after digestion of the glycogen by α-amylase and separation of the sarcoplasmic reticulum fragments by centrifugation (Table 2). Again, a 6-fold increase in phosphorylase activity is observed if the material is diluted 20-fold. Whereas glycogen was readded in order to measure phosphorylase activity, phosphatase

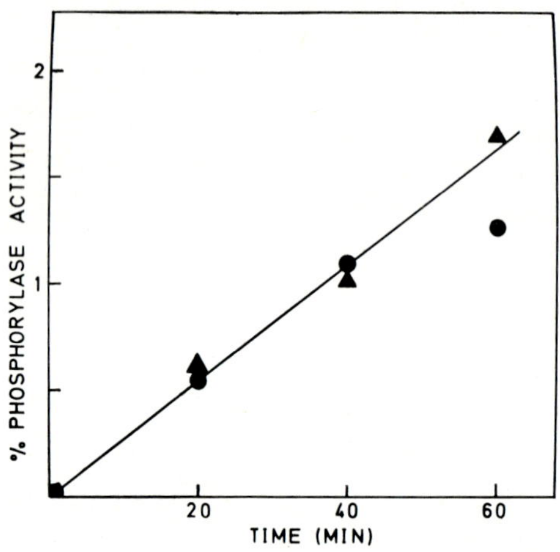

Fig. 9. Activation of phosphorylase *b* by reaction with 6 deamino-6-(3′carboxy-4′-nitro phenylthio) adenosine-5′-phosphate. The reaction is carried out according to [14]. (● complex bound phosphorylase, ▲ purified enzyme). The concentration of the AMP thioether derivative was 4.2 mM and phosphorylase was 15 mg/ml in each fraction. One hundred percent phosphorylase activity was determined in substrate containing 1 mM AMP. The covalently modified phosphorylase *b* was measured in the absence of AMP after a 10-fold dilution

activity was determined in absence of the carbohydrate. Therefore, the dilution effect is independent of glycogen for the phosphatase and, by analogy, for the phosphorylase activity. In this protein glycogen complex, there may be a factor present which suppresses the conformational change induced by the binding of nucleotides to purified phosphorylase.

The concentrated amylase supernatant can be dialyzed for one hour with little loss of this factor, whereas low molecular weight compounds are removed. However, after 20-fold dilution of the same fraction, the factor passes easily through a membrane retaining molecules of a molecular weight higher than 10,000. It appears that this factor is bound to a larger protein in the concentrated fraction but is dissociated by high dilution.

The factor can be enriched from the amylase supernatant fraction by heat denaturation of about 90% of the proteins and ultrafiltration (Diaflo PM 10 membrane). Increasing amounts of this factor added to purified phosphorylase *b* activated by AMP completely abolishes any activity, as shown in Fig. 10. The activity of

Table 2. *Effect of AMP-1-N oxide on phosphorylase b and phosphatase activity. Both experiments were performed with 3 mM effector present. Phosphorylase activity was tested as described in legend to Fig. 8 and phosphatase activity to Fig. 6*

Fraction	Phosphorylase activity %	Phosphatase inhibition %
Amylase supernatant concentrated	16	0
Amylase supernatant 20-fold diluted	100	67

phosphorylase *a* without AMP can also be completely inhibited, but some 5-fold higher concentrations are needed. As yet the factor has not been identified, but the fraction contains sugars from glycogen digestion, inorganic phosphate, IMP (identified by spectral characteristics) and peptides (identified after acid hydrolysis). Except for the last component, control experiments eliminated the other known components as the inhibitor.

Although the nature and the mechanism of the action of this factor are not yet known, all of the evidence suggests that it has a physiologically important role. One has to keep in mind that the overall regulation of phosphorylase is certainly still more complex. Besides Ca ions, epinephrine is also able to activate the phosphorylase system involving c-AMP and the general protein kinase. In addition, the hormone regulates the tension developed during muscle contraction and the duration of a single twitch, thereby acting as a modulator of the Ca effect on both processes, contraction and glycogen degradation. One may hope that studies on such integrated enzyme systems in addition to these on intact cells and purified enzymes, will allow us to analyze such interrelationships.

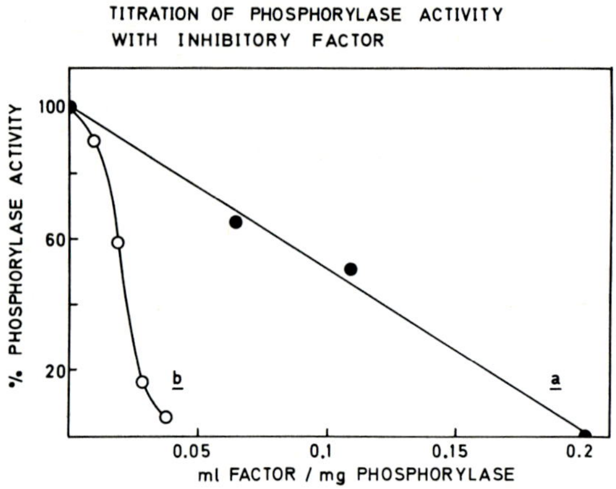

TITRATION OF PHOSPHORYLASE ACTIVITY WITH INHIBITORY FACTOR

Fig. 10. Inhibition of phosphorylase activity with inhibitory factor. Increasing amounts of inhibitor isolated as described in the text were added to a reaction mixture containing 5 mg/ml phosphorylase. Phosphorylase assays were performed as described in the legend to Fig. 8, with the exception that 80 μM AMP was used instead of AMP-1-N oxide. Phosphorylase *a* had the same concentration as the *b* form in the reaction mixture

References

1. CORI, C. F., CORI, G. T., GREEN, A. A.: J. biol. Chem. **151**, 39 (1943).
2. KREBS, E. G., FISCHER, E. H.: Biochim. biophys. Acta (Amst.) **20**, 150 (1956).
3. COHEN, P., DEWER, T., FISCHER, E. H.: Biochemistry **10**, 2683 (1971).
4. HELMREICH, E., CORI, C. F.: Advanc. enzyme Regulat. **3**, 91 (1956).
5. FISCHER, E. H., HEILMEYER, L. M. G., Jr., HASCHKE, R. H.: In: Current topics, in cellular regulations, **4**, p. 211. (HORECKER, B. L., STADTMAN, E. R., Eds.). New York: Academic Press 1971.
6. MORGAN, H. E., PARMEGGIANI, A.: J. biol. Chem. **239**, 2440 (1964).
7. HOYLE, G.: Sci. Amer. **222**, 84 (1970).
8. TATA, J. R.: Biochem. J. **90**, 284 (1964).
9. MEYER, F., HEILMEYER, L. M. G., Jr., HASCHKE, R. H., FISCHER, E. H.: J. biol. Chem. **245**, 6642 (1970).

10. SODERLING, T. R., HICKENBOTTOM, G. P., REIMANN, E. M., HUNKLER, F. L., WALSH, D. A., KREBS, E. G.: J. biol. Chem. **245**, 6317 (1970).
11. HEILMEYER, L. M. G., Jr., MEYER, F., HASCHKE, R. H., FISCHER, E. H.: J. biol. Chem. **245**, 6649 (1970).
12. PORTZEHL, H., CALDWELL, P. C., RÜEGG, J. C.: Biochim. biophys. Acta (Amst.) **79**, 581 (1964).
13. HASCHKE, R. H., HEILMEYER, L. M. G., Jr., MEYER, F., FISCHER, E. H.: J. biol. Chem. **245**, 6657 (1970).
14. HULLA, F. W., FASOLD, H.: Biochemistry (1972) (in press).

Discussion

B. HESS (Dortmund): I would like to ask about the sensitivity of the system with respect to ionic strength variation. First, is there any stabilization of the complex if you work with very low ionic strength and, secondly, can one of the K_m values which is decreased upon dilution be increased again if you work with very low ionic strength?

L. HEILMEYER (Würzburg): We are using only one buffer concentration, i.e. only one ionic strength condition. Since this material is isolated from muscle, low ionic strength conditions are used in order not to solubilize the myofibrils. I do not know what effect a change in the ionic strength has on the stability of the complex or on the K_m values of the participating enzymes.

B. HESS (Dortmund): There are other complexes in the muscle, as Dr. PETTE has shown, but as far as I remember, very low concentrations of divalent ions are needed to obtain the F-actin aldolase complex. Is that right?

D. W. G. PETTE (Konstanz): I am not sure whether the binding of several glycolytic enzymes to F-actin has a physiological role. It is true that this complex forms at relatively low ionic strength which might also be non-physiological. Certain effects have been found with respect, e.g. to the Michaelis constant of aldolase, which was increased by a factor of 10 by binding to the F-actin. However, the low ionic strength was a central factor in finding this complex.

E. HELMREICH (Würzburg): You made the rather sweeping statement that Ca^{++} ions are the signal which trigger simultaneously both muscle contraction and phosphorylase activation. Although I can easily see that Ca^{++} may trigger both processes at the same time, it is more difficult to see that on removal of Ca^{++} phosphorylase kinase and troponin should also be inactivated at the same time. After all, different proteins with presumably very different rates of association/dissociation of the components are involved. Would it not make more sense if phosphorylase activation were to continue even during relaxation? After all, the physiological importance of phosphorylase activation is to provide energy or replenish the energy stores of the muscle cell.

L. Heilmeyer (Würzburg): Since the protein glycogen complex contains variable amounts of enzymes, due to losses during preparation, no answer can be given to your question concerning the time course of phosphorylase activation and inactivation in the cell. It is, however, very probable that a muscle twitch and the corresponding changes in phosphorylase activity occur at different rates, since Ca^{2+} exerts its effect directly on the troponins, whereas in the latter case a cascade of enzyme reactions is involved.

D. W. G. Pette (Konstanz): Did you study the stability of the complex in regard to certain glycolytic metabolites? For example, in the case of aldolase bound to F-actin we found that fructose-di-phosphate dissolves such complexes.

L. Heilmeyer (Würzburg): We did not add metabolites directly to the protein glycogen complex. We added phosphate and then activated the phosphorylase. By the degradation of glycogen, glycolytic intermediates were produced. In this case, a flux of intermediates from glycogen down to the triose phosphates was demonstrated. Under these conditions, the complex was quite active and stable. After complete degradation of glycogen, the enzymes begin to dissociate. But as we have shown, in the α-amylase supernatant fraction the protein/protein interactions still occur. It is only upon dilution that these interactions are lost.

D. W. G. Pette (Konstanz): Could you exclude the possibility that the enzymes were dissociated in the experiment when you just found that triose phosphates are formed?

L. Heilmeyer (Würzburg): We studied this by ultracentrifugation in sucrose gradients and we could show that after digestion of the glycogen all the enzymes stay at the top of the gradient, whereas the enzymes sediment under these conditions if they are bound to the polysaccharide.

B. Hess (Dortmund): I would like to make a comment on what Ernst Helmreich said. With respect to the timing of such a process, we have to keep in mind that we need the high concentration to be quick. This is a very simple relation, even if a decrease of the K_m is observed, which would mean that the transition time of these cycles would be greatly reduced. If the system is treated as a simple first-order system, as I have just shown, the τ-value you observe is related to the K_m or to the enzyme concentration which means, in case of the protein-glycogen complex, the effective enzyme concentration. If your K_m is decreasing or increasing the τ-value might become very large, which would cause the relaxation time immediately to jump from msec to sec. If some enzymes leak out of the complex, probably out of the glycogen matrix, this would again decrease the concentration of the enzyme, which would result in a prolongation of a cycle, may be from 2 sec to 2 min. This refers to all systems in solution in a very general fashion and one has to wonder whether these terms stay constant.

L. **Heilmeyer** (Würzburg): Furthermore, one has to keep in mind that this system is artificial and that the added ATP is used very quickly; in *ca.* 10 sec the added 1 mM ATP is degraded. Certainly time constants, for example for the kinase reaction, are changed by going down several orders of magnitude with the concentration of ATP.

K. **Schnackerz** (Würzburg): Would you care to speculate on the architecture of your glycogen complex?

L. **Heilmeyer** (Würzburg): There are several aspects in relation to this problem. In respect to the size of this particle, it is known that isolated glycogen has S-values of *ca.* 120. The protein glycogen particle has approximately the same sedimentation constant of about 120 S, even though the material is composed of *ca.* 50% protein and 50% carbohydrate. Of course we do not know anything about the size of the glycogen which is integrated in the particle. My feeling is that the isolated glycogen is already an aggregate formed during separation from protein.

Muscle

Protein Assemblies in Muscle

S. Lowey

The Children's Cancer Research Foundation, Boston, Mass. and the Rosenstiel Basic Medical Sciences Research Center, Brandeis University, Waltham, Mass., USA

With 14 Figures

Introduction

The myofibril, the contractile unit of muscle, may be thought of as a highly concentrated suspension of proteins ($\sim 20\%$) enveloped by a membranous sac. Some of these proteins are globular in conformation, such as actin or troponin; some are highly α-helical, such as tropomyosin; while the most abundant protein, myosin, combines both globular and fibrous features in one large molecule. All of these proteins exist as part of highly organized structures comprising several hundred molecules; none function independently as monomers. In the case of the thick filament, the predominant molecular species is myosin. This protein consists of two major chains each with a molecular weight of 200,000 daltons, and four small subunits in the range of 20,000 daltons. A regulatory role for the light chains is suggested by the observation that myosins from fast muscles show a characteristic light-chain pattern which is distinct from that of slower muscles. The thin filament is composed largely of actin, but in addition contains the regulatory proteins troponin and tropomyosin. With the exception of actin, all of these proteins consist of more than one polypeptide chain. Calcium, the activator of muscle contraction, binds to one of the three subunits of troponin, thereby allowing the thin filament to associate with the myosin filament. This paper will summarize the properties of the individual proteins, describe their assembly into filamentous structures, and discuss the kinds of interactions within filaments and between filaments which may lead to a contractile force.

Excitation — Contraction Coupling

Electron microscope studies of fast skeletal muscle fibers have shown a complex membrane network within each cell. One membrane system, known as the sarcoplasmic reticulum, envelops each

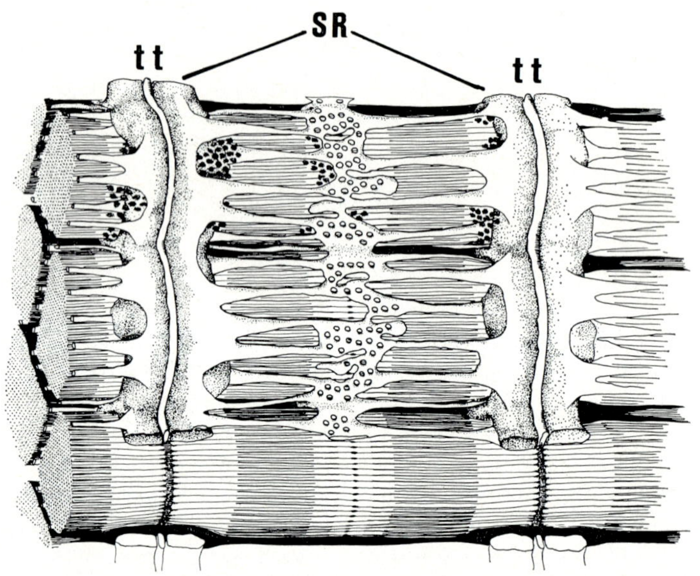

Fig. 1. Three dimensional reconstruction of a small portion of a frog muscle fiber. The contractile filaments are grouped into myofibrils, each of which is surrounded by a system of internal membranes. The sarcoplasmic reticulum (SR) envelops each myofibril from Z to Z-line; the transverse tubules (tt) seen at the Z-line are continuous with the surface membrane. Reprinted from L. D. Peachy, J. Cell Biol. **25**, 209 (1965)

myofibril and serves as an intracellular calcium repository. A second membrane system, the T-system or transverse tubules, is continuous with the plasma membrane but discrete from the sarcoplasmic reticulum, Fig. 1. When a muscle is excited by an action potential at the surface membrane, the T-system serves to carry the current flow into the interior of the cell. At the junction of the transverse tubules and the sarcoplasmic reticulum, some type of electrical transmission probably occurs which induces a permeability change in the sarcoplasmic reticulum with the consequent release of calcium (PEACHY, 1968; COSTANTIN, 1971). The exact details of how potential changes may trigger calcium movement are still not fully understood, but it is clear that the presence of calcium in the myofibril signals the onset of contraction. Relaxation is restored when calcium is transported back into the reticular vesicles. Thus contraction is controlled by a biochemical process within the muscle cell which regulates the concentration of ionized calcium (WEBER, 1966). How calcium transmits its signal to the contractile proteins is a fascinating example of the interplay of different proteins within a polymeric structure.

Structure of the Myofibril

Electron micrographs of muscle reveal two sets of filaments arranged in hexagonal arrays (HANSON and HUXLEY, 1953), Fig. 2. The thick filament is approximately 1.6 μ long and consists of several hundred myosin molecules. Cross bridges project along the entire length of the filament except for a bare region (pseudo-H-zone) of about 0.2 μ at the center of the filament. The unusual bipolar shape of the myosin filament reflects the distinctive, polar shape of the myosin molecule (HUXLEY, 1963). Attached to the bare region of the thick filament is the M-protein(s), which in cross sections of certain muscles, is seen to form bridges between the myosin filaments Fig. 2e (PEPE, 1971).

The thin filament is about 1 μ long and composed chiefly of actin molecules and the less abundant regulatory proteins, tropomyosin and troponin (EBASHI and ENDO, 1968). The thin filaments appear to be anchored in the sarcomere by the Z-line, a structure shown to possess a considerable degree of 2-dimensional order (HUXLEY, 1963). The protein composition of the Z-line is still uncertain, but α-actinin has been identified as one of the constituents (MASAKI, ENDO and EBASHI, 1967).

A necessary condition for contraction is overlap between the thin and thick filaments (the A-band region). Upon stimulation of the fiber, the thin and thick filaments interact with one another and generate the necessary force to move the actin filaments relative to the myosin filaments towards the center of the sarcomere (HUXLEY

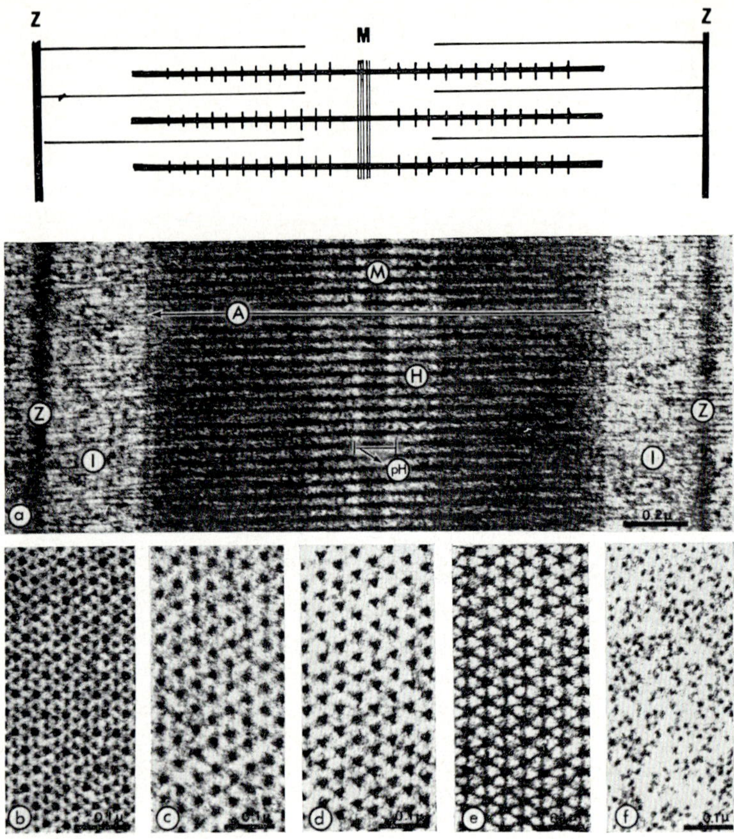

Fig. 2. Top: Schematic representation of the repeating unit (sarcomere) of a myofibril. Bottom: Reprinted from Pepe, F. A. in "Subunits in Biological Systems", Part A, ed. by Timasheff S. N., Fasman, G. D. New York: Marcel Dekker 1971. a) Longitudinal section through the sarcomere of the lateral muscles of the freshwater killifish. Cross section through the A-band in the region of b) overlap; c) the H-zone; d) the pseudo-H-zone and e) the M-band. f) Cross section through the I-band

and Hanson, 1954). One of the roles of the more minor Z- and M-line structures may be to preserve the integrity of these highly ordered arrays of filaments during the disruptive force of the contractile event.

Proteins of the Myofibril

Ever since KÜHNE (1868) discovered that a viscous protein could be extracted from muscle by strong salt solutions, people have been finding increasing numbers of proteins in muscle tissue. KÜHNE's original "myosin" preparation was later shown to contain actin in addition to myosin and was renamed actomyosin (STRAUB, 1942;

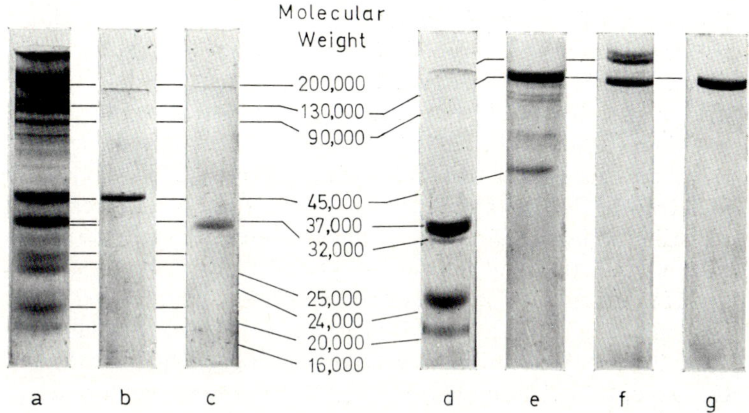

Fig. 3. Characterization of the myofibrillar extract and individual isolated proteins by SDS polyacrylamide gel electrophoresis. a) Total extract from chicken muscle; b) rabbit actin; c) chicken tropomyosin; d) rabbit troponin; e) chicken α-actinin; f) M-line proteins purified by ammonium sulfate; g) M-line protein isolated by ion-exchange chromatography

BANGA and SZENT-GYÖRGYI, 1942). But the description of actomyosin did not end there; some twenty years later, EBASHI and his coworkers (1964) discovered that actomyosin, as extracted from muscle, contained two additional proteins, tropomyosin and troponin. And so the list of isolated myofibrillar proteins has grown to include at least half a dozen proteins whose function and location in the sarcomere are reasonably well-defined.

Fig. 3 gives a comparison of the electrophoretic profile of the total myofibrillar extract with that of the individual proteins in the presence of sodium dodecyl sulfate and a reducing agent. An important conclusion to be drawn from this study is that most of the proteins in the myofibril have probably been identified. The number of "new" proteins to be discovered must surely be very limited. Another interesting feature of these gels is the diversity of subunit molecular weights: the sizes range all the way from 16,000 to 200,000 daltons.

Table 1 lists the molecular weights of the fibrous proteins in solvents chosen to minimize aggregation and yet not introduce any denaturation. Several interesting points emerge from this compilation: with a few exceptions, notably actin [YOUNG, 1967 (1)] and the newly discovered C-protein (OFFER, 1972), the majority of the fibrous proteins consist of more than one polypeptide chain. In the case of the structural proteins, tropomyosin (WOODS, 1967) and α-actinin (GOLL et al., 1971; MARGOSSIAN, 1972), the stable functional unit consists of two chains of identical mass. The same may hold true for the M-proteins, but here there is still some uncertainty regarding which (or if all) of the reported proteins originate in the M-line (MASAKI et al, 1968; KUNDRAT and PEPE, 1971; MORIMOTO and HARRINGTON, 1972; LOWEY and HARRISON, unpublished data). Troponin is a more complex molecule; it appears to have three subunits in about equimolar amounts, Fig. 3d (GREASER and GERGELY, 1971). The largest subunit (37,000 daltons) binds strongly to tropomyosin while the two smaller components are essential for inhibition of actomyosin ATPase (24,000 subunit) and calcium binding (20,000 dalton component) (EBASHI et al., 1971; HARTSHORNE et al., 1969; SCHAUB and PERRY, 1969). The complexity of the troponin molecule reflects its multiple functions: it is the major site for calcium binding in the myofibril and must regulate the interactions of the thin and thick filaments. Finally, the most abundant and also the largest multisubunit protein is myosin. Here the differences in chain weight are particularly striking. The bulk of the myosin molecule is composed of two large polypeptide chains, each with a molecular weight of about 200,000 daltons (GERSHMAN et al., 1969; GAZITH et al., 1970; LOWEY et al., 1969). The remaining 15% of myosin consists of several low molecular weight subunits in the 20,000 dalton size range. The next section will describe the structure and function of this molecule in greater detail.

Table 1. *Properties of Contractile Proteins*[a]

Localization in myofibril	% Total protein	Protein	Sedimentation coefficient	Viscosity ml/g	Molecular weight	Subunit molecular weight	% α-Helix
Thick filament	55	Myosin	6.4	210	470,000	200,000 20,000	57
Thick filament	2	C-Protein	4.6	14	140,000	140,000	< 10
Thin filament	25	G-Actin	3.3	4	46,000	46,000	26
Thin filament	5	Tropomyosin	2.6	34	64,000	32,000	90
Thin filament	5	Troponin	4.0	4	80,000	37,000 24,000 21,000	35
Z-line	tr	α-Actinin	6.2	9	180,000	90,000	60
M-line	tr	M-Protein(s)	9.5 6.0 5.4	— — 4.5	180,000 — 88,000	90,000 140,000 43,000	40 < 10 26

[a] Compiled from references in text.

The Myosin Molecule

Myosin from fast skeletal muscles (Fig. 4a, b, c) contains approximately 4 moles of light chains: 2 moles of an 18,000 dalton

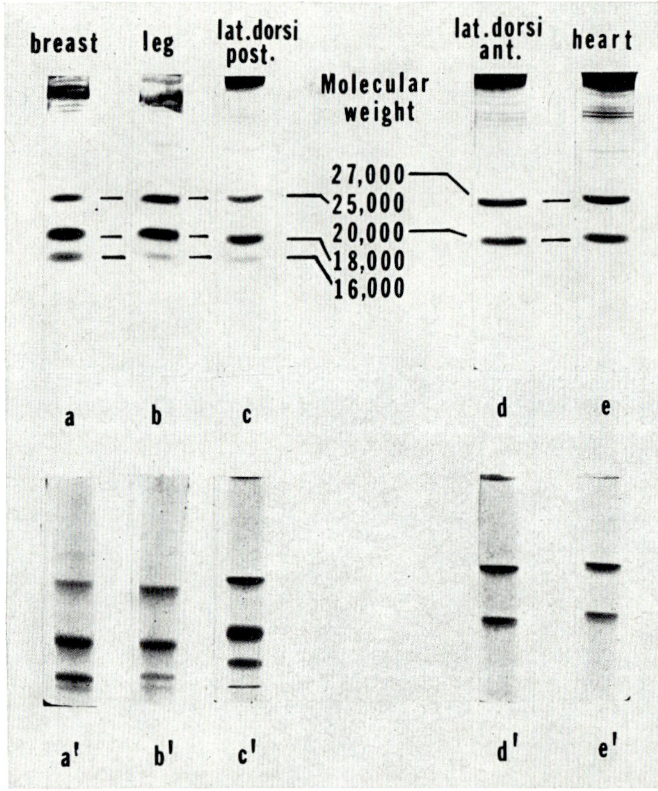

Fig. 4. Light chains from chicken muscles. a) to e) are 10% SDS polyacrylamide gels; a′) to e′) are standard gels of light chains isolated from the myosins depicted in a) to e). Reprinted from Lowey and Risby, Nature (Lond.) **234**, 81 (1971)

subunit and a total of 2 moles of the 25,000 and 16,000 dalton subunits (Lowey and Risby, 1971; Weeds and Lowey, 1971). Myosin from slow muscles (Fig. 4d and e) seems to have equimolar amounts of a 27,000 and 20,000 dalton subunit (LOWEY and RISBY, 1971). The relationship between the light chain pattern and the speed of the muscle adds support to the suggestion that the light chains may be involved in regulating the hydrolytic activity of myosin (STRACHER, 1969; DREIZEN and GERSHMAN, 1970). However, one class of light chains, the 18,000 molecular weight species in fast muscle myosins, do not seem to be essential for ATPase activity. If myosin is reacted with 5,5′-dithio bis-(2 nitrobenzoic acid)

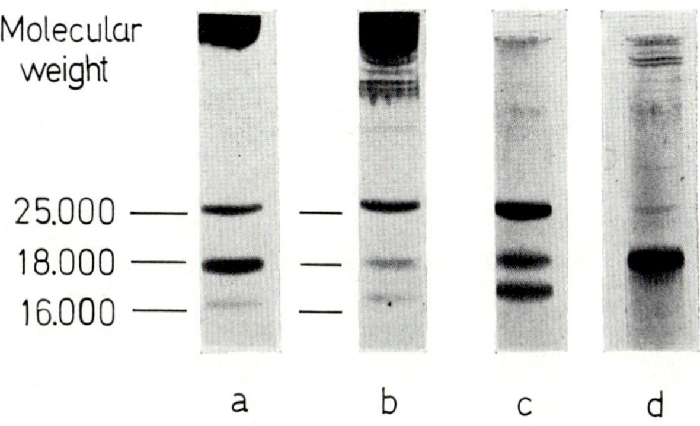

Fig. 5. a) SDS polyacrylamide gel electrophoresis of native myosin; b) myosin after treatment with DTNB; c) light chains removed from DTNB-treated myosin with denaturing solvents; and d) the light chain liberated from myosin by the DTNB reaction

(DTNB), an appreciable portion of the 18,000 dalton subunits are dissociated from myosin (Fig. 5d) without significantly altering the enzymic activity of the myosin (Fig. 5b) (GAZITH et al., 1970). The remaining light chains can only be removed by denaturing the myosin (Fig. 5c). The function of the DTNB-light chains remains unknown. It is important to note that the composition of the DTNB-light chains is chemically quite distinct from that of the 16,000 and 25,000 dalton subunits which are chemically related in sequence (WEEDS and LOWEY, 1971; WEEDS and FRANK, 1972).

A schematic model of myosin based on the studies described above and upon the many studies of native myosin and its proteolytic subfragments is shown in Fig. 6. Myosin is depicted as a polar molecule consisting of two similar halves; each half contains one heavy chain and a minimum of one light chain—probably two in the case of myosins from vertebrate muscles.

The conformation and tertiary structure of myosin has been largely deduced from studies of its proteolytic fragments by hydrodynamic methods and direct visualization by electron microscopy (Fig. 7) (SLAYTER and LOWEY, 1967; LOWEY et al., 1969). By means of the enzyme papain, it is possible to prepare many different types

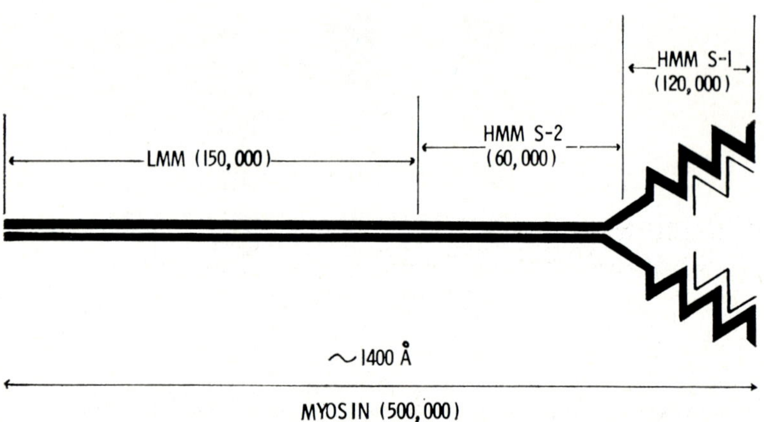

Fig. 6. Schematic representation of the myosin molecule. Reprinted from LOWEY et al., J. molec. Biol. **42**, 1 (1969)

of subfragments from myosin including a single-headed myosin, isolated rods, single globular "heads" (HMM S-1), the water-soluble portion of myosin containing the two heads attached to a 500 Å tail (HMM), the water-soluble portion of the rod (HMM S-2) and the water-insoluble region of the rod (LMM) (Fig. 7). Optical rotatory

---

Fig. 7. A composite of selected myosin molecules and subfragments of myosin. From the top to the bottom are: myosin, single-headed myosin, rods, HMM, HMM S-1, LMM and HMM S-2. Reprinted from LOWEY et al., J. molec. Biol. **42**, 1 (1969)

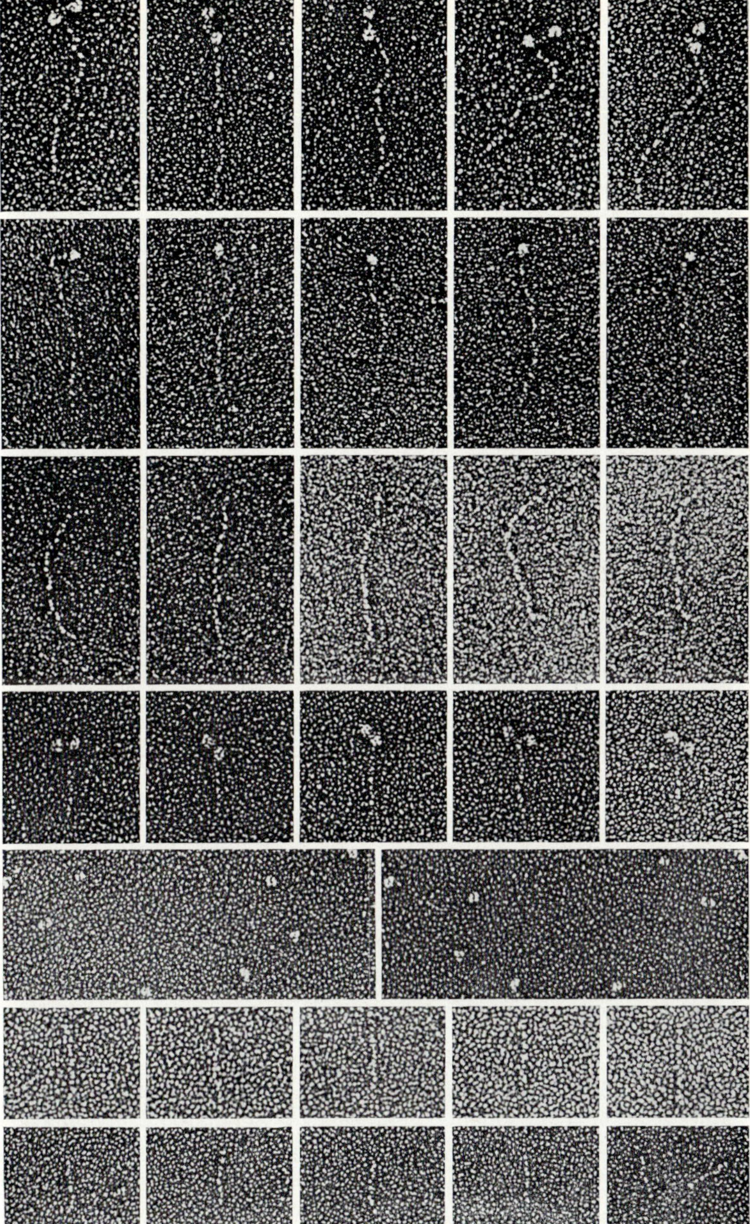

Fig. 7

dispersion measurements of the isolated rods give an unusually high α-helix content ($> 90\%$), consistent with an absence of proline residues in the amino acid composition of the rod. The HMM S-1 subfragments have a helix content of about 30%, a fairly typical value for globular proteins. Thus, each of the two large polypeptide

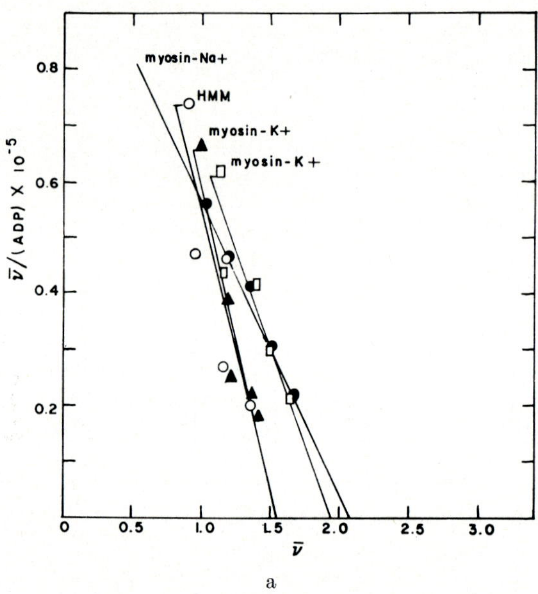

a

Fig. 8. a) Scatchard plot for the equilibrium binding of ADP to myosin and HMM where $\bar{\nu}$ is the average number of moles ADP bound per mole protein [Reprinted from Lowey and Luck, Biochemistry 8, 3195 (1969)]; b) Scatchard plot for the binding of HMM and HMM S-1 to F-actin, where $\bar{\nu}$ is the average number of moles of HMM (or HMM S-1) bound per mole of actin monomer (Margossian and Lowey, J. molec. Biol. [in press]

chains depicted in Fig. 6 is α-helical over a distance of about 1400 Å and terminates in a globular conformation about 100 Å in diameter (Lowey et al., 1969). The α-helical segments stabilize each other by strong side chain interactions along the length of the rod (Cohen and Holmes, 1963), while the globular regions appear to have considerable flexibility.

Each "head" of myosin contains at least one site for binding nucleotides (SCHLISELFELD and BÁRÁNY, 1968) and one site for actin binding [YOUNG, 1967 (2)]. Fig. 8a shows the binding of the competitive inhibitor, Mg-ADP, to myosin and HMM by equilibrium dialysis (LOWEY and LUCK, 1969). At sufficiently high con-

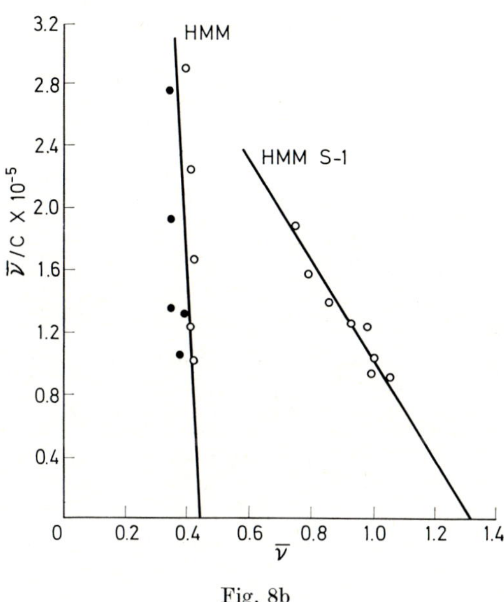

Fig. 8b

centrations of free nucleotide, close to 2 moles of Mg-ADP are bound per mole protein. The binding of F-actin to HMM can be measured in the absence of ATP by centrifuging the acto-HMM complex and determining the concentration of free HMM in the supernatant. These experiments give a value of 2 moles of actin bound per mole HMM and, as might be expected, a value of about 1 mole of actin per HMM S-1 head (Fig. 8b) (LOWEY, 1970; MARGOSSIAN and LOWEY, 1972). It is also possible to determine the rate of Mg-ATP hydrolysis by HMM and HMM S-1 in the presence of increasing amounts of the activator, actin (Fig. 9). From the intercept of double-reciprocal plots of ATPase activity versus actin concentration, it can be shown that the maximum *molar* ATPase of HMM is about twice that of HMM S-1. These results indicate that each of the two heads in myosin behaves relatively independ-

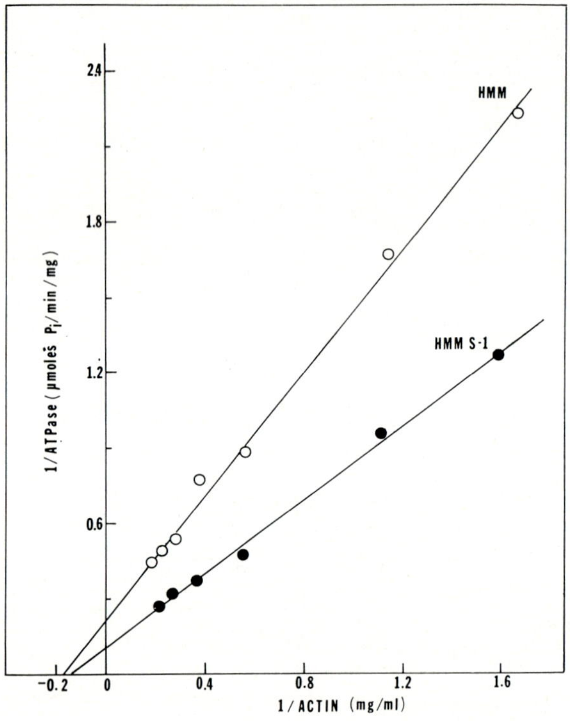

Fig. 9. Double-reciprocal plots of the actin-activated Mg-ATPase activity of HMM and HMM S-1 (Margossian and Lowey, J. molec. Biol. [in press])

ently in its interactions with actin and ATP (Margossian and Lowey, 1972).

The Thick Filament

As the ionic strength of a myosin solution is gradually lowered from about 0.5 M to less than 0.1 M, small aggregates are seen to form at intermediate salt concentrations which have a central bare

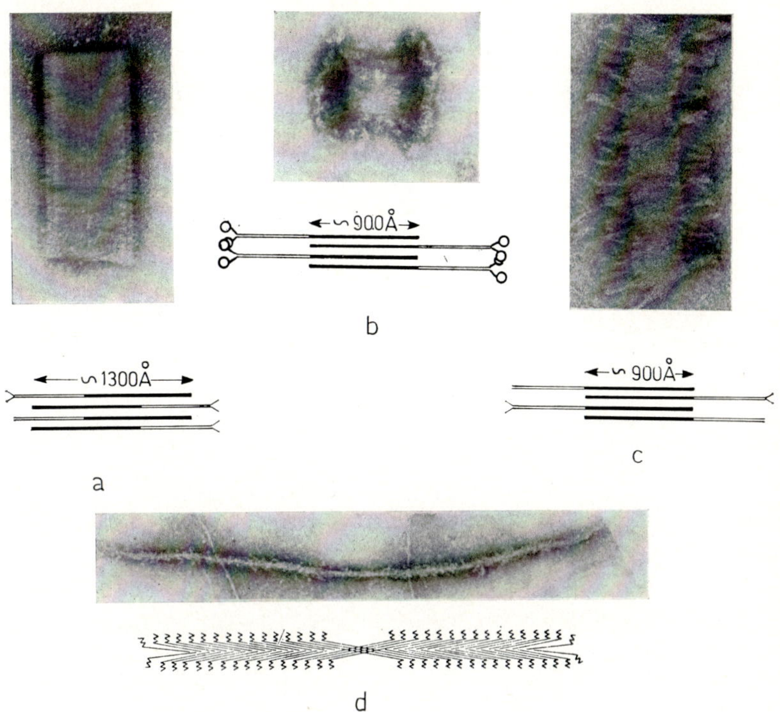

Fig. 10. a) and c) are two types of segments from myosin rods; b) is a segment from myosin. Schematic representation of the molecular overlaps are shown below the electron micrographs. Reprinted from HARRISON et al., J. molec. Biol. **59**, 531 (1971). d) Native thick filament from muscle, and schematic diagram of molecular packing. Reprinted from HUXLEY, Sci. Amer. **213**, 18 (1965)

region of about 2000 Å with globular appendages on either side. It is apparent from the shape of the myosin molecule that filament assembly is initiated by the antiparallel aggregation of the rod portion of myosin. Growth proceeds at either end by the addition of molecules pointing in the same direction (HUXLEY, 1963). These synthetic myosin filaments are very similar in appearance to the native thick filaments (Fig. 10d) except for a more variable length distribution.

The ability of myosin to aggregate into the filamentous state merely by being in an appropriate ionic environment suggests that filament formation is essentially a self-assembly process; meaning, all the information to build a thick filament is contained in the myosin molecule (Harrington and Josephs, 1968). Recently a

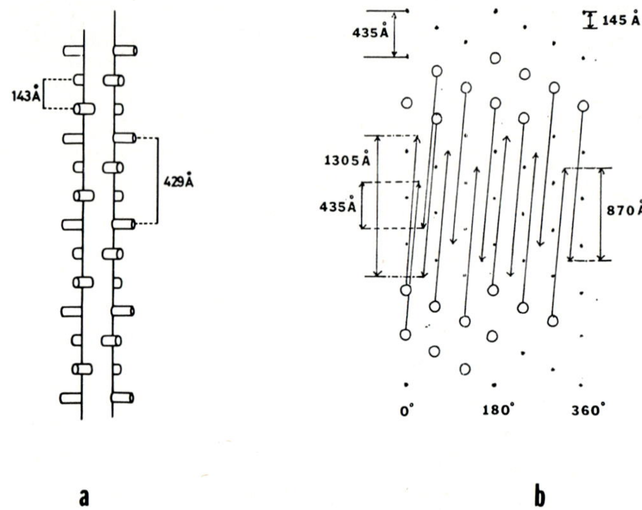

Fig. 11. a) Schematic diagram of a myosin filament showing arrangement of cross bridges on a 6/2 helix. Reprinted from Huxley and Brown, J. molec. Biol. **30**, 383 (1967). b) Modified two-dimensional lattice of the cylindrical surface in a). The length of the rod is taken as 1450 Å, and the molecular axes are tilted relative to the fiber axis (see Fig. 10 d). Reprinted from Harrison et al., J. molec. Biol. **59**, 531 (1971)

second component has been discovered in the thick filament, named the C-protein (Offer, 1972), Table 1. Until the function of this protein is better understood, however, it is premature to speculate on whether it may be involved in a length-determining mechanism.

The arrangement of the myosin molecules in the polar portion of the thick filament can be described as a 6/2 helix (Fig. 11a). The "cross-bridges" (consisting primarily of HMM S-1) project from the filament in pairs related by a two-fold axis; each pair must be rotated 120° and translated 143 Å to generate the helix (Huxley and Brown, 1967). In order that the cross-bridges appear in equivalent positions on the surface of the filament, the rods must be tilted relative to the filament axis, as shown by the schematic drawing in Fig. 10d. Although electron microscopy and X-ray diffraction can tell us about the arrangement of molecules in the polar region of the filament, they do not reveal the molecular packing in the bare region of the filament.

One approach to this problem was provided by the work of Cohen who showed that the simple, rod-shaped muscle proteins such as tropomyosin, paramyosin and the rod fragment of myosin form ordered aggregates *in vitro* which can be related to the native structures (Cohen and Longley, 1966; Cohen et al., 1970, 1971). Two distinct modes of aggregation have been observed with the rod subfragment of myosin: One with an overlap of 1300 Å and another with an overlap of about 900 Å (Fig. 10a and c). Myosin itself can only form an aggregate with a 900 Å overlap (Fig. 10b). Both types of segments are consistent with a value of 1450 Å for the length of the rod. By arranging molecules of this length on the surface lattice indicated by the X-ray results, one can generate the interactions found in the bipolar segments (Fig. 11b). Although a segment with a 435 Å overlap has not been found in myosin from striated muscles, this overlap is observed in rod segments from a smooth muscle myosin (Kendrick-Jones et al., 1971). These results suggest that the surface lattice which defines the packing of myosin molecules in the polar region of the thick filament may be directly related to the antiparallel packing of molecules in the *in vitro* aggregates (Harrison et al., 1971).

The Thin Filament

In a salt-free solvent, actin exists as a monomeric unit with a molecular weight of 45,000 daltons. Upon the addition of salt, G-actin-ATP is polymerized to F-actin-ADP and inorganic phosphate is released in the process. Negatively stained actin polymers appear indistinguishable from the isolated, native thin filaments except for their length: the native filaments are uniformly 1 μ long, whereas the *in vitro* filaments are of indefinite length (Fig. 12a). Electron microscopy (Hanson and Lowey, 1963) and X-ray diffraction (Huxley and Brown, 1967) have shown the thin filaments to consist of actin monomers, each about 55 Å in diameter, arranged in a helix whose pitch is about 2 × 360 Å (Fig. 13). The specific orientation of the actin globules in the chains is revealed by the attachment of the HMM S-1 subfragments to actin. The distinctive

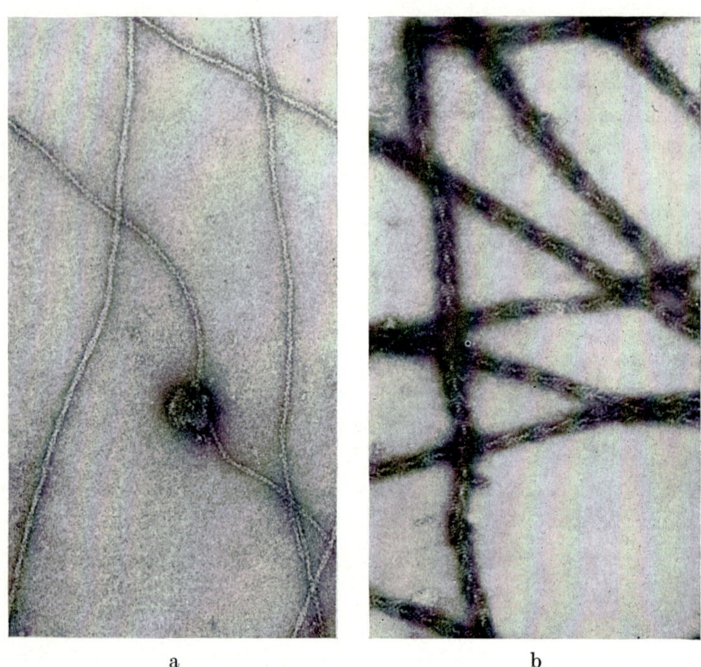

Fig. 12. a) Purified F-actin filaments. b) Actin filaments "decorated" with HMM S-1. Reprinted from Moore et al., J. molec. Biol. **50**, 279 (1970)

"arrowhead" appearance of the "decorated" actin filaments emphasizes the underlying structural polarity of the thin filaments: meaning each actin monomer must be related to its four neighbors in such a manner that all the actin units are oriented in the same direction (Fig. 12b) (Moore, Huxley and DeRosier, 1970).

Tropomyosin appears to lie in the two grooves of the actin helix (O'Brien et al., 1971), where it positions a troponin molecule about every 400 Å along the thin filament (Fig. 13). Evidence in support of this model comes from a periodicity of 400 Å observed in I-bands stained with antibodies prepared against troponin (OHTSUKI et al., 1967). This periodicity is also evident in the I-band of untreated muscles (HUXLEY and BROWN, 1967), and has been identified with the end-to-end bonding of tropomyosin in its *in vitro* aggregates (COHEN and LONGLEY, 1966). The tropomyosin filaments must undergo a periodic bending in order to fit into the helical grooves of the actin filament. This variable bending or "supercoiling" of the tropomyosin filament is consistent with the data obtained from crystals of tropomyosin by electron microscopy and X-ray diffrac-

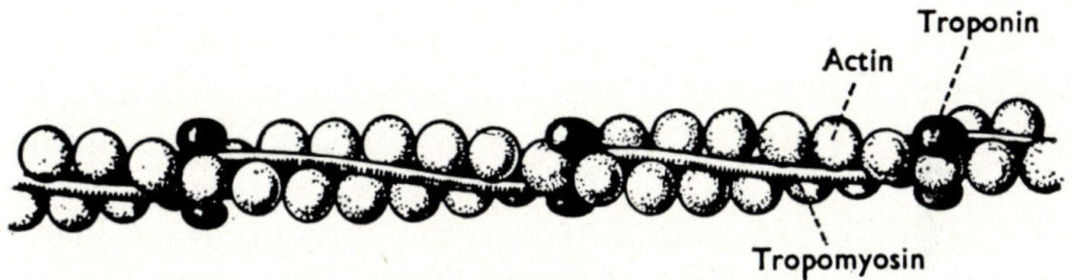

Fig. 13. A model for the thin filament. Reprinted from EBASHI et al., Quart. Rev. Biophys. **2**, 351 (1969)

tion (CASPAR et al., 1969; COHEN et al., 1971). The "kite-shaped" appearance of the nets in negatively stained preparations of tropomyosin crystals is a direct consequence of the bending imposed by the pattern of cross-connections between tropomyosin filaments (CASPAR et al., 1969). When troponin is co-crystallized with tropomyosin, the troponin molecule(s) can be seen attached to the middle of each long arm in the tropomyosin net (HIGASHI and OOI, 1968). The presence of troponin has also been observed to produce changes in the unit cell dimensions of the tropomyosin crystal. An implication of these variations is that troponin may be able to influence the conformation and interactions of the tropomyosin filament (COHEN et al., 1971).

The functional role of troponin-tropomyosin in vertebrate striated muscle is to regulate the interactions of actin with myosin. At free calcium concentrations below 10^{-8} M, actin is unable to associate with myosin and the muscle is at rest. Upon activation, the calcium level rises to $>10^{-5}$ M and contraction occurs. By binding calcium, a conformational change may be induced in each troponin molecule which is somehow transmitted to the 7 to 8 actin units under its control (Ebashi and Endo, 1968; Weber and Bremel, 1971; Spudich and Watt, 1971). Since tropomyosin interacts strongly with both troponin and actin, it is most likely the mediator through which the "ON-OFF" messages are communicated. One can envisage a mechanism whereby small conformational changes are introduced into tropomyosin by troponin; these changes may in turn affect the interactions between tropomyosin and actin and ultimately lead to a modification in the reactivity of the actin molecules. Alternatively, the regulation by troponin-tropomyosin may involve actual movement by the tropomyosin molecule from a position where it can sterically block the interaction of actin and myosin to one where the interaction is uninhibited (Huxley, 1972). A considerable body of X-ray diffraction evidence is now accumulating to show that tropomyosin moves in the thin filaments during contraction (Vibert, Haselgrove, Lowy and Poulsen, 1972; Huxley, 1972).

Interactions between Filaments

It is reasonable to assume that during relaxation the helical portion of myosin interacts with the backbone of the thick filament and the HMM S-1 globules are arranged on its surface. Upon stimulation of the muscle, a "bridge" is formed between the thin and thick filaments. The attachment of HMM S-1 to actin must generate a sufficient force to move the actin filament relative to the myosin filament. One of the puzzling features of this interaction has been the problem of how a link can be maintained despite the increasing interfilament distance caused by contraction[1]. A plausible solution to this problem can be found by considering the solubility properties of the myosin molecule: LMM is the least soluble region of myosin, and hence is primarily responsible for anchoring myosin in the core of the thick filament. HMM S-2 is much more soluble than LMM and, correspondingly, has weaker interactions with the

[1] It is well known that muscle contracts at "constant volume"; i.e. when the fiber shortens, the filaments move further apart, and when the muscle is stretched, the filaments move closer together (Huxley, 1969).

backbone of the filament. HMM S-1 is water-soluble and its interactions are probably limited to the actin filament. Thus it appears highly likely that HMM S-2 can bend reversibly away from the axis of the thick filament; the degree of bending will depend on the interfilament distance HMM S-1 must bridge to link with actin (HUXLEY, 1969; LOWEY et al., 1967, 1969; PEPE, 1966). X-ray diffraction studies on whole muscle have provided strong evidence for the existence of such a mechanism (HUXLEY and BROWN, 1967; HUXLEY, 1968).

Although the precise details of the interaction between myosin and actin are still largely unknown, a general picture of the sequence of events occuring during the cyclical attachment of myosin and actin is beginning to emerge. The clarification of the chemical events accompanying the structural changes has been provided largely through the kinetic studies of TAYLOR and his coworkers. For many years it was thought that actin activates the Mg-ATPase activity of myosin by forming an active complex. It now appears from rapid stopped-flow measurements that the activation of myosin ATPase is due to the displacement of the products of the hydrolysis reaction by actin. This conclusion followed in part from the observation that the rate of acto-HMM dissociation was much faster than the initial rate of ATP hydrolysis by HMM (LYMN and TAYLOR, 1971). The cycle of events can be summarized as follows: ATP rapidly dissociates the actomyosin complex by binding to the ATPase sites of myosin; free myosin then splits ATP and forms a relatively stable myosin-products complex; actin recombines with this complex and dissociates the products, thereby reforming the actin-myosin bridge which is presumably the force generating step (Fig. 14). Although this model provides an important conceptual framework for future studies of the contractile cycle, the individual steps and rate constants will undoubtedly undergo modification: for instance, recent spin-label and fluorescence experiments (GERGELY and SEIDEL, 1972; WERBER et al., 1972) have shown that the steady state hydrolysis of ATP does not lead to a simple complex of myosin and products (M · Pr), but rather to some intermediate conformation (M · ADP · P) with different spectral properties from (M · Pr).

Two different conformations of the cross-bridges of myosin (HMM S-1) have actually been seen in electron micrographs of sectioned insect flight muscle (Reedy et al., 1965): In the relaxed state, the dissociated bridge lies approximately perpendicular to the thick filament axis, whereas in the rigor state the bridges are attached at an angle of about 45° to the actin filaments. Independent evidence for two such distinct conformations has come from X-ray diffraction studies of living muscle (Huxley and Brown, 1967) and from three-dimensional image reconstructions of electron

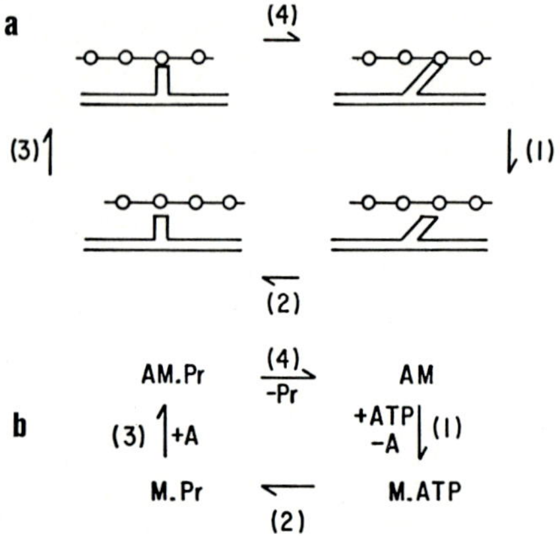

Fig. 14. a) Heuristic diagram of a contractile cycle for a single cross-bridge. b) Steps in chemical mechanism for ATP hydrolysis by actomyosin. Pr stands for reaction products, ADP and phosphate. Reprinted from R. W. Lymn and E. W. Taylor, Biochemistry **10**, 4617 (1971)

micrographs of actin filaments decorated with HMM S-1 (Moore et al., 1970) (see Fig. 12b). Although these studies all suggested that HMM S-1 was in a tilted or "slewed" configuration when attached to actin, the question remained whether this conformation depended on the presence of actin. By means of an analogue of ATP, which has the property of being able to dissociate actomyosin but not be hydrolysed by it, Holmes and his coworkers have been able to show that the "slewed" conformation persists even when myosin is dissociated from actin (Mannherz et al., 1972). They concluded from their studies that myosin can exist in two different conformational states, and that the transition between states required the hydrolysis of ATP. These structural findings complement very nicely the model of the cross-bridge cycle proposed by LYMN and TAYLOR (1971) on the basis of their kinetic data, Fig. 14.

Finally, I would like to comment briefly on the regulation of the actin-myosin interaction. As discussed previously, existing evidence implies that control of the interaction resides largely in the thin filament through the action of the calcium binding troponin. This mechanism implies a relatively passive role for myosin and the thick filament in so far as regulation is concerned. Although data favoring a more active role for myosin in vertebrate muscles are scant, it has been shown that molluscan muscles regulate contraction-relaxation primarily through the thick filament (KENDRICK-JONES et al., 1970). Molluscan myosin preparations contain a light chain which binds calcium, and have a calcium-dependent ATPase activity in the presence of pure actin (SZENT-GYÖRGYI et al., 1972). These invertebrate muscles do not have a troponin regulatory system, even though the thin filaments do contain tropomyosin. The finding of a regulatory subunit of molecular weight about 20,000 daltons in molluscan myosin raises the question of whether the light chains of rabbit myosin, in particular the DTNB-light chains, might not also have some hitherto undiscovered regulatory function in the myosin-actin interaction.

Another indication of myosin's participation in the control mechanism comes from the observation that at low concentrations of ATP, natural actomyosin from vertebrate muscles can hydrolyse ATP even in the absence of calcium (BREMEL and WEBER, 1971). It is well known that in the complete absence of ATP myosin forms "rigor links" with actin, irrespective of the calcium level. It has now been proposed that at low ATP concentrations, the fraction of myosin molecules present in rigor links can "switch ON" the neighboring unoccupied actin monomers, thereby enabling them to combine with free myosin and activate ATP hydrolysis (BREMEL and WEBER, 1972). Thus even in vertebrate muscles, myosin can influence the flow of information in the thin filament.

Although our knowledge concerning the structure and interactions of the myofibrillar proteins has greatly increased in the last decade, clearly much remains to be discovered before we can fully understand the intricate series of events culminating in a contractile force.

Acknowledgements. This work was supported by Public Health Service Grant AM-04762, Public Health Service Research Career Program Award K3-AM-10630, National Science Foundation Grant GB-8616, and a grant from the MuscularDystrophy Associations of America.

References

Banga, I., Szent-Györgyi, A.: Studies **1**, 5 (1941—42).

Bremel, R. D., Weber, A.: Nature (Lond.) **238**, 97 (1972).

Caspar, D. L. D., Cohen, C., Longley, W.: J. molec. Biol. **41**, 87 (1969).

Cohen, C., Longley, W.: Science **152**, 794 (1966).

— Lowey, S., Harrison, R. G., Kendrick-Jones, J., Szent-Györgyi, A. G.: J. molec. Biol. **47**, 605 (1970).

— Szent-Györgyi, A. G., Kendrick-Jones, J.: J. molec. Biol. **56**, 223 (1971).

— Caspar, D. L. D., Parry, D. A. D., Lucas, R. M.: Cold Spr. Harb. Symp. quant. Biol. **36**, 205 (1971).

Constantin, L. L.: Contractility of muscle cells and related processes, p. 69 and 89. (Podolsky, R. J., Ed.). New Jersey: Prentice-Hall, Inc. 1971.

Dreizen, P., Gershman, L. C.: Biochemistry **9**, 1688 (1970).

Ebashi, S., Ebashi, F.: J. Biochem. (Tokyo) **55**, 604 (1964).

— Endo, M.: Progr. Biophys. molec. Biol. **18**, 123 (1968).

— Wakabayashi, T., Ebashi, F.: J. Biochem. (Tokyo) **69**, 441 (1971).

Gazith, J., Himmelfarb, S., Harrington, W. F.; J. biol. Chem. **245**, 15 (1970).

Gergely, J., Seidel, J. C.: Cold Spr. Harb. Symp. quant. Biol. **37** (1972) (in press).

Gershman, L. C., Stracher, A., Dreizen, P.: J. biol. Chem. **244**, 2726 (1969).

Goll, D. E., Suzuki, A., Singh, I.: Biophys. Soc. Abstr. **107** a (1971).

Greaser, M. L., Gergely, J.: J. biol. Chem. **246**, 4224 (1971).

Hanson, J., Huxley, H. E.: Nature (Lond.) **173**, 530 (1953).

— Lowy, J.: J. molec. Biol. **6**, 46 (1963).

Harrington, W. F., Josephs, R.: Devel. Biol. Suppl. **2**, 21 (1968).

Harrison, R. G., Lowey, S., Cohen, C.: J. molec. Biol. **59**, 531 (1971).

Hartshorne, D. J., Theiner, M., Mueller, H.: Biochem. biophys. Acta (Amst.) **175**, 320 (1969).

Higashi, S., Ooi, T.: J. molec. Biol. **34**, 699 (1968).

Huxley, H. E., Hanson, J.: Nature (Lond.) **173**, 973 (1954).

— J. molec. Biol. **7**, 281 (1963).

— Brown, W.: J. molec. Biol. **30**, 383 (1967).

— J. molec. Biol. **37**, 507 (1968).

— Science **164**, 1356 (1969).

— Cold Spr. Harb. Symp. quant. Biol. **37**, (1972) (in press).

Kendrick-Jones, J., Lehman, W., Szent-Györgyi, A. G.: J. molec. Biol. **54**, 313 (1970).

— Szent-Györgyi, A. G., Cohen, C.: J. molec. Biol. **59**, 527 (1971).

KÜHNE, W.: Physiol. Chem. Leipzig 1868.
KUNDRAT, E., PEPE, F. A.: J. Cell Biol. **48**, 340 (1971).
LOWEY, S., GOLDSTEIN, L., COHEN, C., LUCK, S. M.: J. molec. Biol. **23**, 287 (1967).
— LUCK, S.: Biochemistry **8**, 3195 (1969).
— SLAYTER, H. S., WEEDS, A. G., BAKER, H.: J. molec. Biol. **42**, 1 (1969).
— 8th Intl. Cong. Biochem. Abstracts, **28** (1970).
— RISBY, D.: Nature (Lond.) **234**, 81 (1971).
— STEINER, L.: J. molec. Biol. **65**, 111 (1972).
LYMN, R. W., TAYLOR, E. W.: Biochemistry **10**, 4617 (1971).
MANNHERZ, H. G., BARRINGTON-LEIGH, J., HOLMES, K. C., ROSENBAUM, G.: Cold Spr. Harb. Symp. quant. Biol. **37** (1972) (in press).
MARGOSSIAN, S. S.: Fed. Proc. **31**, 866 Abs. (1972).
— LOWEY, S.: J. molec. Biol. (1972) (in press).
MASAKI, T., ENDO, M., EBASHI, S.: J. Biochem. (Tokyo) **62**, 630 (1967).
— TAKAITI, O., EBASHI, S.: J. Biochem. (Tokyo) **64**, 909 (1968).
MOORE, P. B., HUXLEY, H. E., DEROSIER, D. J.: J. molec. Biol. **50**, 279 (1970).
MORIMOTO, K., HARRINGTON, W. F.: J. biol. Chem. **247**, 3052 (1972).
O'BRIEN, E. J., BENNETT, P. M., HANSON, J.: Phil. Trans. B **261**, 201 (1971).
OFFER, G.: Cold Spr. Harb. Symp. quant. Biol. **37** (1972) (in press).
OHTSUKI, I., MASAKI, T., NONOMURA, Y., EBASHI, S.: J. Biochem. (Tokyo) **61**, 817 (1967).
PEACHY, L. D.: Ann. Rev. Physiol. **30**, 401 (1968).
PEPE, F. A.: J. Cell Biol. **28**, 505 (1966).
— In: Subunits in biological systems, p. 323. (TIMASHEFF, S. N., FASMAN, G. D., Eds.). New York: Marcel Dekker, Inc. 1971.
REEDY, M. K., HOLMES, K. C., TREGEAR, R.: Nature (Lond.) **207**, 1276 (1965).
SCHAUB, M. C., PERRY, S. V.: Biochem. J. **115**, 993 (1969).
SCHLISELFELD, L. H., BÁRÁNY, M.: Biochemistry **9**, 3206 (1968).
SLAYTER, H. S., LOWEY, S.: Proc. nat. Acad. Sci. (Wash.) **58**, 1611 (1967).
SPUDICH, J. A., WATT, S.: J. biol. Chem. **246**, 4866 (1971).
STRACHER, A.: Biochem. biophys. Res. Commun. **35**, 519 (1969).
STRAUB, F. B.: Studies **2**, 3 (1942).
SZENT-GYÖRGYI, A. G., SZENTKIRALYI, E. M., KENDRICK-JONES, J.: J. molec. Biol. (1972) (in press).
VIBERT, P. J., HASELGROVE, J. C., LOWY, J., POULSEN, F. R.: Nature (Lond.) New Biol. **236**, 182 (1972).
WEBER, A.: In: Current topics in bioenergetics, p. 203. (SANADI, D. R., Ed.). New York: Academic Press 1966.
— BREMEL, R. D.: Contractility of muscle cells and related processes, p. 37. (PODOLSKY, R. J., Ed.). New Jersey: Prentice-Hall, Inc. 1971.
WEEDS, A. G., LOWEY, S.: J. molec. Biol. **61**, 701 (1971).
— FRANK, G.: Cold Spr. Harb. Symp. quant. Biol. **37** (1972) (in press).

Werber, M., Szent-Györgyi, A. G., Fasman, G. D.: Biochemistry **11**, 2872 (1972).
Woods, E. F.: J. biol. Chem. **242**, 2859 (1967).
Young, M.: (1) J. biol. Chem. **242**, 4449 (1967).
— (2) Proc. nat. Acad. Sci. (Wash.) **58**, 2393 (1967).

Discussion

H. K. Schachman (Berkeley): What is the extent of overlap in the antiparallel aggregation of myosin molecules?

S. Lowey (Boston): The length of the bare zone in the native myosin filaments is reported to be between 1300 and 1700 Å. The extent of overlap between the rod portion of the myosin molecules can only be inferred from studies of the aggregates formed in the presence of divalent cations: myosin forms bipolar segments with an overlap of 900 Å while the rod forms segments with a 900 Å or a 1300 Å overlap.

U. Gröschel-Stewart (Würzburg): Are the light chains necessary for myosin-actin interaction?

S. Lowey (Boston): No detailed information is available at present on the role of the light chains in myosin-actin interactions.

U. Gröschel-Stewart (Würzburg): Do the light chains of smooth-muscle myosin (uterus, or chicken gizzard, as described by Kendrick-Jones) resemble slow-muscle myosin light chains in structure?

S. Lowey (Boston): I think preliminary data show (Kendrick-Jones) that smooth-muscle myosin from chicken gizzard also contains two classes of light chains.

W. Hasselbach (Heidelberg): Is it a proven fact that G-actin does not react with L-myosin?

S. Lowey (Boston): The activation of the ATPase activity of myosin by F-actin is very much greater (about 100-fold) than that by G-actin.

Multienzyme Complexes

The Pyruvate Dehydrogenase Complex of E. coli K-12 Structure and Synthesis

U. HENNING, O. VOGEL, W. BUSCH and J. E. FLATGAARD

Max-Planck-Institut für Biologie, Tübingen, Germany

With 10 Figures

Since the pioneering experiments of GUNSALUS and HAGER [1] and of KOIKE, REED, and CARROLL [2, 3] α-ketoacid dehydrogenase complexes from a number of sources have been studied from many aspects. These enzymes catalyze an oxidative decarboxylation according to:

$$R\text{—}\overset{\overset{\displaystyle O}{\|}}{C}\text{—}CO_2H + CoA + NAD^+ \rightarrow$$

$$CO_2 + R\text{—}\overset{\overset{\displaystyle O}{\|}}{C}\text{—}S\text{—}CoA + NADH + H^+ \qquad (1)$$

This overall reaction, proceeding with enzyme bound intermediates, requires three different enzyme components present in the complex: an α-ketoacid dehydrogenase (enzyme 1, E 1), a dihydrolipoamide transacetylase (E 2), and a dihydrolipoamide dehydrogenase (E 3). In the case of the pyruvate dehydrogenase complex they effect reaction (1) by the following sequence of reactions:

$$\text{Pyruvate} + \text{TPP-E1} \rightarrow CO_2 + \text{Hydroxyethyl-TPP-E1}, \qquad (2)$$

$$\text{Hydroxyethyl-TPP-E1} + \text{Lipoyl-E2} \rightarrow \text{S-acetyl-dihydrolipoyl-E2} + \text{TPP-E1}, \qquad (3)$$

$$\text{S-Acetyl-dihydrolipoyl-E2} + \text{CoA} \rightarrow \text{Acetyl-CoA} + \text{Dihydrolipoyl-E2}, \qquad (4)$$

$$\text{Dihydrolipoyl-E2} + \text{FAD-E3} \rightarrow \text{Lipoyl-E2} + \text{reduced FAD-E3}, \qquad (5)$$

$$\text{Reduced FAD-E3} + \text{NAD}^+ \rightarrow \text{NADH} + \text{H}^+ + \text{FAD-E3}. \qquad (6)$$

Some time ago we were attracted by the *E. coli* pyruvate dehydrogenase complex from a genetic point of view because it appeared at that time that the molar ratio of the constituent polypeptide chains was far from unity [4], in other words, one had to assume the existence of some sort of a control mechanism which would allow for the production of the right number of polypeptide chains. We therefore began to study the regulation of the synthesis of the enzyme complex and it became clear during these studies [5 to 7] that one would be hampered in further such experiments without a precise knowledge about the composition of this multienzyme complex. It was for this reason that we began, and by now essentially completed, an investigation on this composition.

The solution to the question on the molecular structure of the enzyme complex has been much facilitated by the development of a certain technology of purification [8] and by the isolation of regulatory mutants [7] producing the complex in large quantities; about 5% of all soluble protein in some such mutants is pyruvate dehydrogenase complex.

The final outcome of the structural studies has shown that the above mentioned control mechanism is not required. It was the proposal of this mechanism which led us to study the regulation of the synthesis of the complex. It turned out that the mechanism of this regulation appears to be a rather unusual one and it thus seems somewhat ironical that studies undertaken on a wrong conceptual basis did reveal an interesting situation.

Structure of the Pyruvate Dehydrogenase Complex

Components. We first have established that the three enzyme components in the complex correspond to three different polypeptide chains. The pyruvate dehydrogenase component when dissociated from the complex [3] exists as an enzymatically active dimer of two identical polypeptide chains with a molecular weight of about 10^5 per chain [9]. There is an N-terminal serine residue and a C-terminal sequence ...arg-leu-ala-CO_2H.

The transacetylase component also consists of identical polypeptide chains with a molecular weight of about 8×10^4; an N-terminus was not found and the C-terminal sequence is ...arg-arg-(leu, val)-met-CO_2H [8]. This component in the complex can also consist mainly of polypeptide chains with a molecular weight of 3.8×10^4 because the 8×10^4 daltons chain can be cleaved enzymatically. The cleavage process will be discussed in more detail below.

The dihydrolipoamide dehydrogenase component when dissociated from the complex has been shown to have a molecular weight of 1.12×10^5 and to contain two moles of FAD per 1.12×10^5 daltons protein [3, 10]. We found that, as expected, the enzyme is a dimer of two identical polypeptide chains with a molecular weight of 5.6×10^4 per chain [8, 11]. The C-terminal sequence found is ...lys-lys-CO_2H.

Enzyme complex. Having established these facts we asked for the number of each polypeptide chain in the enzyme complex. At first sight it would appear a simple matter to determine the polypeptide chain proportions in the complex by merely dissociating a known quantity of complex and measuring the quantities of component enzymes liberated. However, there are intolerable sources of error. On the one hand the classical dissociation procedure [3] is not sufficiently quantitative for such a purpose [8]. On the other hand separation of the components occurs, of course, with certain yields and these yields cannot be measured with sufficient precision. We have therefore chosen another strategy [12]. The component polypeptide chains can be separated by dodecylsulfate polyacrylamide gel electrophoresis ([8, 9] Fig. 1). The intensity of staining of the resolved bands can be measured with high accuracy. The great advantage of the procedure is that for each amount of enzyme complex applied to a gel column completely identical experimental conditions are achieved for each protein component.

In initial experiments concerning this question two difficulties showed up. First, as mentioned already, in several instances the amount of transacetylase polypeptide chains moving as a 8×10^4 daltons protein began to decrease in a time dependent process. One major and at least three minor faster moving bands appeared instead with the major band behaving as a 3.8×10^5 daltons subunit (Fig. 2). Second, the relative amount of pyruvate dehydro-

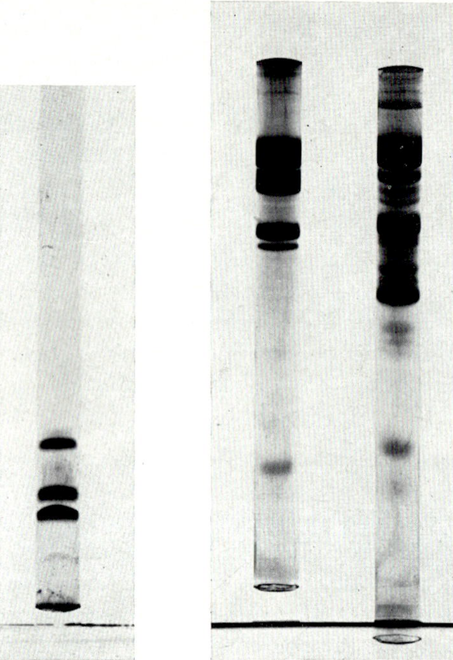

Fig. 1 Fig. 2

Fig. 1. Dodecylsulfate polyacrylamide gel electrophoresis of pyruvate dehydrogenase complex. The bands from top to bottom are pyruvate dehydrogenase, transacetylase, and dihydrolipoamide dehydrogenase subunits (30 μg enzyme complex). For experimental conditions see [9]

Fig. 2. Fragmentation of the transacetylase subunit. Left column: enzyme complex containing only the 8×10^4 daltons transacetylase subunit. The band just below the dihydrolipoamide dehydrogenase subunit stems from contamination of this preparation with some α-ketoglutarate dehydrogenase complex, the band is the transsuccinylase subunit. Right column: enzyme complex with the *same specific activity* as that of the left column; the transacetylase subunit has been fragmented to yield a series of smaller polypeptide chains. The main fragment (the most prominent fastest moving band) has a molecular weight of 3.8×10^4

genase polypeptide chains varied rather considerably (up to 30%) from one enzyme complex preparation to another.

Cleavage of the transacetylase. It could be shown that the fragmentation of the transacetylase chain is an enzymatic one [11].

The only way we have found so far to prevent fragmentation is storage of the protein in dodecylsulfate. In the presence of the detergent the transacetylase can be boiled for hours without any cleavage would occur. We have obtained evidence that in fact it may be the transacetylase itself which displays "autolytic" proteolytic activity. We have used preparative dodecylsulfate polyacrylamide gel electrophoresis for the isolation of the polypeptide chains from the enzyme complex [8]. In this procedure the subunits of the enzyme complex elute, of course, in sequence according to their molecular weight: the 8×10^4 daltons transacetylase subunit appears after the 5.6×10^4 daltons dihydrolipoamide dehydrogenase subunit. In an attempt to avoid the laborious removal of the detergent from the protein fractions we have used in several cases elution buffer not containing dodecylsulfate. In these cases the transacetylase subunit was found, upon subsequent analytical dodecylsulfate polyacrylamide gel electrophoresis, to have been cleaved with the main fragment having the 3.8×10^4 molecular weight. Clearly, the 3.8×10^4 daltons fragment must have arisen after the preparative separation. Such cleavage has never been observed when the elution buffer contained dodecylsulfate. The simplest explanation would be that upon a decrease in detergent concentration the transacetylase subunit can partially renature and become active again in proteolysis. It can, of course, not be excluded that a small amount of a protease migrates exactly as the transacetylase does and that it renatures upon removal of the detergent.

We have also compared tryptic fingerprints from the 8×10^4 daltons chain with such fingerprints from the 3.8×10^4 daltons fragment. It was found that a few peptides are missing in the pattern of the 3.8×10^4 daltons fragment and that a few others are present which are not seen in the pattern of the 8×10^4 daltons chain. The cleavage therefore actually is analogous to a chymotrypsinogen to chymotrypsin type conversion [13].

Fragmentation of the transacetylase does not ruin the enzymatic activity of the pyruvate dehydrogenase complex nor does it lead to any dissociation of the complex. The two complexes from which the analytical electrophoresis patterns of Fig. 2 were obtained had the same specific activity.

It has remained entirely unclear what the conditions are which promote fragmentation. We had suspected [8] that the degree of transacetylase fragmentation is different in different strains. While this may be true, we have since found that it can also be quite different in enzyme complexes isolated from the same strain. Although we have seen enzyme complexes, immediately upon isolation, containing almost only cleaved transacetylase we do not know whether or not the cleavage occurs *in vivo*, i.e., has physiological significance. As has been observed independently by PERHAM and THOMAS [14], cleavage can definitely occur *in vitro*. In summary, pyruvate dehydrogenase complex can exist in a number of states with respect to the fragmentation of the transacetylase subunits.

The pyruvate dehydrogenase "core" complex. As has been mentioned above it was observed that the amount of pyruvate dehydrogenase component in the complex did not appear to be constant from one preparation to another. We have described a purification procedure for the enzyme complex consisting of essentially two steps: chromatography on Biogel A-50-m followed by chromatography on calcium phosphate gel [8]. It is possible to purify the complex by chromatography on calcium phosphate gel alone [11]. Enzyme complex obtained this way proved to be of the type containing less pyruvate dehydrogenase component. Therefore, the behavior of the complex during chromatography on this gel was followed more closely. All fractions eluted from the gel column were assayed for pyruvate dehydrogenase activity and it was found that a peak of such activity elutes just before the complex (Fig. 3). The amount of material exhibiting pyruvate dehydrogenase activity and which is separated from the complex on calcium phosphate gel was found to vary; in some cases it was almost absent while in other cases it comprised up to 20% of the activity which eluted as enzyme complex. We have not yet investigated this variation systematically; at present it seems that it is due to differences in growth conditions.

The material from the peak exhibiting pyruvate dehydrogenase activity has been subjected to tryptic digestion and a fingerprint of these tryptic peptides was indistinguishable from the pattern we had found [9] for the tryptic peptides obtained from pyruvate dehydrogenase dissociated by the classical procedure [3] from the enzyme complex.

The important point now is that when pyruvate dehydrogenase

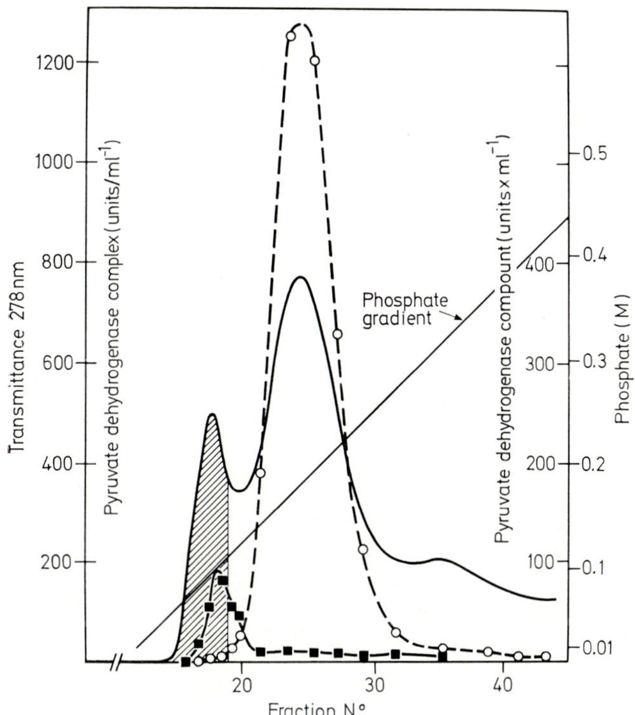

Fig. 3. Calcium phosphate gel chromatography of pyruvate dehydrogenase complex. The enzyme complex had first been purified by chromatography on Biogel A-50-m [8]. It was then applied to a calcium phosphate gel column (about 1 l with ca. 80 mg protein onto a 8 × 6 cm column). The column was then subjected to a linear pH 7 potassium phosphate gradient (300 ml 0.01 M in the mixing flask and 300 ml 1 M in the reservoir). ○, activity in reaction (1); —, transmittance 278 nm; ■, activity in reaction (1) appearing only upon adding, to the fractions, the subcomplex transacetylase-dihydrolipoamide dehydrogenase

complex purified by chromatography on calcium phosphate gel is rechromatographed on this gel then absolutely *no* pyruvate dehydrogenase (or any protein) is separated from the complex. The molar proportions of polypeptide chains in some 20 enzyme complex preparations obtained by the gel chromatography have invariably been the same. We therefore have called pyruvate de-

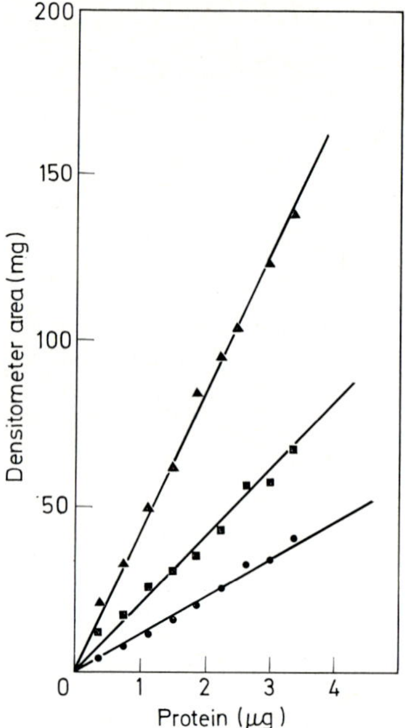

Fig. 4. Staining intensities of the components of the pyruvate dehydrogenase core complex. ▲, pyruvate dehydrogenase, ■, transacetylase; ●, dihydrolipoamide dehydrogenase. Each vertical series of symbols corresponds to one dodecylsulfate polyacrylamide gel column. The staining intensity per unit weight is different for each component, therefore, standard curves for each isolated component have been prepared. The slopes of curves such as those of this Fig. divided by the slopes of the standard curves yield the weight proportions of the chains in the complex, and division of these by the molecular weight of the polypeptide chains yields the molar proportions of the drains in the complex

hydrogenase complex obtained by gel chromatography "core" complex. This clearly is an operational designation and we have chosen, in addition, to define core complex as that multienzyme complex which contains the 8×10^4 daltons transacetylase subunit.

Polypeptide chain composition of the core complex. After finally having the core complex as a reproducible entity we could proceed to determine its polypeptide chain composition [12]. It was performed by the procedure outlined above; an example demonstrating the accuracy of the method is shown in Fig. 4. The results are summarized in Table 1. It is clear that a) there is a 1:1:1 molar ratio of the three chains in the core complex, that b) the amount of pyruvate dehydrogenase subunit in the complex can increase to yield a ratio of 1.35:1:1, and that c) core complex has also been

Table 1. *Molar ratios of polypeptide chains in pyruvate dehydrogenase complexes*

	Pyruvate dehydrogenase	Transacetylase	Flavoprotein
7 Core	1.07	0.95	1.00
2 "Complete"	1.35	1.01	1.00
2 Core "native"	1.00	1.02	1.00

All values were calculated by setting the data for the flavoprotein subunit 1.00. *Core complex* is enzyme complex from which excess pyruvate dehydrogenase component had been separated during chromatography on calcium phosphate gel. It is the complex which upon rechromatography on this gel loses no further components. "*Complete complex*" has been obtained in only two instances; it is unknown why in these cases excess pyruvate dehydrogenase was not removed by the preceding gel chromatography. "*Native*" *core complex* was found in two cases. In one it was complex which had been purified from commercially available cells. These cells (*E. coli* K-12) were grown by the manufacturer in an enriched medium (not specified) to 2/3 log phase. In the other case it was complex we have purified from strain YMel grown to about 2×10^8 cells/ml in complete medium. Chromatography of these preparations on calcium phosphate gel did not result in separation of any pyruvate dehydrogenase component.

found as a "native" one, i.e., in two cases the cells apparently did synthesize an enzyme complex with the 1:1:1 molar ratio of subunits.

To answer the question of the absolute numbers of polypeptide chains in the core complex knowledge of its molecular weight is required. It was determined [11] by measuring the sedimentation and diffusion coefficients and it was calculated to be

$$3.75 \pm 0.2 \times 10^6.$$

With the 1:1:1 molar ratio of the different polypeptide chains in the core complex an excellent fit for the molecular weight is reached when 16 copies of each chain are present in the complex:

$16 \times 100{,}000$	(pyruvate dehydrogenase subunit)	$= 1.6 \times 10^6$
$16 \times 80{,}000$	(transacetylase subunit)	$= 1.28 \times 10^6$
$16 \times 56{,}000$	(dihydrolipoamide dehydrogenase subunit)	$= 0.896 \times 10^6$
	Sum	$= 3.776 \times 10^6$

The number 16 has been confirmed by measuring the FAD content of the core complex [12]. It was found in two different preparations that 16.5 moles FAD are present per mole 3.75×10^6 daltons core complex.

A polypeptide chain ratio of 1.35:1:1 (see Table 1) implies the presence of about 3 additional pyruvate dehydrogenase dimers per enzyme complex. This "excess" pyruvate dehydrogenase may not participate in enzymatic catalysis in reaction (1), i.e., at least the removal of this "excess" enzyme does not cause a corresponding loss of activity in reaction (1). It therefore remained obscure what purpose the excess enzyme serves, if it does have any physiological significance.

Assembly

It was mentioned in the Introduction that we had started to investigate the regulation of the synthesis of the enzyme complex. In the course of these studies a fairly large number of mutants were isolated which lack enzymatic activity in reaction (1). They were used to establish the genetics of the system (Fig. 5). Among missense mutants in the pyruvate dehydrogenase structural gene a considerable fraction showed up which appeared to exhibit an impaired inducibility. That is, normally the synthesis of the enzyme complex is inducible (the inducing metabolite is pyruvate [6]) and in such mutants the defective enzyme complex is inducible to a lesser degree than is the case in wild type. This is only true for mutants in the pyruvate dehydrogenase gene, it has not been found for mutants in the transacetylase gene.

Our first suspicion was that the assembly and the synthesis of the complex are somehow coupled, specifically that the synthesis of the pyruvate dehydrogenase component is necessary for correct assembly and that this may be a requirement for the synthesis of the rest of the component polypeptide chains. Therefore, a study was undertaken concerning the *in vitro* assembly of pyruvate dehydrogenase and the subcomplex consisting of transacetylase and dihydrolipoamide dehydrogenase [15]. It is long known [3] that the enzyme complex can be reassociated *in vitro* from the component enzymes. Our question, in context with the mutants mentioned, was more specific and it can best be understood in considering Fig. 6.

The subcomplex mentioned has, of course, no activity in reaction (1) and this activity appears when pyruvate dehydrogenase

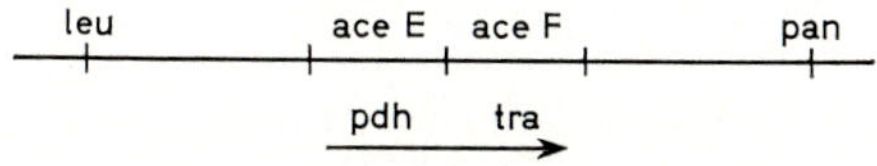

Fig. 5. Section of the *E. coli* chromosome. *leu*, leucine locus; *pan*, pantothenate locus; *ace*E and *ace*F, structural genes for the pyruvate dehydrogenase and transacetylase components, respectively. Arrow: direction of transcription

component is added. A given amount of subcomplex can thus be titrated with pyruvate dehydrogenase by measuring catalytic activity in reaction (1). The question posed was what happens in the area of the titration curve where the amount of pyruvate dehydrogenase is limiting: is this component bound statistically to the subcomplex molecules or is the binding highly cooperative so that essentially only complete complexes and subcomplex without pyruvate dehydrogenase are present?

Analytical treatment of the data of Fig. 6 revealed that the appearance of activity in the overall reaction exhibited a Hill-coefficient of 1.8. Thus either the binding is cooperative—certainly not highly cooperative—or, more likely, for each pyruvate dehydrogenase dimer bound two active sites are manifested. Binding studies with radioactive pyruvate [15] showed the latter explanation to be true. A precise determination of the number of binding sites was difficult because substrate binding showed a rather strong

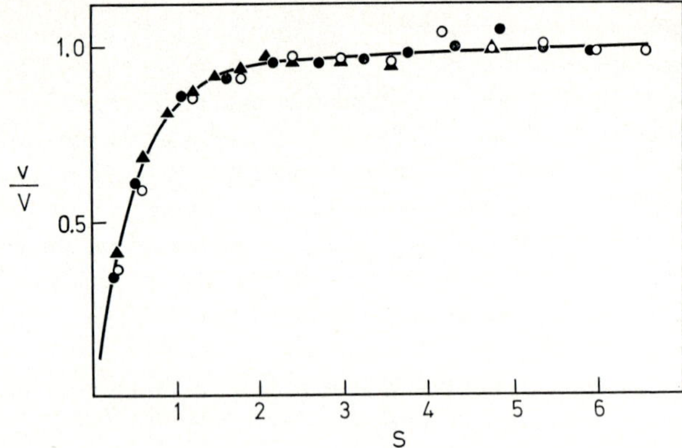

Fig. 6. Titration of subcomplex with pyruvate dehydrogenase component. S, units (enzymatic activity) pyruvate dehydrogenase added per unit subcomplex present; v/V, units enzymatic activity (reaction 1) appearing per unit subcomplex. The different symbols represent different sets of experiments performed with widely different amounts of subcomplex; ○ contained 30 times and ▲ about 300 times more subcomplex than ●

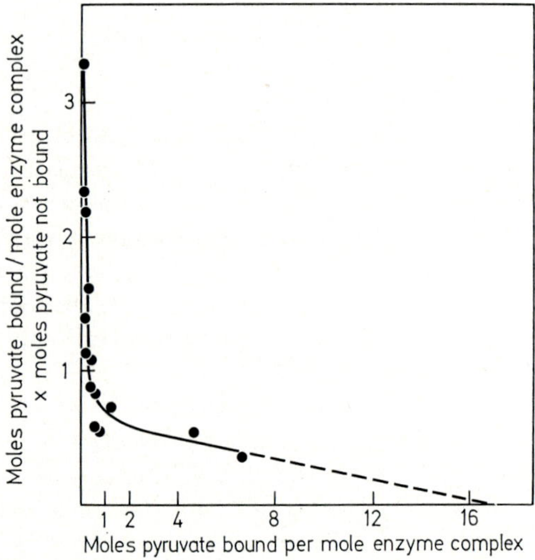

Fig. 7. Binding of 1-^{14}C-pyruvate to enzyme complex, Scatchard plot. Binding was measured, using equilibrium dialysis, to purified core complex (mol.-weight 3.75×10^6)

negative cooperative effect over a fairly wide range of substrate concentration. Fig. 7 shows a *Scatchard* plot of such binding data. Although the extrapolation cannot be very accurate it is clear that there is reasonable agreement with our structural studies, i.e., there are about 16 binding sites for pyruvate on the enzyme complex.

We have gone one step further and have analyzed by sucrose density gradient centrifugation the enzyme complexes formed along the titration curve [15]. Using, from a number of such experiments, the molecular weights calculated from the apparent sedimentation coefficients it turned out that the increase of these molecular weights—with increasing amounts of pyruvate dehydrogenase added—exhibits a Hill coefficient around 1 (Fig. 8). Clearly then, no cooperativity could be detected for the *in vitro* binding of pyruvate dehydrogenase to the subcomplex. Although obvious it should be pointed out that these experiments also show that a complete pyruvate dehydrogenase complex is not at all required for its enzymatic activity.

Regarding the mutant behavior discussed in the beginning of this section it certainly appears that one does not have to look for any coupling between assembly and synthesis. What then is the connection between the apparent involvement of the pyruvate dehydrogenase component in the synthesis of the enzyme complex?

Synthesis

As mentioned before we have isolated mutants which are defective in their mechanism of regulation of pyruvate dehydrogenase complex synthesis; they produce the enzyme complex constitutively [7]. We have begun to genetically localize these mutant alleles. Orienting experiments first showed that all (10 studied so far) are very closely linked to the pyruvate dehydrogenase complex structural genes. To date, a more detailed localization has been performed with two constitutive mutants and the result is schematically shown in Fig. 9. It seems that mutant alleles leading to this constitutive synthesis can reside *within* the pyruvate dehydrogenase structural gene.

It should be emphasized that these genetic mapping experiments have so far been done in detail with only two of the constitutive mutants. Recombination analyses can suffer from inherent

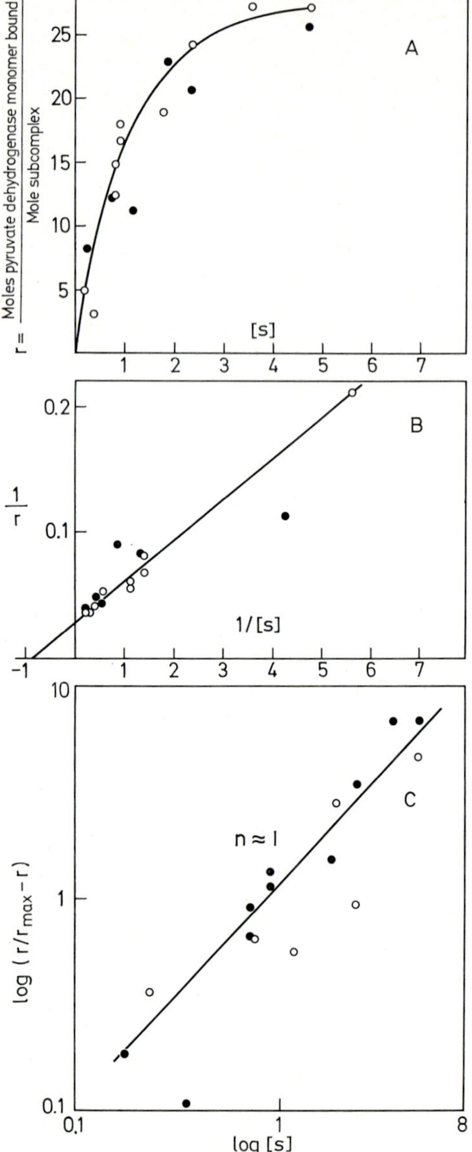

Fig. 8. Binding of pyruvate dehydrogenase component to subcomplex. A. S: units (enzymatic activity) pyruvate dehydrogenase added per unit subcomplex present. B. LINEWEAVER-BURK plot of the data from A. It has been calculated that the K_M for: pyruvate dehydrogenase + subcomplex ⇌ pyruvate dehydrogenase complex is around 10^{-12} M. The extrapolated number of maximally bound pyruvate dehydrogenase monomers (r_{max}) is about 32, an amount which has never been found in isolated enzyme complexes. C. HILL plot of the data from A and B

difficulties (cf. 16) and the evidence summarized in Fig. 9 therefore needs to be taken as preliminary. If, however, in fact mutation to constitutivity does occur in the structural gene under discussion we would have to conclude that the product of that gene not only is to become an enzyme but also that it plays a role a regulator gene product would, i.e., in the "simplest" classical case a repressor.

There is one straight-forward prediction from such an assumption. Mutants not producing any pyruvate dehydrogenase should be defective in regulation of the synthesis of the pyruvate dehydrogenase complex. Moreover, if so, this defect should be *trans*-recessive. We have found this to be true in three cases one of which is illustrated in Fig. 10. Mutant *ace*E66 is a nonsense (amber) mutant in the pyruvate dehydrogenase structural gene [17]. As a nonsense mutant it does not produce pyruvate dehydrogenase but does synthesize subcomplex consisting of transacetylase and dihydrolipoamide dehydrogenase. *Ace*F10 is a mutant (type unknown) in the transacetylase which does not produce this gene product but does synthesize, normally inducibly, enzymatically active pyruvate dehydrogenase component which, in crude extracts of the mutant, is found as a dimer [9]. *Ace*E66 apparently exerts the well known phenomenon of polarity, i.e., when the cells are grown under inducing conditions only about 10% of the amount

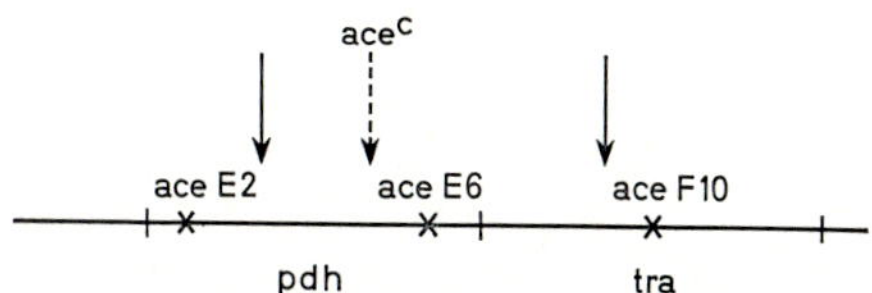

Fig. 9. Localization of *ace*^{c} mutant alleles. *ace*^{c}, mutant allele conferring constitutivity to the synthesis of the enzyme complex: *ace*E2 and *ace*E6, missense mutants in the pyruvate dehydrogenase structural gene; *ace*F10, mutant in the transacetylase structural gene. Two different constitutive alleles have been found to reside to the right of *ace*E2 and to the left of *ace*F10, i.e., somewhere between the solid arrows. For one constitutive allele strong evidence exists that it is located to the right of *ace*E2 and to the left of *ace*E6, i.e., somewhere between the left solid and the broken arrow, i.e., *within* the pyruvate dehydrogenase structural gene

of subcomplex is made which one would expect to be produced when the system is induced. Fig. 10 shows that pyruvate dehydrogenase synthesis in *ace*F10 is regulated normally while subcomplex synthesis in *ace*E66 is not; *ace*E66 could be designated a low level constitutive mutant because the specific activity of the subcomplex

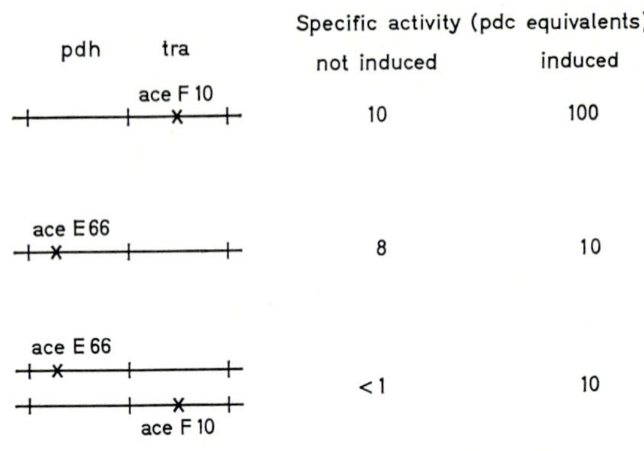

Fig. 10. *Cis-trans* behavior of a nonsense mutant (*ace*E66). Specific activity is expressed in terms of reaction (1) which, of course, cannot proceed in mutant *ace*F10 or *ace*E66. It can, however, simply be measured by adding, to crude extracts, saturating amounts of the component missing, i.e., pyruvate dehydrogenase in case of *ace*E66 and subcomplex transacetylase-dihydrolipoamide dehydrogenase in case of *ace*F10. Thus the specific activities refer to pyruvate dehydrogenase component in *ace*F10 and to subcomplex in *ace*E66. In case of the diploid strain *ace*E66/F′ *ace*F10 pyruvate dehydrogenase complex is produced by the association of the subcomplex coming from the *ace*E66 and of the pyruvate dehydrogenase component coming from the *ace*F10 genomes. The amount of subcomplex synthesized is limiting, therefore, the corresponding specific activities in this case indicate the degree of expression of the *ace*E66 locus only. "Not induced" refers to acetate and "induced" to glucose as carbon sources (cf. [6])

is about the same regardless as to whether the cells are grown under inducing or non-inducing conditions. If this low level constitutivity were due to the absence of pyruvate dehydrogenase one would of course expect that the presence of this enzyme component would allow again a normal regulation. Therefore the diploid strain shown in Fig. 10 was constructed which carries the mutant allele *ace*E66 on the chromosome and the *ace*F10 allele on an episome [6]. Clearly, in the diploid subcomplex synthesis under non-inducing conditions is no longer constitutive. Removal of the episome led to the reappearance of the low level constitutivity.

The simplest, although not necessarily correct interpretation of the experiments presented in this section would be that the pyruvate dehydrogenase component acts as a repressor. Other explanations are possible. We can hardly, however, avoid an interpretation which would *not* use the pyruvate dehydrogenase component as a regulatory element.

Discussion and Summary

Structure of the pyruvate dehydrogenase complex

1. The enzyme complex is not a unique entity. It can be synthesized containing 16 copies of each of the three different polypeptide chains but this number can increase to at least 22 for the pyruvate dehydrogenase subunit. The complex consisting of 48 polypeptide chains has been called core complex. Its molecular weight calculated from the molecular weights of the polypeptide chains is 3.776×10^6; the molecular weight found by measuring the sedimentation and diffusion coefficients is 3.75×10^6.

2. The transacetylase component can exist, in the pyruvate dehydrogenase complex in a number of states concerning the size of its polypeptide chains. The 8×10^4 daltons chain can be cleaved enzymatically to yield several fragments with the main one having a molecular weight of 3.8×10^4. The fragmentation of the chain need not influence the structural stability of the enzyme complex nor measurably its catalytic activity. It has remained unknown whether or not the cleavage can occur *in vivo*.

3. If more than 8 pyruvate dehydrogenase dimers are present in the enzyme complex then this "excess" enzyme is removed rather easily under conditions where none of the remaining 8 dimers is lost. The structural implication of this fact is the following. There is convincing evidence from electron microscopic studies ([18] to [21]) that the transacetylase has cubic shape. The simplest way of constructing a cube from 16 chains would, of course, be to use dimers as morphological subunits. A cube consisting of 16 identical polypeptide chains (or 8 dimers) naturally possesses

16 equivalent binding sites (or 8 when looking at dimers); and the transacetylase *does* bind, in the core complex, 8 dimers each of the other two components. The very fact that the pyruvate dehydrogenase dimers exceeding the 8 ones in the core complex are removed so easily shows that the "excess" dimers are bound differently—which fact constitutes indirect evidence for the equivalent positions mentioned.

Assembly. For the binding of the pyruvate dehydrogenase component to the subcomplex transacetylase-dihydrolipoamide dehydrogenase no cooperativity could be found. A complete pyruvate dehydrogenase complex is not required for catalytic activity in the overall reaction.

Synthesis. Evidence is briefly presented which strongly suggests that the pyruvate dehydrogenase component is involved in the regulation of the synthesis of the enzyme complex. The simplest interpretation of the data available to date would be that this protein can act as a repressor type protein.

A few more general conclusions may be drawn from the results presented. First, the information required to determine the structure of the pyruvate dehydrogenase complex is still completely but *not uniquely* present in the enzyme complex. We have seen that in the cell the amount of pyruvate dehydrogenase in the enzyme complex can be varied, and we have seen that much larger variations can be achieved *in vitro*. In other words, the structural information of the corresponding structural genes does *not* suffice to determine the structure of the final product. An additional information parameter comes in, namely that of the time; it depends on how much of a given gene product is made as to how the final structure looks like. Since it is a time parameter it can no longer be present in the final product. One thus has to at least modify the idea of the law that the primary structure determines all other structures of a protein (cf. 22).

Second, we have shown that the pyruvate dehydrogenase component participates in the mechanism of regulation of enzyme complex synthesis. (We have not discussed here the fact, derived from various experimental approaches, that this participation has nothing to do with the enzymatic activity of the component; *see* also [6].) Whether or not the interpretation is correct that the pyruvate dehydrogenase can act as a repressor it appears clear that the product of a structural gene for an enzyme has properties so far in such systems ascribed only to regulatory genes. It could, in fact, well be that in our case a separate regulatory gene does not exist.

References

1. Gunsalus, I. C.: In: A symposium on the mechanism of enzyme action, p. 545 (McElroy, W. D., Glass, B., Eds.). Baltimore: The Johns Hopkins Press 1954.
2. Koike, M., Reed, L. J., Carroll, W. R.: J. biol. Chem. **235**, 1924 (1960).
3. — — — J. biol. Chem. **238**, 30 (1963).
4. Henning, U., Herz, C., Szolyvay, K.: Z. Vererbungsl. **95**, 236 (1964).
5. — Dietrich, J., Murray, K. N., Deppe, G.: In: Molecular genetics, p. 223 (Schuster, H., Wittmann, G., Ed.). Berlin-Heidelberg-New York: Springer 1968.
6. Dietrich, J., Henning, U.: Europ. J. Biochem. **14**, 258 (1970).
7. Flatgaard, J. E., Hoehn, B., Henning, U.: Arch. Biochem. **143**, 461 (1971).
8. Vogel, O., Beikrich, H., Müller, H., Henning, U.: Europ. J. Biochem. **20**, 169 (1971).
9. — Henning, U.: Europ. J. Biochem. **18**, 103 (1970).
10. Williams, C. H.: J. biol. Chem. **240**, 4793 (1965).
11. Vogel, O., Höhn, B., Henning, U.: Europ. J. Biochem. (in press).
12. — — — Proc. nat. Acad. Sci. (Wash.) **69**, 1615 (1972).
13. Desnuelle, P.: In: The enzymes, 2nd Ed., Vol. **4**, p. 93 (Boyer, P., Lardy, H., Myrbäck, K., Eds.). New York: Acad. Press 1960.
14. Perham, R. N., Thomas, J. O.: FEBS Letters **15**, 8 (1971).
15. Busch, W., Henning, U.: In preparation.
16. Drapeau, G. R., Brammar, W. J., Yanofsky, C.: J. molec. Biol. **35**, 357 (1968).
17. Henning, U., Dennert, G., Hertel, R., Shipp, W. S.: Cold Spr. Harb. Symp. quant. Biol. **31**, 227 (1966).
18. Fernándes-Morán, H., Reed, L. J., Koike, M., Willms, C. R.: Science **145**, 930 (1964).
19. Willms, C. R., Oliver, R. M., Henney, H. R., Mukherjee, B. B., Reed, L. J.: J. biol. Chem. **242**, 889 (1967).
20. Reed, L. J., Oliver, R. M.: Brookhaven Symp. Biol. **21**, 397 (1968).
21. — In: Current topics in cellular regulation, p. 233 (Horecker, B. L., Stadtman, E. R., Eds.). New York: Acad. Press 1969.
22. Rosenbaum, M.: Nature (Lond.) New Biology **230**, 12 (1971).

Discussion

F. Lynen (München): Have you also determined the lipoic acid content of your enzyme? As far as I remember, Reed found 32 molecules per enzyme complex, which would mean that each of your transacetylase peptide chains (m.w. 80,000) carries two lipoic acids.

U. Henning (Tübingen): Reed and associates [Henney, H. R., Willms, C. R., Muramatsu, T., Mukherjee, B. B., Reed, L. J.: J. biol. Chem. **242**, 898 (1967)] have reported that the transacetylase polypeptide chain has a molecular weight of about 36,000 and that there is one lipoyl residue per chain; they proposed the presence of 24 such chains in the transacetylase. Unless E. Coli Crookes (used by Reed) produces a transacetylase different from that of K-12, these authors were probably working with the fragment of the transacetylase. If so, the 80,000 daltons chain should indeed contain 2 lipoyl residues. We have not measured the lipoic acid content but at present are not very much attracted by the possibility of one transacetylase chain possessing 2 lipoyl residues. There are 16 moles of FAD and about 16 binding sites for pyruvate in the core complex. With a 1:1:1 ratio of the three different polypeptide chains in the complex, two lipoyl residues per chain would result in a 1:2:1 ratio of active centers in the complex. This, of course, is not impossible but would at least be rather strange. I may add that, as you have heard this morning, Dr. Green would obviously very much prefer to have 24 subunits in a protein exhibiting cubic symmetry (as the transacetylase does). Clearly, however, with a 1:1:1 molar ratio of polypeptide chains we cannot accomodate 24 polypeptide chains of m.w. $8 \cdot 10^4$ daltons in the transacetylase. From a crystallographic point of view, this cube would therefore be fairly ugly—unless our data are wrong. One should perhaps remember that proteins with an odd number of subunits exist, a fact which appeared rather unacceptable several years ago.

H. Holzer (Freiburg): Does the breakdown of PDH (eventually dependent on TPP) play any part in the control of the synthesis of the PDC complex?

H. Henning (Tübingen): We have no evidence whatsoever to indicate that under any growth condition the enzyme complex or the pyruvate dehydrogenase component would be degraded—at least as long as this component is not mutationally altered. TPP does have, indirectly, a very pronounced effect on this regulation; this effect, however, certainly has no physiological significance. There are conditions under which thiamine-requiring strains can be grown in the absence of thiamine. If this is done on glucose as carbon source, synthesis of pyruvate dehydrogenase complex becomes fully induced and the complex made is apoenzyme regarding TPP. The reason is that pyruvate is the inducing metabolite and pyruvate is accumulated when the enzyme complex is inactive [for experimental details *see* Dietrich, J., Henning, U.: Europ. J. Biochem. **14**, 258 (1970)].

M. Lazdunski (Marseille): How many classes of sites did you find in binding studies with pyruvate or analogs of pyruvate? Has the experiment been done with the isolated pyruvate dehydrogenase dimer?

U. Henning (Tübingen): Binding studies have been performed only with radioactive (1-C^{14}) pyruvate and not with analogues. Also, so far only the isolated complex has been used and not the pyruvate dehydrogenase component alone. Dissociation of this component is performed at pH 9.5 and we are not sure just how "native" the component still is after this treatment.

We cannot yet distinguish classes of binding sites. All I can state is that the binding of pyruvate to the enzyme complex exhibits a Hill coefficient of 0.6 and this coefficient changes to 1 at pyruvate concentrations *lower* than the K_m (2×10^{-4} M). Most likely there are interesting consequences for the mechanism of this negative cooperativity because a very similar phenomenon is observed for pyruvate in the catalytic activity [Bisswanger, H., Henning, U.: Europ. J. Biochem. **24**, 376 (1971)]. Here the substrate shows *positive* cooperativity (n = 1.8) and n also changes to 1 at the same pyruvate concentration below the same K_m.

C. Veeger (Wageningen): Work by Dr. van den Broek in my department has shown that transhydrogenase from *Azotobacter vinelandii* starts to aggregate to a molecular weight of 30 to 50 × 10^6 daltons after removal of PDC from purified extracts. Further work shows that the transhydrogenase can form specific complexes with transacetylase [van den Broek et al.: Europ. J. Biochem. **24**, 31 (1971)]. Recent work by Bresters et al. (FEBS Letters, in press) shows that PDC from Azotobacter shows similar regulatory properties as the E. Coli complex. It could be demonstrated that the Azotobacter complex contains phosphotransacetylase, even in preparations of high purity, which is inseparable from the complex. Furthermore we were able to demonstrate the anaerobic formation of ATP by this complex from pyruvate via the sequence: acetylCoA → acetylP → ATP. The efficiency of this complex is 60% of the rate of oxidative phosphorylation. Since E. Coli also contains transhydrogenase and phosphotransacetylase, it cannot be excluded that the complex as isolated is part of a similar complex as the Azotobacter PDC. In addition, preliminary studies have shown that a similar ATP production is present in flight-muscle mitochondria, which points to the idea that acetylphosphate-mediated anaerobic ATP synthesis from pyruvate could be present in obligatory aerobic cells.

U. Henning (Tübingen): When isolated, the E. coli pyruvate dehydrogenase complex does not contain any measurable phosphotransacetylase activity. We have not looked for transhydrogenase activity; however, this enzyme cannot be present in stoichiometric amounts because we would have found any polypeptide chain which might have been in the complex in addition to the chains of the three enzyme components. I cannot exclude, of course, that the enzymes you mentioned are lost from the complex during purification. I am somewhat doubtful about this, however. If you grow K 12 on glucose it does not produce very much phosphotransacetylase (as compared to growth on acetate) and pyruvate dehydrogenase complex activity in crude extracts is much stimulated by the addition of phosphotransacetylase.

N. M. Green (London): If we accept Dr. Henning's value of 16 subunits in the lipoate transacetylase, we can arrange them to give the appearance of a cube in the electron microscope, but alternate subunits would be in non-equivalent positions and the symmetry would be D^4 rather than octahedral. This would be consistent with the electron microscopic data. However there is a conflict with the X-ray diffraction results of de Rosier and Reed on the closely related lipoate transsuccinylase. The symmetry of this enzyme is truly octahedral and in the electron microscope it appears to be similar to the transacetylase.

U. Henning (Tübingen): Yes, the transsuccinylase and transacetylase really have a very similar appearance in the electron microscope, and de Rosier, Oliver, and Reed's X-ray evidence for cubic symmetry of the crystalline transsuccinylase did not make us very happy. As you just have pointed out, however, electron micrographs can reveal shape but not symmetry, and there is no *a priori* cogent reason why the transacetylase should not possess different symmetry from the transsuccinylase. The issue may also be even more complicated. Perham and Thomas have shown that the transsuccinylase polypeptide chain can be degraded. They and we found that the transacetylase can be cleaved somewhere near the middle of the polypeptide chain. Who knows whether such cleavage could cause a rearrangement of the structure of the transacetylase, maybe even more so if one looks at crystalline preparations of the isolated enzyme?

The Multienzyme Systems of Fatty Acid Biosynthesis

M. Sumper and F. Lynen

Max-Planck-Institut für Zellchemie, München, Germany

With 18 Figures

The Multienzyme Systems of Fatty Acid Biosynthesis

The synthesis of long-chain fatty acids from acetyl-CoA is catalyzed by two enzyme systems. Both systems are multienzyme assemblies; the acetyl-CoA carboxylase with just two partial enzyme activities is one of the simplest, and the fatty acid synthetase with at least seven partial enzyme activities is one of the more complex multienzyme systems.

Acetyl-CoA carboxylase from yeast was isolated in homogeneous form and compared in its properties with pyruvate carboxylase from yeast. Both enzymes have very similar sedimentation coefficients and molecular weights. Both enzyme are composed of four protomers. Acetyl-CoA carboxylase and pyruvate carboxylase split under identical conditions into a variety of aggregates: besides the protomer, dimeric, trimeric and polymeric forms are found. Patterns of the dissociated enzymes obtained by sedimentation in sucrose density gradients and by electrophoresis in polyacrylamide are almost identical.

Within the limits of sensitivity of the immunochemical techniques used in this study, no cross-reaction could be observed between antiacetyl-CoA carboxylase and pyruvate carboxylase. This indicates that the substructures catalyzing the ATP-dependent carboxylation of biotin, common to both enzymes, are not based on identical primary structures. From these results it is proposed that the genes for acetyl-CoA carboxylase and pyruvate carboxylase may have been derived from a common ancestor.

Regarding the second multienzyme system involved in fatty acid biosynthesis, fatty acid synthetase, it has become possible to gain a preliminary insight in the structural organization of the complex by means of a simple method recently found to promote reversible dissociation of the complex. The molecular weights of the subunit proteins are between 200,000 and 250,000 which may indicate that the dissociation did not result in single enzyme components. After dissociation, the activity of the condensing enzyme and both reductases disappeared, but the activity of all other component enzymes did not decrease. Reactivation occurs on decrease of ionic strength by dilution or dialysis, suggesting that the catalytically active conformation of the reductase enzymes requires some protein-protein interactions which occur only in more complex structures.

Fatty acid synthetase produces palmityl-CoA and stearyl-CoA in equal amounts under standard conditions. A model is proposed, based on the known enzymatic properties of the fatty acid synthetase, which rationalizes the chain termination at the level of C_{16} and C_{18}-acid. The model assumes that the fatty acyltransferase involved experiences hydrophobic interaction with the growing carbon chain, starting at the level of the C_{13}-acid. The intensity of this interaction increases by an energy increment of -0.9 kcal per additional CH_2 group. To calculate the probability of the product release at a particular chain length, an equation was derived from the model for the quantitative description of the observed product distribution. The formula suggests the conditions under which either short acyl-CoA derivatives or exclusively stearyl-CoA can be produced. Synthesis under these conditions was tested experimentally and results indicated that the formula can be applied to a wide range of experimental conditions.

Die Multienzymsysteme der Fettsäurebiosynthese

Ausgehend vom Acetyl-CoA wird die Synthese von langkettigen Fettsäuren durch zwei Enzymsysteme, nämlich die Acetyl-CoA-Carboxylase und die Fettsäuresynthetase, katalysiert. Beide Enzymsysteme gehören zu den Multienzymkomplexen, wobei die Acetyl-CoA-Carboxylase mit zwei enzymatischen Teilfunktionen zu den einfachsten, die Fettsäuresynthetase mit wenigstens sieben enzymatischen Teilfunktionen zu den komplizierten Multienzymkomplexen zu rechnen ist.

Der erste Teil des Referates berichtet über Untersuchungen an der Acetyl-CoA-Carboxylase, die sich neben der biochemischen Charakterisierung dieses Enzyms mit der Frage einer möglichen phylogenetischen Beziehung der Biotinenzyme untereinander beschäftigt. Der zweite Teil behandelt ein Modell des Kettenabbruches in der Fettsäurebiosynthese sowie Versuche zur Zerlegung des Multienzymkomplexes Fettsäuresynthetase.

1. Die Acetyl-CoA-Carboxylase aus Hefe und ihre strukturelle Verwandtschaft zur Pyruvatcarboxylase

Die Acetyl-CoA-Carboxylase ist ein Biotinenzym. Wie bei allen Biotinenzymen läuft auch diese Carboxylierungsreaktion (Acetyl-CoA → Malonyl-CoA) in einer zweistufigen Folge ab. Eine zentrale Rolle im Reaktionsablauf spielt das carboxylierte Derivat des enzymgebundenen Biotinrestes. Im ersten Reaktionsschritt wird dieser Biotinrest zum N-Carboxybiotin carboxyliert (Abb. 1). Nach der Herkunft dieser Carboxylgruppe kann man die Biotinenzyme in zwei Gruppen einteilen. Die Enzyme der ersten Gruppe verwenden Hydrogencarbonat und benötigen ATP um die neue N-Carboxy-Bindung zu knüpfen. In diese Gruppe von Biotinenzymen gehören z. B. Acetyl-CoA-Carboxylase, β-Methylcrotonyl-CoA-Carboxylase und Pyruvatcarboxylase. Die zweite Gruppe von Biotinenzymen verwendet nicht Hydrogencarbonat, sondern eine Carbonsäure und zwar entweder eine β-Ketosäure oder ein Malonyl-CoA-Derivat als Carboxylgruppendonor. Diese Transcarboxylierungsreaktion verläuft ohne die Beteiligung von ATP. In diese Gruppe von Biotinenzymen gehört z. B. die Methylmalonyl-CoA: Pyruvat Transcarboxylase.

Im zweiten Schritt des Reaktionsablaufes wird dann die Carboxylgruppe vom Carboxybiotin auf das jeweilige Substrat übertragen. Diesen Schritt haben alle Biotinenzyme gemeinsam.

Neuerdings wurde für die Acetyl-CoA-Carboxylase aus E. coli in den Arbeitskreisen von Vagelos [1—3] und Lane [4, 5] gezeigt, daß die beiden Teilschritte der Enzymkatalyse durch zwei verschiedene Untereinheiten der Acetyl-CoA-Carboxylase katalysiert werden. Das Enzymsystem läßt sich leicht in eine Biotincarboxylase, in eine Carboxyltransferase und in ein Protein, das als Träger der prosthetischen Gruppe Biotin fungiert, auftrennen. Diese Trennung in Teilenzyme ist bei Biotinenzymen aus eukaryonten Organismen bisher nicht gelungen: Hier liegen wesentlich stabilere Enzymkomplexe vor.

In Abb. 2 sind die wichtigsten Teilreaktionen, die von Biotinenzymen katalysiert werden, zusammengestellt. Teilreaktion A, die

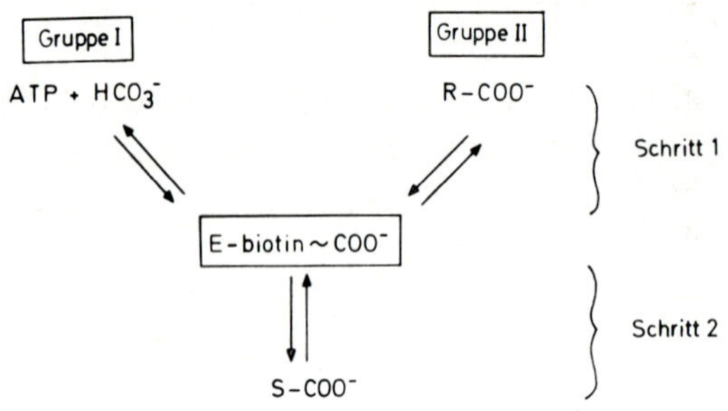

Gruppe I: Acetyl-CoA-Carboxylase, Propionyl-CoA-Carboxylase, ß-Methylcrotonyl-CoA-Carboxylase, Pyruvat-Carboxylase

$$CH_3\text{-}\overset{O}{\overset{\|}{C}}\text{-SCoA} + ATP + HCO_3^- \xrightleftharpoons{Mg^{++}} HOOC\text{-}CH_2\text{-}\overset{O}{\overset{\|}{C}}\text{-SCoA} + ADP + P$$

Gruppe II: Methylmalonyl-CoA: Pyruvat-Transcarboxylase

$$\boxed{HOOC}\text{-}\underset{CH_3}{\underset{|}{CH}}\text{-}\overset{O}{\overset{\|}{C}}\text{-SCoA} + CH_3\text{-}\overset{O}{\overset{\|}{C}}\text{-COO}^- \rightleftharpoons \underset{CH_3}{\underset{|}{CH_2}}\text{-}\overset{O}{\overset{\|}{C}}\text{-SCoA} + \boxed{HOOC}\text{-}CH_2\text{-}\overset{O}{\overset{\|}{C}}\text{-COO}^-$$

Abb. 1. Reaktionsfolge bei Biotinenzymen

ATP-abhängige Carboxylierung des Biotins, wird von allen Biotinenzymen der ersten Gruppe katalysiert. Teilreaktionen des Typs B umfassen die reversiblen Carboxylierungen von Acyl-CoA-Derivaten, die als Homologe oder Vinyloge des Acetyl-CoA eine chemisch nahestehende Substratgruppe bilden. Teilreaktion C ist die reversible Carboxylierung von Pyruvat durch Oxalacetat. Die von den verschiedenen bekannten Biotinenzymen katalysierten Gesamtreaktionen setzen sich aus diesen wenigen Grundtypen von Teilreaktionen zusammen. So haben z. B. die Biotinenzyme Acetyl-CoA-Carboxylase, Propionyl-CoA-Carboxylase, β-Methylcrotonyl-CoA-Carboxylase und Geranyl-CoA-Carboxylase die Teilaktivitätenkombination AB (AB_1, AB_2 ...), die Pyruvatcarboxylase hat die Teilaktivitätenkombination AC und die Methylmalonyl-CoA: Pyruvat Transcarboxylase die Teilaktivitätenkombination BC.

A		E-biotin + HCO_3^- + ATP	$\rightleftharpoons$	E-biotin-CO_2^- + ADP + P
B	(1)	E-biotin-CO_2^- + $CH_3COSCoA$	$\rightleftharpoons$	E-biotin + $^-OOC\text{-}CH_2COSCoA$
	(2)	E-biotin-CO_2^- + $CH_3CH_2COSCoA$	$\rightleftharpoons$	E-biotin + $CH_3CH(COO^-)COSCoA$
	(3)	E-biotin-CO_2^- + $CH_3\text{-}C(CH_3){=}CHCOSCoA$	$\rightleftharpoons$	E-biotin + $CH_3\text{-}C(CH_2COO^-){=}CHCOSCoA$
	(4)	E-biotin-CO_2^- + $R\text{-}CH_2\text{-}C(CH_3){=}CHCOSCoA$	$\rightleftharpoons$	E-biotin + $R\text{-}CH_2\text{-}C(CH_2COO^-){=}CHCOSCoA$
C		E-biotin-CO_2^- + $CH_3\text{-}C(=O)\text{-}COOH$	$\rightleftharpoons$	E-biotin + $^-O_2C\text{-}CH_2\text{-}C(=O)\text{-}COOH$

Abb. 2. Zusammenstellung der Teilreaktionen, die von Biotinenzymen katalysiert werden

Uns interessierte die Frage, ob dieses „Baukastensystem" der Teilaktivitäten sowie die Ähnlichkeit der Teilreaktionen der Gruppe B die Folge einer phylogenetischen Beziehung unter den Biotinenzymen ist. Es wäre vorstellbar, daß die Biotinenzyme (besonders die des Typs AB und AC) sich aus einem gemeinsamen Vorfahren entwickelt haben. Das Baukastensystem von Teilaktivitäten könnte das Ergebnis einer Neukombination von Genen sein, die die entsprechenden Teilenzyme codieren. Alle Biotinenzyme, die Hydrogencarbonat als Carboxylgruppendonor benützen, katalysieren Teilreaktion A, die ATP-abhängige Caboxylierung des Biotins. Ist die Annahme einer phylogenetischen Verwandtschaft richtig, so sollten die für Teilreaktion A verantwortlichen Proteinteilstrukturen in den betreffenden Biotinenzymen eine strukturelle Ähnlichkeit untereinander aufweisen. Dies vorausgesetzt, kann man weiter erwarten, daß auch diejenigen Teilstrukturen, die den zweiten Reaktionsschritt katalysieren, eine gewisse Ähnlichkeit untereinander aufweisen. Zumindest derjenige Bereich, der die Assoziation mit der ersten Teilstruktur (ATP-abhängige Biotincarboxylierung) ermöglicht, muß dann analog aufgebaut sein.

Unter diesem Gesichtspunkt haben wir zwei Biotinenzyme, nämlich Acetyl-CoA-Carboxylase und Pyruvatcarboxylase, in homogener Form aus einem eukaryonten Organismus (Saccharomyces cerevisiae) isoliert und geprüft, ob sich Hinweise für eine strukturelle Ähnlichkeit dieser Biotinenzyme nachweisen lassen [6].

Zur experimentellen Prüfung einer strukturellen Verwandtschaft haben wir die Molekulargewichte, das Dissoziationsverhalten und die Untereinheiten der beiden Enzyme verglichen sowie immunochemische Methoden angewandt.

Der Sedimentationskoeffizient der Pyruvatcarboxylase war von Young et al. [7] mit 15,6 S, das Molekulargewicht mit 600000 bestimmt worden. Wir haben mit der Zuckergradientenzentrifugationsmethode von Martin und Ames [8] für die Acetyl-CoA-Carboxylase einen Mittelwert von 15,5 S für die Sedimentationskonstante gefunden, also einen Wert der praktisch mit demjenigen für die Pyruvatcarboxylase übereinstimmt. Viel besser ließ sich das Sedimentationsverhalten beider Enzyme durch direkten Vergleich prüfen. Durch Mischen der reinen Enzyme und Zentrifugation in einen Zuckergradienten zeigte sich, daß nach einer 15stündigen Zentrifugation bei 40000 Upm keinerlei Auftrennung der beiden Enzyme erfolgte. Die Maxima der Enzymaktivitäten für Acetyl-CoA-Carboxylase und Pyruvatcarboxylase lagen vielmehr in ein und derselben Fraktion (Abb. 3). Man darf aus diesem Experiment den Schluß ziehen, daß beide Enzyme sehr ähnliche Molekulargewichte haben, wenn sie sich nicht stark in ihrer Gestalt unterscheiden. Ein starker Unterschied der Gestalt wird durch die noch zu beschreibenden Experimente sehr unwahrscheinlich gemacht.

Bei der Untersuchung der Stabilitätseigenschaften der Acetyl-CoA-Carboxylase zeigte sich, daß bei pH-Werten über 8 ein rascher Verlust der enzymatischen Aktivität eintritt. So ist z. B. in 0,1 M Tris · HCl vom pH 8,5 die Enzymaktivität innerhalb von 7 Std auf 5% des Ausgangswertes abgesunken. Eine teilweise Reaktivierung (etwa bis zu 45%) ließ sich durch Überführen des Enzyms in Phosphatpuffer hoher Ionenstärke und pH-Werten um 6,5 erreichen. Um zu prüfen, ob diese Instabilität in einer Dissoziation des Enzyms bei alkalischen pH-Werten zu suchen ist, wurde das inaktivierte Enzym im Zuckergradienten untersucht. Wie in Abb. 4 A zu

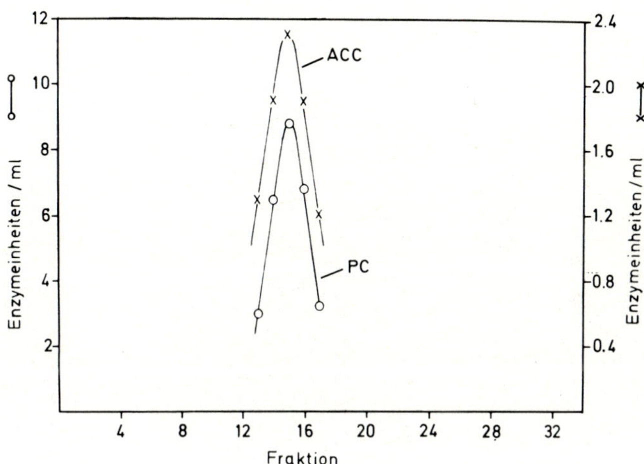

Abb. 3. Vergleich des Sedimentationsverhaltens von Acetyl-CoA-Carboxylase (ACC) und Pyruvatcarboxylase (PC) [6]. 2 mg Pyruvatcarboxylase und 3 mg Acetyl-CoA-Carboxylase wurden gemischt und zusammen auf einen linearen Zuckergradienten (5—20%) in 0,3 M Kaliumphosphat pH 6,5 aufgetragen. Zentrifugationsdauer 15 Std, Temperatur 8 °C, 39000 UpM (SW 40 Rotor). Der Gradientenbecher wurde unten angestochen und fraktioniert

erkennen ist, beobachtet man ein stark verändertes Sedimentationsverhalten. Der Enzymkomplex ist dissoziiert, es treten Spaltstücke mit Sedimentationskoeffizienten von etwa 12 S, 9 S und 6 S auf. Es war nun interessant zu vergleichen, ob Pyruvatcarboxylase unter diesen Bedingungen ebenfalls dissoziiert. Wir inkubierten beide Enzyme unter identischen Bedingungen in 0,1 N NH_3-Lösung für etwa 4 Stunden und unterwarfen sie dann einer Zuckergradientenzentrifugation. Das Resultat zeigt Abb. 4 B. Bemerkenswerterweise zer-

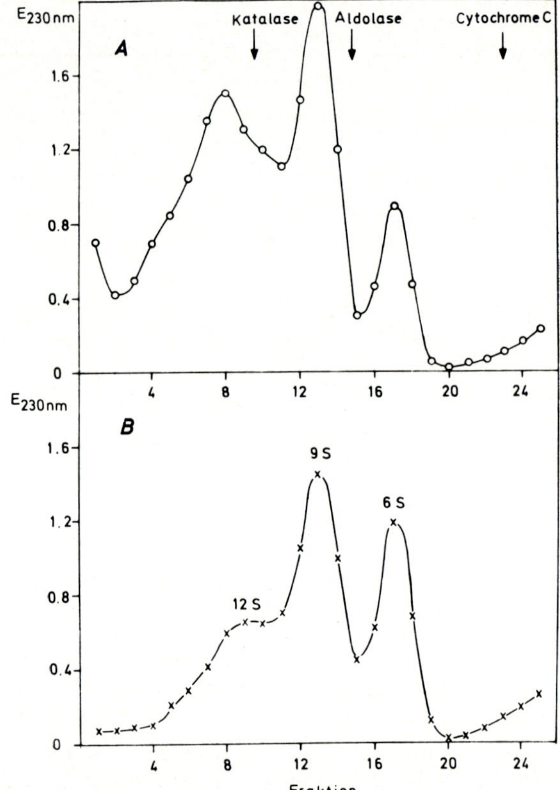

Abb. 4. Zuckergradientenzentrifugation dissoziierter Acetyl-CoA-Carboxylase (A) und Pyruvatcarboxylase (B) [6]. Die Enzyme wurden 5 Std gegen 0,1 M NH_3 dialysiert (20 °C) und dann auf einen linearen Zuckergradienten (5—20%) in 0,1 M TRIS · HCl pH 9,0 aufgetragen. Zentrifugationsdauer 16,5 Std, Temperatur 6 °C, 39000 UpM (SW 40 Rotor). Die Referenzenzyme wurden in separaten Bechern zentrifugiert. Die Gradientenbecher wurden unten angestochen und fraktioniert

fällt die Pyruvatcarboxylase in ganz analoger Weise, es treten ebenfalls Bruchstücke mit Sedimentationskoeffizienten von 12 S, 9 S und 6 S auf.

Neben dem Sedimentationsverhalten verglichen wir das elektrophoretische Verhalten der Spaltstücke von Acetyl-CoA-Carboxylase und Pyruvatcarboxylase in der Diskelektrophorese. Wiederum wurden dazu beide Biotinenzyme in gleicher Weise in 0,1 N NH_3-Lösung vorinkubiert und dann in einem 6%igen Polyacrylamidgel elektrophoretisch aufgetrennt. Abb. 5 gibt das Elektrophorese-

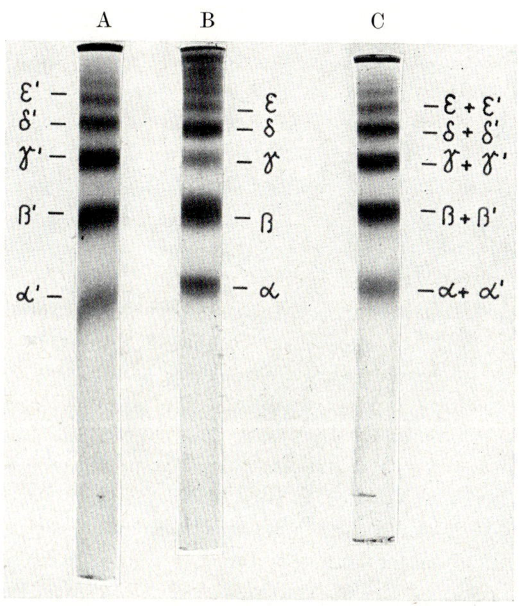

Abb. 5. Polyacrylamidgel-Elektrophorese von Acetyl-CoA-Carboxylase und Pyruvatcarboxylase [6]. Die Enzyme wurden durch Dialyse gegen 0,1 N NH_3-Lösung (+ 0,01 M Thioglycol) dissoziert (20 °C). A. Pyruvatcarboxylase; B. Acetyl-CoA-Carboxylase; C. Pyruvatcarboxylase und Acetyl-CoA-Carboxylase zusammengemischt. 6%iges Gel; Trennung bei pH 10

muster wieder. Auch hier zeigen die Spaltstücke von Acetyl-CoA-Carboxylase und Pyruvatcarboxylase ein äußerst ähnliches Muster. Die Beweglichkeiten der entsprechenden Banden α und α', β und β' usw. sind innerhalb der Genauigkeit dieser Methode ununterscheidbar, wie das Elektrophoresemuster C beweist: Hier wurden die dissoziierten Enzyme zusammengemischt und gemeinsam der Elektrophorese unterzogen. Es erfolgte keine Auftrennung der entsprechenden Banden von Acetyl-CoA-Carboxylase und Pyruvatcarboxylase. Die Elektrophoresemuster der dissoziierten Enzyme sind lediglich durch die relative Intensitätsverteilung der einzelnen Proteinbanden zu unterscheiden. So ist bei der Acetyl-CoA-Carboxylase die Bande δ immer intensiver als Bande γ, während bei der Pyruvatcarboxylase das umgekehrte Intensitätsverhältnis beobachtet wird.

Der Zusammenhang zwischen dem Elektrophoresemuster und dem Sedimentationsmuster konnte leicht durch einen entsprechenden Vergleich der durch Zuckergradientenzentrifugation isolierten 12 S, 9 S und 6 S Spaltstücke hergestellt werden. Dabei zeigte sich, daß Bande α mit dem 6 S Spaltstück, Bande β mit dem 9 S Spaltstück und Bande γ mit dem 12 S Spaltstück identisch ist. (Bezüglich der in der Elektrophorese zusätzlich beobachteten Banden mit geringer Beweglichkeit [δ, ε, . . .] s. u.).

Als nächstes interessierte uns die nähere Charakterisierung der 12 S, 9 S und 6 S Spaltstücke. Aufschlußreich war die Beobachtung, daß die Proteinverteilung in den einzelnen Komponenten des Sedimentationsmusters nicht konstant festgelegt ist. Diese Proteinverteilung ist vielmehr stark von den gewählten Bedingungen abhängig, besonders von der gewählten Temperatur während der Dissoziation. Beim Übergang zu tieferen Temperaturen während der Dissoziation verschwindet bei beiden Enzymen die 12 S-Komponente zunehmend zugunsten der 9 S und 6 S-Komponenten. Besonders stark ist dieser Temperatureffekt bei der Pyruvatcarboxylase ausgeprägt: Wird die Dissoziation bei ca. 2 bis 4 °C vorgenommen, so ist sowohl die 12 S wie auch die 9 S-Komponente nahezu vollständig zugunsten der 6 S-Komponente verschwunden.

Auch das Vorhandensein oder Nichtvorhandensein des Coenzyms Biotin hat starken Einfluß auf die Proteinverteilung im Spaltmuster, wie für das Apoenzym der Acetyl-CoA-Carboxylase gefunden wurde. Abb. 6 zeigt das elektrophoretische Spaltmuster von Apo- und Holoenzym der Acetyl-CoA-Carboxylase, die in identischer Weise vorbehandelt wurden. Offensichtlich ist die Dissoziation in die 9 S und 6 S-Bruchstücke beim Apoenzym erleichtert. Offenbar ist die Quartärstruktur des Enzyms in Abwesenheit des Coenzyms labiler.

Diese Befunde zeigen, daß sich die schweren 12 S und 9 S-Bruchstücke in die 6 S-Untereinheit überführen lassen. Weil also das gesamte Proteinmaterial in die 6 S-Komponente überführbar ist, muß diese 6 S-Untereinheit die kleinste Einheit sein, die noch über die gesamte Primärstruktur des Enzyms verfügt. Die 12 S und 9 S-

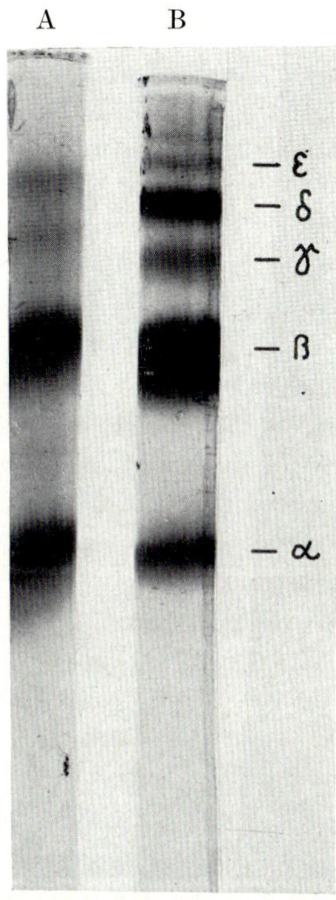

Abb. 6. Polyacrylamidgel-Elektrophorese von dissoziiertem Apoenzym (A) und Holoenzym (B) der Acetyl-CoA-Carboxylase [6]. Apo- und Holoenzym wurden durch dreistündige Dialyse gegen 0,1 N NH_3-Lösung (+ 0,01 M Thioglycol) bei 20 °C dissoziert. 6 %iges Gel; Trennung bei pH 10

Spaltstücke müssen Aggregate dieser 6 S-Untereinheit sein. Für die Richtigkeit dieser Annahme gibt es weitere experimentelle Hinweise. Beide Biotinenzyme können durch Wachstum der Hefe auf ^{14}C-Biotin-enthaltenden Nährmedium radioaktiv markiert werden. Spaltet man die markierten Enzyme in die 12 S, 9 S und 6 S-Bruchstücke und untersucht die Radioaktivitätsverteilung im Spaltmuster, so muß man eine konstante spezifische Radioaktivität über das gesamte Spaltmuster erwarten, wenn die gemachte Annahme zutrifft. Das entsprechende Experiment ergab, daß für beide Spaltmuster, also für Acetyl-CoA-Carboxylase und Pyruvatcarboxylase, die Verteilung von Protein und Radioaktivität sich entsprechen [6].

Eine weitere Stütze für die Annahme, daß die schwereren Bruchstücke beider Enzyme Aggregate einer 6 S-Untereinheit sind, liefern immunochemische Methoden. Es wurden Kaninchenantisera gegen native Acetyl-CoA-Carboxylase und native Pyruvatcarboxylase hergestellt. Mit Hilfe der Doppeldiffusionstechnik von OUCHTERLONY wurden die durch Gradientenzentrifugation isolierten 12 S, 9 S und 6 S-Bruchstücke zusammen mit den nativen Enzymen immunochemisch verglichen. Abb. 7 zeigt das Ergebnis für beide Enzyme. In beiden Fällen bildet sich in der Immunodiffusion zwischem nativen Enzym, dem 12 S, 9 S und dem 6 S-Bruchstück eine Präzipitatlinie kompletter Fusion aus. Die Abwesenheit von Spornbildungen zeigt, daß jedes Spaltstück, also auch das kleinste 6 S-Spaltstück, noch über alle antigenen Stellen des nativen Enzymkomplexes verfügt.

Mit Hilfe der Näherungsbeziehung (1) [9] wurden die Sedimen-

$$\frac{s_1}{s_2} = \left(\frac{M_1}{M_2}\right)^{2/3} \qquad (M = \text{Molekulargewicht}) \qquad (1)$$

tationskoeffizienten (s) eines Dimeren, eines Trimeren und eines Tetrameren der 6 S-Untereinheit berechnet. Die errechneten Werte sind für ein Dimeres 9,5 S, für ein Trimeres 12,0 S und für ein Tetrameres 15,1 S. Diese Werte stehen mit den experimentell gefundenen Werten in guter Übereinstimmung. Demnach wären also die nativen Enzyme Tetramere aus 6 S-Untereinheiten, die bei der Spaltung in Trimere (12 S), Dimere (9 S) und Monomere (6 S) zerfallen.

In diesem Zusammenhang ist es interessant, daß eine Auftragung der Beweglichkeiten der Proteinbanden α, β, γ, ... der Diskelektrophorese gegen den Logarithmus ihrer Molekulargewichte (M-, 2 M-, 3 M-, wobei M- das Molekulargewicht der 6 S-Untereinheit ist) eine lineare Beziehung ergibt [6]. Die Auswertung der Elektrophoresen in dieser Weise deutet darauf hin, daß die beobachteten, langsamer wandernden Proteinbanden ε, ζ, ... höhere Assoziate mit pentamerer, hexamerer und höherer Anordnung der 6 S-Unterein-

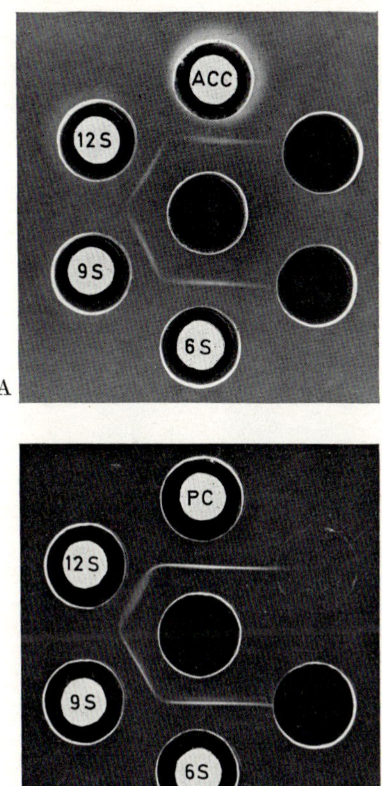

Abb. 7. Immunodiffusion der Spaltstücke von Acetyl-CoA-Carboxylase und Pyruvatcarboxylase [6]. A. Agarose-Gel (0,6 %) in 0,1 M Kaliumphosphatpuffer pH 7,5. Die Platte wurde 24 Std bei 20 °C entwickelt. Loch 1: native Acetyl-CoA-Carboxylase; Loch 2: 12 S Spaltstück; Loch 3: 9 S Spaltstück; Loch 4: 6 S Spaltstück. Im Zentrum befand sich Antiacetyl-CoA-Carboxylase-Serum. B. Agarose-Gel (0,6 %) in 50 mM Tris · HCl pH 7,5. Die Platte wurde 48 Std bei 20 °C entwickelt. Loch 1: native Pyruvatcarboxylase; Loch 2: 12 S Spaltstück; Loch 3: 9 S Spaltstück; Loch 4: 6 S Spaltstück. Im Zentrum befand sich Antipyruvatcarboxylase-Serum

heit darstellen. In dieser Assoziationstendenz dieser Hefeenzyme spiegelt sich bereits eine Beziehung zu den Acetyl-CoA-Carboxylasen aus Säugetiergeweben wieder, die ja als aktive Enzyme polymere Strukturen ausbilden. Dieses Assoziationsbestreben der Acetyl-CoA-Carboxylase und der Pyruvatcarboxylase aus Hefe beobachtet man besonders stark, wenn man die Enzyme mit $2{,}5 \cdot 10^{-4}$ M p-Chlormercuribenzoat in 0,1 M Trispuffer pH 8,5

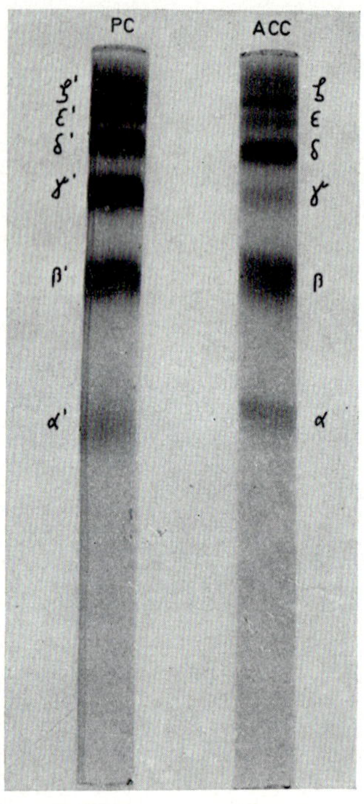

Abb. 8. Polyacrylamidgel-Elektrophorese von Pyruvatcarboxylase (PC) und Acetyl-CoA-Carboxylase (ACC), die in Gegenwart von p-Chlormercuribenzoat dissoziiert wurden [6]. Beide Enzyme wurden 3 Std bei 20 °C gegen 0,1 M Tris · HCl-Puffer pH 8,5 und $2{,}5 \cdot 10^{-4}$ M p-Chlormercuribenzoat dialysiert. 6 %iges Gel, Trennung bei pH 10

inkubiert. Abb. 8. zeigt das Aufspalten beider Enzyme unter diesen Bedingungen. Besonders die Pyruvatcarboxylase bildet hier bevorzugt höhere Assoziate aus.

Die bisher geschilderten Ergebnisse beweisen die strukturelle Ähnlichkeit der beiden Biotinenzyme aus Hefe. Wie weit geht nun diese Verwandtschaft? Geht sie soweit, daß diejenigen Proteinstrukturen, die die Carboxylierung des Biotins katalysieren (diese Teilreaktion ist ja beiden Enzymen gemeinsam), identische Primärstruktur besitzen. Diese Möglichkeit ließ sich immunochemisch prüfen. Anti-Acetyl-CoA-Carboxylase, hemmte die enzymatische Aktivität der Acetyl-CoA-Carboxylase vollständig, ebenso hemmte Anti-Pyruvatcarboxylase Pyruvatcarboxylase vollständig. Wie für den Fall der Pyruvatcarboxylase genauer untersucht wurde, wird speziell die erste Teilreaktion, also die Carboxylierung des Biotins, gehemmt, während die zweite Teilreaktion kaum beeinflußt wurde. Anti-Acetyl-CoA-Carboxylase zeigte selbst in großem Überschuß eingesetzt keine inhibierende Wirkung auf Pyruvatcarboxylase, d. h. es war keine Kreuzinhibition zu beobachten. Auch in der Immunelektrophorese ließ sich innerhalb der Nachweisgrenze dieser Methode keine Kreuzreaktion zwischen Anti-Acetyl-CoA-Carboxylase und Pyruvatcarboxylase feststellen.

Man muß aus diesem Befund den Schluß ziehen, daß die beiden Enzymen gemeinsame Teilaktivität durch Proteinteilstrukturen katalysiert werden, die eine nicht identische Primärstruktur haben und damit von unabhängigen Genen codiert werden. Wie läßt sich die trotzdem gefundene strukturelle Analogie beider Biotinenzyme verstehen? Wir spekulieren, daß beide Gene, die die Sequenz von Acetyl-CoA-Carboxylase und Pyruvatcarboxylase kontrollieren, von einem gemeinsamen Vorfahren abstammen. Durch einen Prozeß einer Genverdopplung entstanden zwei Gene, die dann eine unabhängige Evolution durchlaufen konnten, um schließlich für zwei Biotinenzyme zu codieren, die zwar verschiedene, aber immer noch ähnliche Funktionen ausüben. Ganz analog könnte man sich die Entstehung der anderen Biotinenzyme vorstellen, die z. T. noch geringere strukturelle Unterschiede aufweisen sollten (z. B. Acetyl-CoA-Carboxylase und Propionyl-CoA-Carboxylase) als das hier untersuchte Paar.

Beispiele dieser Art von Genverdopplungen sind bekannt [10]. Besonders interessant ist der Fall von α-Lactalbumin und Lysozym. Es besteht für diese beiden Proteine kaum ein Zweifel, daß sie aus einem gemeinsamen Ur-Gen hervorgegangen sind. Diese beiden Proteine üben heute verschiedene Funktionen aus, haben aber noch eine in ca. 35% aller Positionen übereinstimmende Aminosäuresequenz. Auch ist sehr wahrscheinlich die Tertiärstruktur trotz der Aminosäuresubstitutionen weitgehend unverändert geblieben [11, 12].

Ein letztes Argument für die strukturelle Verwandtschaft der Biotinenzyme liefert die Biotin:Apoenzym-Ligase. Dieses Enzym katalysiert die kovalente Bindung des Coenzyms an eine definierte Lysinseitenkette des Apoenzyms. Die Ligase ist in der Lage, die aktiven Zentren von verschiedenen Biotinenzymen zu erkennen, selbst von verschiedenen Biotinenzymen aus verschiedenen Organismen [13].

2. Der Multienzymkomplex Fettsäuresynthetase aus Hefe

Die Fettsäuresynthetase aus Hefe ist ein sehr stabiler Multienzymkomplex mit wenigstens sieben katalytischen Teilaktivitäten. Das Molekulargewicht des Enzymkomplexes wurde mit $2{,}3 \times 10^6$ bestimmt. Zwei verschiedene Typen von SH-Gruppen erwiesen sich für den Reaktionsablauf von großer Bedeutung. Die sog. zentrale SH-Gruppe ist die Thiolgruppe eines proteingebundenen 4′-Phosphopantetheinmoleküls (Abb. 9). Hauptaufgabe dieses Moleküls ist es, die als Thioester gebundenen Zwischenprodukte der Fettsäuresynthese den verschiedenen aktiven Zentren der Teilenzyme zuzuführen. Das 4′-Phosphopantethein wirkt also als ein mobiler Substratcarrier innerhalb des Enzymkomplexes, ein Prinzip, das bei allen Multienzymsystemen verwirklicht zu sein scheint, um ein unkontrolliertes Wegdiffundieren der zahlreichen, im Reaktionscyclus auftretenden Zwischenprodukte zu verhindern. Die zweite SH-Gruppe, die sog. periphere SH-Gruppe ist Bestandteil eines Cysteinrestes im aktiven Zentrum des kondensierenden Teilenzyms. Die Fettsäuresynthese beginnt mit der Übertragung eines Acetylrestes vom Acetyl-CoA über die zentrale SH-Gruppe zur peripheren SH-Gruppe. Auf die zentrale SH-Gruppe wird ein Malonylrest übertragen. Der Acetylrest kondensiert nun mit dem Malonylrest: Das Ergebnis ist ein Acetacetylrest, der als Thioester am Pantethein des „Schwingarmes“ gebunden ist. Durch Reduktion zum β-Hydroxybutyrylrest, Dehydratisierung zum Crotonylrest und abschließende Hydrierung zum Butyrylrest wird der erste Cyclus der Synthese beendet. Der gesättigte C_4-Acylrest wird nun wieder auf die periphere SH-Gruppe übertragen und kann nun mit einem weiteren, auf die zentrale SH-Gruppe übertragenen Malonylrest kondensieren.

Der Acetylrest wird also sukzessive durch C_2-Einheiten verlängert. Theoretisch könnten dabei Fettsäuren beliebiger Kettenlänge entstehen. Tatsächlich werden jedoch nur Palmitinsäure und Stearinsäure in Form ihrer CoA-Thioester als Endprodukte abgegeben. Im folgenden soll ein Modell beschrieben werden, das diesen spezifischen Kettenabbruch erklären kann [14]. Die naheliegende Deutung wäre die Annahme einer hohen Kettenlängenspezifität des transferierenden Teilenzyms, das die synthetisierten Acylreste auf

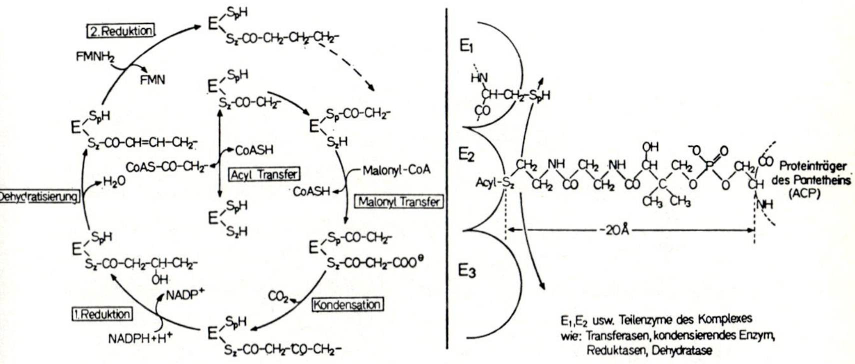

Abb. 9. Reaktionsschema der Fettsäuresynthese am Multienzymkomplex. Links: Fettsäuresynthesecyclus, rechts: chemische Natur der „peripheren" und „zentralen" SH-Gruppe

CoA zurücküberträgt. SCHWEIZER [15] konnte diese Möglichkeit ausschließen, indem er zeigte, daß diese Transferase in der Lage ist, Acylreste mit Kettenlängen von 6 C-Atomen bis 18 C-Atomen mit vergleichbaren Geschwindigkeiten zu übertragen.

Ausgangspunkt der Betrachtung ist die Reaktionsmöglichkeit eines Acylrestes, der als Thioester an der zentralen SH-Gruppe gebunden ist:

$$\begin{array}{ccccc} & \overset{(K^0_T\,K'_T)}{\underset{}{K_T}} & & K_K & \\ \text{Transferase} & \rightleftarrows & \text{E-SH}_z & \rightleftarrows & \text{E-SH}_p \\ k_T \swarrow & & & & \searrow k_K \\ \text{Produkt} & & & & \text{Kondensation.} \end{array}$$

In dieser Situation hat dieser Acylrest zwei Möglichkeiten weiter zu reagieren. Entweder wird er auf die periphere SH-Gruppe übertragen, so kann er mit einem weiteren Malonylrest kondensieren und im geschilderten Reaktionscyclus zu einem um 2 C-Atome verlängerten Acylrest umgewandelt werden. Oder aber er kann auf die Acyltransferase übertragen werden und dann nach Übertragung auf CoA als Endprodukt den Enzymkomplex verlassen. Die Wahrscheinlichkeit, daß ein gegebener Acylrest den Komplex als Endprodukt verläßt, ist ganz einfach vom Verhältnis der Geschwindigkeiten von Kondensations- und Transferreaktion abhängig $\left(W = \frac{V_T}{V_K + V_T}\right)$. Wir nehmen an, daß die Gleichgewichtsverteilung der Acylreste zwischen Transferase, zentraler und peripherer SH-Gruppe ein vorgelagertes Gleichgewicht ist. Die Geschwindigkeit der Produktbildung ist proportional der Beladung der Transferase, die Kondensationsgeschwindigkeit ist proportional der Beladung der peripheren SH-Gruppe. Unter diesen Voraussetzungen hätte jeder Acylrest die gleiche Wahrscheinlichkeit Produkt zu bilden. Ein Modell des Syntheseabbruchs muß aber eine Änderung dieser Wahrscheinlichkeit mit wachsender Kettenlänge beschreiben, mit anderen Worten, das Enzymsystem muß die Fähigkeit besitzen, verschieden lange Alkanketten unterscheiden zu können. Ein solches Unterscheiden ist durch eine Wechselwirkung der Alkanketten mit einer Art „Meßstrecke" möglich. Als Arbeitshypothese haben wir einen hydrophoben Bereich an der Transferase angenommen (Abb. 10). Neben dem aktiven Zentrum, an dem der Acylrest kovalent gebunden wird, soll der hydrophobe Bereich eine Art „zweite Bindungsstelle" für die Alkankette sein. Setzt die Wechselwirkung beim Erreichen einer bestimmten Kettenlänge des Acylrestes ein, so trägt jede hinzukommende Methylengruppe mit einem konstanten Beitrag zur freien Energie dieser Wechselwirkung bei. Durch die zunehmende Wechselwirkungsenergie mit wachsender Kettenlänge wird die Gleichgewichtsverteilung der Acylreste zugunsten der Transferase verschoben, die Geschwindigkeit der Produktbildung nimmt somit zu. Formuliert man diese Modellvorstellung mathematisch, so läßt sich in einfacher Weise der folgende Ausdruck für die Wahrscheinlichkeit der Produktbildung eines gegebenen Acylrestes der Kettenlänge C_n ableiten:

$$W_n = \frac{1}{1 + A \exp(m\Delta F/RT)}$$

mit $$A \sim \frac{k_K}{k_T} \times \frac{K_K}{K^0_T} \times \frac{[\text{Mal} - \text{CoA}]}{[\text{Mal} - \text{CoA} + \text{Ac} - \text{CoA}]} .$$

Dabei bedeutet A eine aus den oben eingeführten Gleichgewichtskonstanten und Geschwindigkeitskonstanten zusammengesetzte Größe, die für gegebene Synthesebedingungen konstant, d. h. kettenlängenunabhängig ist. Der Korrekturfaktor der Kondensa-

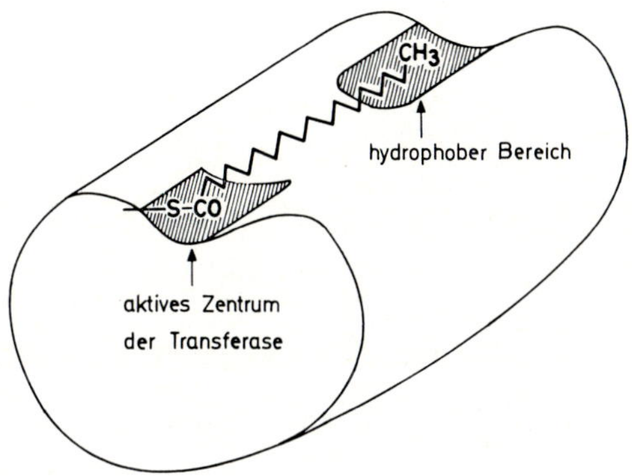

Abb. 10. Schematische Darstellung des postulierten hydrophoben Bereiches am transferierenden Teilenzym

tionsgeschwindigkeit $\frac{[\text{Mal} - \text{CoA}]}{[\text{Mal} - \text{CoA} + \text{Ac} - \text{CoA}]}$ wird durch die Tatsache bedingt, daß sowohl Acetyl-CoA als auch Malonyl-CoA um die zentrale SH-Gruppe konkurrieren, jedoch nur der Malonylrest die Kondensationsreaktion ermöglicht. ΔF beschreibt die Wechselwirkungsenergie pro Methylengruppe, m ist die Anzahl der Methylengruppen, die zur Wechselwirkung mit dem hydrophoben Bereich beitragen.

Mit diesem Ausdruck kann man nun Produktverteilungen berechnen. Nimmt man als freie Energie der Wechselwirkung einen Betrag von $\Delta F = -0{,}4$ kcal pro Methylengruppe an, so ergibt das Modell eine recht breite Endproduktverteilung (Abb. 11). Berechnet man die Verteilung mit einem Energieinkrement von $\Delta F = -1{,}0$ kcal, so würde man eine Endproduktverteilung erwarten, die im wesentlichen nur noch zwei Fettsäuren — in Abb. 11 allgemein mit C_n und C_{n+2} bezeichnet — auftreten läßt. Tatsächlich sind die Energieinkremente für hydrophobe Wechselwirkungen von Methylengruppen von NEMETHY und SCHERAGA [16] in dieser Größenordnung abgeschätzt worden. Um den speziellen Fall der Produktverteilung der Fettsäuresynthetase berechnen zu können, ist noch folgende Information notwendig: Wann, d. h. ab welcher Kettenlänge setzt die Wechselwirkung zwischen Transferase und wachsender Alkankette ein. Um eine Aussage hierüber machen zu können, muß man die Endproduktanalyse der Fettsäuresynthese betrachten. Man findet mit hoher Reproduzierbarkeit, daß jede Fettsäure im Kettenlängenbereich von C_6 bis C_{12} mit einem konstanten Anteil von 1 bis 2% im Produktmuster vertreten ist. Offensichtlich kann in diesem Kettenlängenbereich die Wechselwirkung noch nicht bestehen. Erstmals die C_{14}-Säure, die Myristinsäure zeigt einen erhöhten Anteil im Produktmuster [14]. Aus diesem Grund nehmen wir an, daß die Wechselwirkung mit dem Erreichen einer Kettenlänge von 13 C-Atomen einsetzt. Unter dieser Voraussetzung läßt sich nun die theoretische Produktverteilung für die Fettsäure-

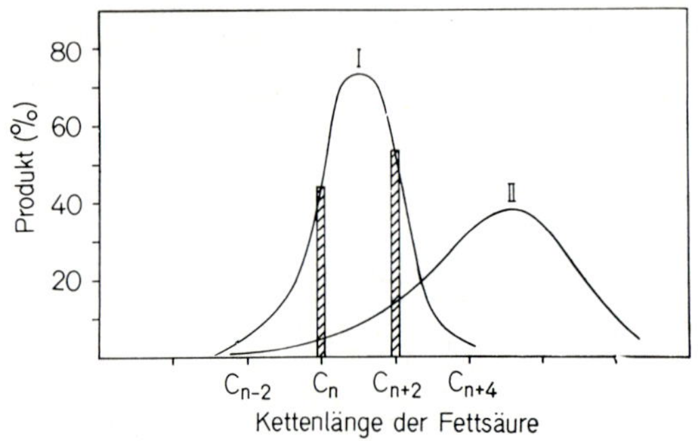

Abb. 11. Berechnete Endproduktverteilungen für zwei verschiedene Energieinkremente ΔF [14]. Kurve I: $\Delta F = -1{,}0$ kcal pro Methylengruppe. Kurve II: $\Delta F = -0{,}4$ kcal pro Methylengruppe

synthetase errechnen. Die Säure C_n entspricht dann der C_{16}-Säure, also der Palmitinsäure, C_{n+2} der C_{18}-Säure, also der Stearinsäure. Das Maximum der Produktverteilungskurve liegt bei der C_{17}-Säure, die jedoch als ungeradzahlige Fettsäure nicht gebildet wird. Ersetzt man den Starter Acetyl-CoA durch Propionyl-CoA, so wird im Synthesecyclus die homologe Reihe der ungeradzahligen Fettsäuren durchlaufen. Mit diesem Trick läßt sich die Produktverteilungskurve experimentell bestimmen. Das Ergebnis zeigt Abb. 12. Mit Propionyl-CoA als Starter tritt Margarinsäure (C_{17}) fast ausschließ-

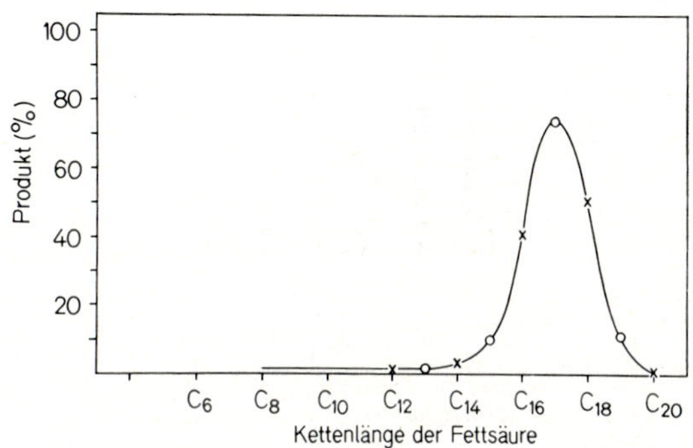

Abb. 12. Prozentuale Verteilung der Endprodukte in der Fettsäuresynthese [14]. Die Reaktionsansätze enthielten im Volumen 2 ml: 200 μMol Kaliumphosphat pH 6,5; 0,8 mg Serumalbumin; 1,2 μMol NADPH; 200 μg Fettsäuresynthetase und entweder 0,25 μMol 1-^{14}C-Acetyl-CoA oder 0,25 μMol Propionyl-CoA. T = 25 °C. Die Reaktion wurde durch Zugabe von 0,5 μMol 2-^{12}C- oder 2-^{14}C-Malonyl-CoA gestartet. Abgestoppt wurde mit 1 ml 10 %iger KOH in Methanol. Die Fettsäuren wurden als Methylester radiogaschromatographisch analysiert. x——x Acetyl-CoA als „Starter"; o——o Propionyl-CoA als „Starter"

lich als Endprodukt auf, die experimentell gefundene Produktverteilung entspricht der berechneten.

Mit dem Modell lassen sich eine Reihe von Voraussagen machen, wie das Produktspektrum der Fettsäuresynthetase verändert werden kann. Allgemein muß jede Änderung des Geschwindigkeitsverhältnisses von Kondensation und Transfer eine Verschiebung des Produktspektrums bewirken. Eine relative Verlangsamung der

Kondensation muß bereits den kürzeren Kettenlängen die Chance geben, noch vor Einsetzen der Wechselwirkung Produkt zu bilden. Ein experimentell geprüfter Extremfall dieser Art ist der folgende: Senkt man die Malonyl-CoA-Konzentration unter den Sättigungswert für Fettsäuresynthetase, so läßt sich die Kondensationsreaktion beliebig verlangsamen. Eine konstante, sehr niedrige Malonyl-CoA-Konzentration ließ sich experimentell durch Zusatz definierter

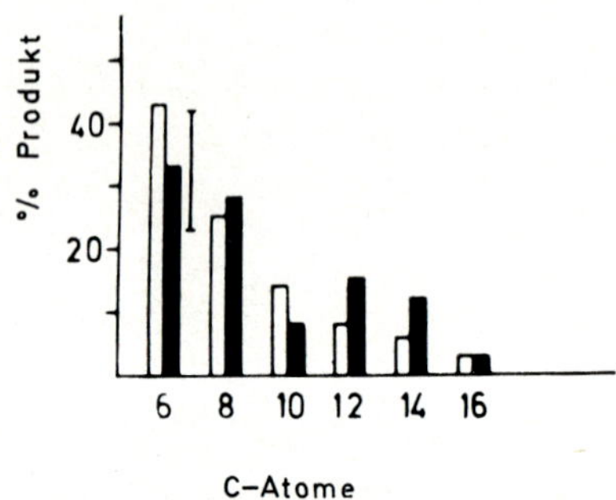

Abb. 13. Synthese kurzkettiger Fettsäuren durch limitierende Mengen Malonyl-CoA (Verlangsamung der Kondensationsreaktion) [14]. Die Reaktionsmischung enthielt im Volumen 2 ml: 100 μMol Tris · HCl pH 7,5; 20 μMol $MgCl_2$; 20 μMol K-Citrat; 7 μMol GSH; 50 μMol $KHCO_3$; 1,5 mg Serumalbumin; 0,4 μMol 1-^{14}C-Acetyl-CoA; 1,2 μMol NADPH; 600 mE Fettsäuresynthetase und 11 mE Acetyl-CoA-Carboxylase (aus Hühnerleber). Die Reaktion wurde durch Zugabe von 7 μMol ATP gestartet. Analyse der Fettsäuren wie in Abb. 12. Weiße Balken: berechnete Produktverteilung; schwarze Balken: experimentell gefundene Produktverteilung

Mengen Acetyl-CoA-Carboxylase einstellen. Wir haben auf diese Weise die Kondensationsgeschwindigkeit um einen Faktor 56 verkleinert. Berechnet man die Produktverteilung, so zeigt sich, daß unter diesen Bedingungen eine extreme Verschiebung zu kurzkettigen Fettsäuren auftreten muß. Abb. 13 vergleicht die Berechnung mit dem Experiment, eine weitgehende Übereinstimmung ist gegeben. Natürlich lassen sich, neben dem hier geschilderten Extremfall, durch entsprechende Geschwindigkeitsverschiebungen von Kondensation und Transfer viele, verschiedene Produktmuster erzielen. Die Übereinstimmung von berechneten und gefundenen Produktmustern war immer hinreichend erfüllt.

Untersuchungen über die strukturelle Organisation des Multienzymkomplexes Fettsäuresynthetase wurden durch seine Stabilität sehr erschwert. Drastische Methoden zur Dissoziation von Quartärstrukturen, wie etwa 6 M Guanidinhydrochlorid, führten zwar zur Spaltung des Komplexes aber gleichzeitig wurden auch sämtliche katalytischen Teilaktivitäten zerstört.

Neuerdings gelang es, einen ersten Schritt in Richtung kompletter reversibler Zerlegung des Multienzymkomplexes zu erzielen [17, 18]. Fettsäuresynthetase kann in 1 M NaCl oder LiCl gelöst werden, ohne ihre enzymatische Aktivität zu verlieren und ohne ein verändertes Sedimentations- und Elektrophoreseverhalten zu zeigen. Friert man aber diese salzhaltigen Enzymlösungen zweimal auf ca. –70 °C ein und läßt bei Raumtemperatur auftauen, so haben sich die enzymatischen und physikalischen Eigenschaften drastisch geändert. Die Dissoziation des Multienzymkomplexes in Untereinheiten unter diesen Bedingungen wird durch die Diskelektrophorese bewiesen (Abb. 14).

Nach zweimaligem Einfrieren und Auftauen ist die enzymatische Aktivität zur Fettsäuresynthese vollständig verschwunden. Eine Reaktivierung läßt sich jedoch bei Verringerung der Ionenstärke durch Verdünnen oder Dialyse bei 20 °C erreichen. Der Reaktivierungsgrad liegt in der Regel um 50%, jedoch wurden Reaktivierungen bis zu 80% der Ausgangsaktivität erreicht. Die reaktivierte Fettsäuresynthetase erweist sich hinsichtlich ihres Sedimentations- und Elektrophoreseverhaltens als ununterscheidbar vom nativen Enzymkomplex. Abb. 15 dokumentiert dies für das elektrophoretische Verhalten. Diese Elektrophoresen wurden in einem sehr großporigem Polyacrylamidgel (3,4%) durchgeführt, um eine Wanderung des intakten Multienzymkomplexes zu ermöglichen; die Spaltstücke werden in diesem großporigen Gel nicht sauber getrennt.

Die Fähigkeit zur enzymatischen Fettsäuresynthese bedingt die intakte enzymatische Aktivität aller Teilenzyme. Der Verlust der Gesamtaktivität hat also nicht notwendigerweise den Verlust aller Teilaktivitäten zur Folge. Mit geeigneten Modellsubstraten [27] können alle Teilaktivitäten separat gemessen und so ihr Verhalten während der Dissoziation und Reaktivierung verfolgt werden. Diese Messungen wurden von Weithmann [19] vorgenommen. Es stellte sich dabei folgendes heraus: Die Transferaseaktivitäten, d. h. die Malonyl-Transferase und die Acetyl-Transferase sowie die Dehydratase sind nach der Spaltung nicht oder nur kaum in ihrer katalytischen Aktivität verändert. Völlig inaktiv sind die beiden Reduktasen, sowie das kondensierende Teilenzym. Diese drei Teilaktivi-

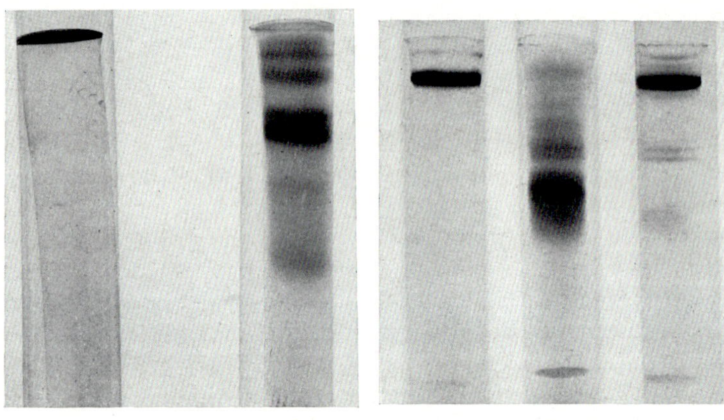

Abb. 14 Abb. 15

Abb. 14. Dissoziation der Fettsäuresynthetase durch Einfrieren und Auftauen in 1 M LiCl-Lösung [18]. Links: Fettsäuresynthetase in 1 M LiCl-Lösung; rechts: gleiches Präparat nach zweimaligem Einfrieren und Auftauen. 5%iges Polyacrylamidgel, elektrophoretische Trennung bei pH 9,5

Abb. 15. Elektrophoretischer Vergleich von nativer und rekonstituierter Fettsäuresynthetase [17]. Links: Fettsäuresynthetase in 1 M LiCl; Mitte: gleiche Präparation nach zweimaligem Einfrieren und Auftauen; rechts: gleiche Präparation nach Reassoziation durch fünffaches Verdünnen mit 0,1 M Kaliumphosphatpuffer pH 7,5, der 40 mM an Cystein war

täten lassen sich unter Bedingungen, die zur Reassoziation des Multienzymkomplexes führen, reaktivieren. In Abb. 16 sind die Reaktivierungskinetiken der Gesamtaktivität und der beiden Reduktasen wiedergegeben. Man erkennt, daß die beiden Reduktasen schneller reaktivieren als die Gesamtaktivität. Die Reaktivierung der Kondensationsreaktion ist offenbar der geschwindigkeitsbestimmende Schritt, da diese Teilaktivität nahezu parallel zur

Gesamtaktivität reaktiviert. Das ist plausibel, denn die kondensierende Teilaktivität benötigt die exakte räumliche Orientierung von mindestens zwei Proteinuntereinheiten, nämlich der Acyl-Carrier-Proteinkomponente, an der die zentrale SH-Gruppe des 4'-Phosphopantetheins gebunden ist und der Proteinkomponente, die die periphere SH-Gruppe trägt. Da die Reduktasen nur unter

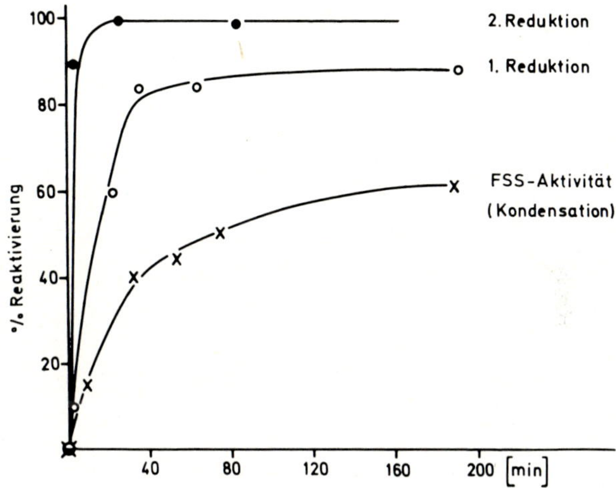

Abb. 16. Kinetik der Reaktivierung der Gesamtaktivität sowie der Aktivität der Reduktasen. Fettsäuresynthetase wurde durch zweimaliges Einfrieren und Auftauen in 1 M LiCl-Lösung dissoziiert. Zum Zeitpunkt Null wurde die Präparation fünffach mit 0,1 M Kaliumphosphatpuffer pH 7,5 der 13 mM an Dithiothreitol war, verdünnt, bei 20 °C inkubiert und dann die Reaktivierung der Enzymaktivitäten zu verschiedenen Zeitpunkten gemessen

Bedingungen reaktivieren, die zur Assoziation des Multienzymsystems führen, ist offenbar für die katalytische Aktivität dieser Teilenzyme eine spezifische Protein-Protein-Wechselwirkung notwendig, die nur nach Assoziation mit einem oder mehreren „Nachbarenzymen" gegeben ist.

Um eine grobe Abschätzung der Molekulargewichte der Spaltstücke zu bekommen, wurde die gespaltene Fettsäuresynthetase mit Hilfe der Zuckergradientenzentrifugation untersucht. Abb. 17 zeigt das Resultat, man beobachtet zwei Gipfel A und B. Gipfel A kann je nach den angewandten Spaltmethoden mehr oder weniger zugunsten von Gipfel B verschwinden. Gipfel B sedimentiert annähernd gleich schnell wie Katalase. Diese Spaltstücke haben also Molekulargewichte in der Größenordnung von 250000. Offenbar führt diese Spaltmethode noch nicht zu den vollkommen getrennten

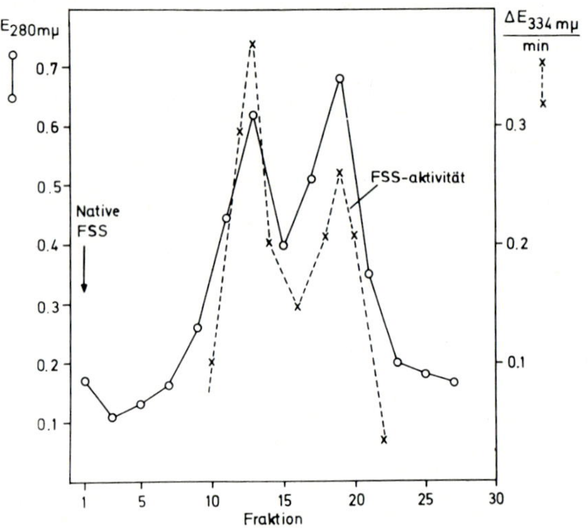

Abb. 17. Zuckergradientenzentrifugation dissoziierter Fettsäuresynthetase und Reaktivierung der enzymatischen Aktivität in den einzelnen Fraktionen [18]. Ca. 7 mg dissoziierter Fettsäuresynthetase (durch Einfrieren und Auftauen einer Lösung, die 1 M NaCl enthielt) wurden auf 30 ml eines Zuckergradienten (5—20%) in 0,1 M Tris · HCl pH 7,5 und 0,5 M NaCl aufgetragen. Zentrifugationsdauer 36 Std, Temperatur ca. 0 °C, 24000 UpM. Der Gradientenbecher wurde am Boden angestochen und Fraktionen von 1 ml wurden gesammelt. Nach fünffachem Verdünnen aller Fraktionen mit 0,1 M Kaliumphosphatpuffer (40 mM Cystein) und dreistündiger Inkubation bei 20 °C wurde der Reaktivierungsgrad der Fettsäuresynthetaseaktivität in allen Fraktionen bestimmt

Teilenzymen, sondern vielmehr zu Bruchstücken, die noch Assoziate darstellen.

Ein überraschendes Ergebnis liefern Reaktivierungsexperimente mit den Proteinkomponenten der Gipfel A und B. Nach Verdünnen und Inkubation bei Raumtemperatur sind beide Fraktionen wieder zu enzymatisch aktiver Fettsäuresynthetase reassoziiert (Abb. 17). Offensichtlich verfügen beide Fraktionen A und B über alle Teilkomponenten, die zur Rekombination des intakten Multienzymkomplexes notwendig sind. Proteingipfel A besteht, wie mit der

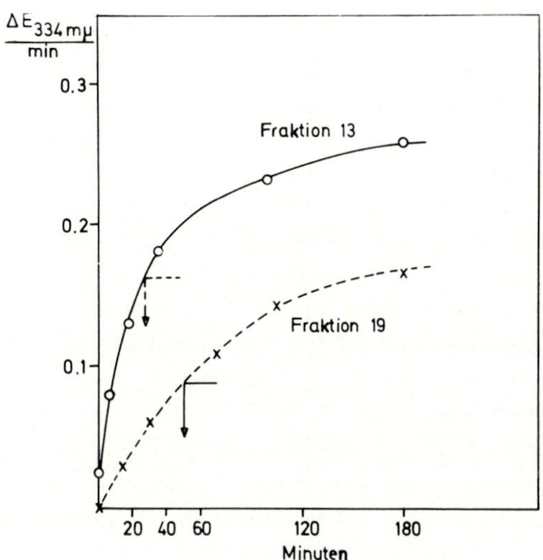

Abb. 18. Kinetik der Reaktivierung der Fettsäuresynthetaseaktivität für Fraktion 13 und 19 der Abb. 17 [18]

Diskelektrophorese untersucht wurde, im wesentlichen aus einer Proteinkomponente, Gipfel B dagegen aus mehreren Komponenten. Auf Grund dieser Befunde nehmen wir an, daß der Multienzymkomplex nach folgendem Schema dissoziiert und reassoziiert:

Fettsäuresynthetase
(M. W. $2{,}3 \times 10^6$)

↑ ↓

n größere Untereinheiten
(M. W. $\approx 5 \times 10^5$) } Gipfel A

↑ ↓

m kleinere Untereinheiten
(M. W. $\approx 2{,}5 \times 10^6$) } Gipfel B

Das würde bedeuten, daß im Multienzymkomplex alle Teilenzyme mehrfach vorhanden sind. Einige quantitative Belege gibt es: Bezogen auf das Molekulargewicht $2{,}3 \times 10^6$ findet man z. B. 4 bis 5 FMN-Moleküle [20], etwa 5 Pantothensäurereste [21], also 5 zentrale SH-Gruppen sowie 5 bis 6 nichtthioesterartige Malonyl-Acceptorstellen [18].

Die Messung der Reaktivierungskinetik beider Proteinfraktionen A und B zeigt ebenfalls, daß Fraktion A eine bereits höher organisierte Unterstruktur darstellt. Ihre Reaktivierungs-Halbwertszeit ist um einen Faktor 2 kürzer als für Fraktion B (Abb. 18).

Zusammenfassend läßt sich sagen, daß ein Aufbau der Fettsäuresynthetase aus Untereinheiten, die für sich noch alle Teilenzyme der Fettsäuresynthetase besitzen, wahrscheinlich gemacht werden konnte, exakte quantitative Aussagen über die Stöchiometrie des Komplexes jedoch noch nicht gemacht werden können. Besonders auffallend ist die Tatsache, daß sowohl die Fettsäuresynthetase als auch die Acetyl-CoA-Carboxylase aus Hefe trotz Anwendung verschiedener Methoden bisher nicht in die Einzelenzyme aufgetrennt werden konnten. Die 6 S-Untereinheit der Acetyl-CoA-Carboxylase ließ sich z. B. durch Einwirkung von 8 M Harnstoff und 100 mM Thioglycol nicht weiter aufspalten. Möglicherweise bestehen bei diesen Multienzymsystemen kovalente Verknüpfungen zwischen den funktionellen Einheiten.

References

1. ALBERTS, A. W., VAGELOS, P. R.: Fed. Proc. **27**, 647 (1968).
2. — — Proc. nat. Acad. Sci. (Wash.) **59**, 561 (1968)
3. — NERVI, A. M., VAGELOS, P. R.: Proc. nat. Acad. Sci. (Wash.) **63**, 1319 (1969).
4. DIMROTH, P., GUCHHAIT, R. B., STOLL, E., LANE, M. D.: Proc. nat. Acad. Sci. (Wash.) **67**, 1353 (1970).
5. GUCHHAIT, R. B., MOSS, J., SOKOLSKI, W., LANE, M. D.: Proc. nat. Acad. Sc. (Wash.) **68**, 653 (1971).
6. SUMPER, M., RIEPERTINGER, C.: Zur Veröffentlichung im Europ. J. Biochem. eingereicht (1972).
7. YOUNG, M. R., TOLBERT, B., UTTER, M. F.: Meth. Enzymol. **13**, 250 (1969).
8. MARTIN, R. G., AMES, B. N.: J. biol. Chem. **236**, 1373 (1961).
9. SCHACHMAN, H. K.: Ultracentrifugation in Biochemistry. New York: Academic Press 1959.
10. NOLAN, C., MARGOLIASH, E.: Ann. Rev. Biochem. **37**, 727 (1968).
11. BREW, K., VANAMAN, T. C., HILL, R. L.: J. biol. Chem. **242**, 3747 (1967).

12. Browne, W. J., North, A. C. T., Phillips, D. C., Brew, K., Vanaman, T. C., Hill, R. L.: J. molec. Biol. **42**, 65 (1969).
13. McAllister, H. C., Coon, M. J.: J. biol. Chem. **241**, 2855 (1966).
14. Sumper, M., Oesterhelt, D., Riepertinger, C., Lynen, F.: Europ. J. Biochem. **10**, 377 (1969).
15. Schweizer, E.: Dissertation, Universität München 1963.
16. Nemethy, G., Scheraga, H. A.: J. chem. Phys. **36**, 3401 (1962).
17. Sumper, M., Riepertinger, C., Lynen, F.: FEBS Letters **5**, 45 (1969).
18. — Dissertation, Universität München, 1970.
19. Weithmann, U.: Unveröffentlichte Versuche.
20. Oesterhelt, D., Bauer, H., Lynen, F.: Proc. nat. Acad. Sci. (Wash.) **63**, 1377 (1969).
21. Winnewisser, W.: Unveröffentlichte Versuche.
22. Lynen, F.: Fed. Proc. **20**, 941 (1961).

Discussion

H. Holzer (Freiburg): Sind Acetyl-CoA-Carboxylase und Pyruvatcarboxylase vollständig spezifisch?

N. Sumper (München): Acetyl-CoA-Carboxylase war nicht in der Lage, Pyruvat zu carboxylieren. Das entsprechende Kontrollexperiment hätte noch eine Carboxylierungsgeschwindigkeit von etwa 0,5% der Acetyl-CoA-Carboxylierungsgeschwindigkeit erfassen können. Dasselbe Ergebnis lieferte der Versuch, Acetyl-CoA mit Hilfe von Pyruvatcarboxylase zu carboxylieren.

H. Holzer (Freiburg): Enthält jede Untereinheit der beiden Enzyme je ein Biotin?

M. Sumper (München): Die Pyruvatcarboxylase aus Hefe enthält nach Untersuchungen im Arbeitskreis von Utter vier Biotinreste pro Enzymmolekül. Für die Acetyl-CoA-Carboxylase aus Hefe ist noch keine quantitative Biotinbestimmung ausgeführt worden. Im Gegensatz zur induzierbaren Pyruvatcarboxylase ist die Acetyl-CoA-Carboxylase ein konstitutives Enzym, das selbst bei völligem Biotinausschluß im Zuchtmedium synthetisiert wird (und dann natürlich als Apoenzym vorliegt). Bei der Acetyl-CoA-Carboxylase besteht deshalb die Schwierigkeit, daß die Präparationen z. T. Apoenzym (Rominger, K. L.: Dissertation, Universität München 1964) enthalten, wodurch zu niedrige Biotinwerte gemessen würden.

W. Gruber (Tutzing): Man sollte evtl. auch Antiseren gegen Untereinheiten der ACC machen und länger immunisieren, um komplettere Antikörper zu erhalten und damit evtl. Kreuzreaktionen mit der Pyruvatcarboxylase.

M. Sumper (München): Es wurde insgesamt 78 Tage immunisiert, wobei alle 14 Tage Enzym gespritzt wurde. Möglicherweise könnte man bei längeren Immunisierungszeiten doch kreuzreagierende Antikörper erhalten.

U. Henning (Tübingen): Können Sie die Untereinheiten von der Pyruvat- und Acetyl-CoA-Carboxylase hybridisieren?

M. Sumper (München): Das haben wir bisher nicht versucht.

Antigen-Antibody Interactions

Conformational Changes in Protein Antigens Induced by Specific Antibodies: Sperm-whale Myoglobin

M. J. Crumpton

National Institute for Medical Research, Mill Hill, London NW7 1AA, Great Britain

With 7 Figures

Introduction

It is well recognized that the conformations of globular proteins are particularly sensitive to environmental changes and that they may be altered by the binding of specific ligands. As a result, it seems likely that interaction between proteins or between the polypeptide chains of multisubunit proteins will also be associated with localized or more extensive conformational alterations.

Antigen-antibody interaction is one system in which attempts have been made to correlate specific binding with changes in the conformations of the reactants. However, contrary to some claims, investigations of the effect of the antigen on the conformation of the antibody have failed to provide convincing evidence in support of the view that binding of antigen by the antibody combining-site causes conformational changes elsewhere in the antibody molecule (for a critical review of the evidence see Metzger, 1970). This interpretation of the data is supported by the results of more recent studies of the effect of bound ligand on the circular dichroism spectra, the rate of exchange of labile hydrogens and the susceptibility to proteolysis of a homogeneous human macroglobulin [Ashman et al., 1971; Ashman and Metzger, 1971]. In contrast, data have been presented by a number of investigators suggesting that specific antibodies may either induce a conformational change in globular protein antigens or, alternatively, select and stabilize a particular conformation of the antigen. For example, mutants producing defective enzymes have been activated to full enzymic activity by antibodies to the wild-type enzyme [penicillinase; Pollock et al (1967); β-D-galactosidase; Rotman and Celada (1968); Melchers and Messer (1970); catalase, Feinstein et al. (1971)]. Similarly, the restoration of the activity of heat-denaturated acetylcholinesterase by antibodies to the native enzyme [Michaeli et al., 1967] and the enhancement of enzymic activity by certain homologous antisera [cephalosporin β-lactamase, Chong and Goldner (1970); ribonuclease, Cinader et al. (1971); glutamate dehydrogenase, Lehmann (1971)] has been accounted for in terms of a modification of the conformation of the antigen. Conformational changes are most probably also the cause for the observations that form the basis of this report; namely, the release of ferrihaem from sperm-whale metmyoglobin by antibodies to apomyoglobin (i.e. the haem-free protein). An account of this phenomenon has been published previously [Crumpton, 1966].

Comparison of Metmyoglobin and Apomyoglobin

A molecule of sperm-whale myoglobin is composed of one polypeptide chain containing 153 amino acid residues; no cyst(e)ine is present [Edmundson, 1965]. X-ray crystallographic analysis [Kendrew et al., 1961] of the oxidised protein (metmyoglobin) showed that the amino acid residues are arranged in helical segments which are separated by non-helical regions and that the polypeptide chain is folded into a bag-like structure containing a non-polar pocket which is occupied by the haem group (Fig. 1).

It is generally accepted that the structure of metmyoglobin in solution resembles closely that of the crystalline protein [Stryer, 1968; Hugli and Gurd, 1970 (1), (2)]. Evidence has, however, been presented in support of the view that the conformation in solution is more motile than in the crystal [Stryer, 1968; Hugli and Gurd, 1970 (2)], and that dissolved proteins are in a state of dynamic equilibrium with localized unfolding constantly occurring [Schechter et al., 1968]. For example, difference Fourier studies of metmyoglobin and azide-metmyoglobin showed no detectable change in the conformation of the protein when the azide ion is bound at the sixth coordination position of the haem [Stryer et al., 1964]. However, if the conformation of the protein was fixed, azide would not be bound because access to the haem is blocked by neigh-

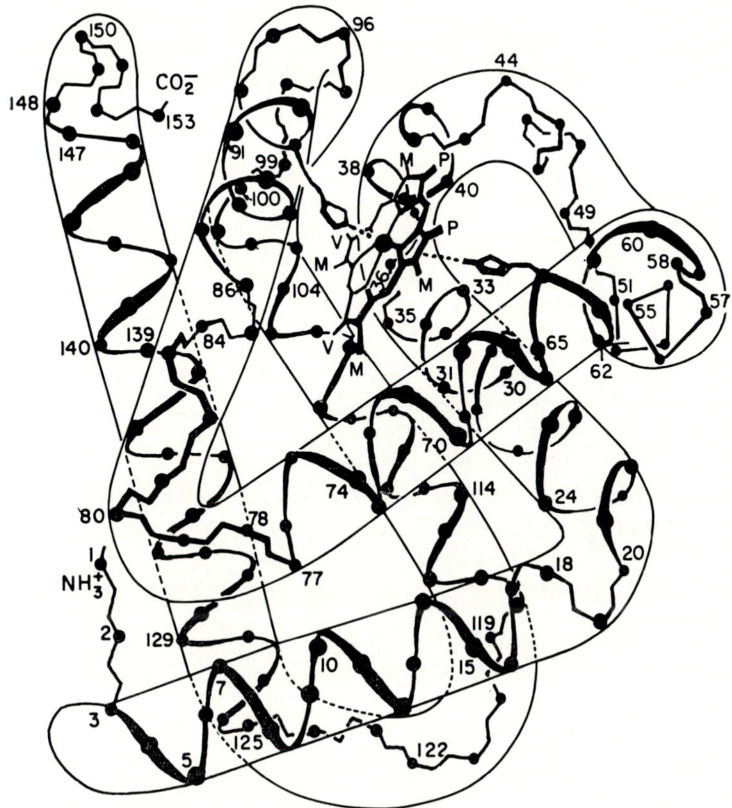

Fig. 1. A two-dimensional representation of the conformation of a molecule of crystalline sperm-whale metmyoglobin [DICKERSON, 1964]. The amino acid residues are numbered from the N- to the C-termini

bouring side-chains. As a result, binding of azide requires a transitory conformation in which the site is more exposed to solvent. The binding of azide ion is, moreover, 21-fold faster in solution than in the crystal indicating that localized unfolding occurs much more readily when the protein is dissolved.

The haem group of myoglobin is noncovalently bound to the protein moiety and may be removed by treatment with acid-acetone in the cold to give the haem-free protein (apomyoglobin). A comparison of the physical properties of apomyoglobin with those of metmyoglobin (Table) indicates that removal of the haem is associated with a small decrease in helical content and an over-all swelling or slight increase in asymmetry of the protein moiety. As shown in Fig. 2 the immunological reactivity of apomyoglobin also differed from that of metmyoglobin. Similar results have been obtained using a variety of antisera to metmyoglobin; some antisera, however, gave less precipitate with metmyoglobin than with apomyoglobin [see Fig. 1; Crumpton and Wilkinson, 1965] whereas other antisera failed to distinguish the two proteins [see Fig. 4; Crumpton, 1966]. The different immunological reactivities of apomyoglobin and metmyoglobin are undoubtedly related to the different conformations of the proteins. This interpretation is consistent with other observations which show that subtle conformational changes can cause alterations in immunological reactivity [e.g. deoxy- and oxy-haemoglobin; Reichlin et al., 1964]. Furthermore, the results of these experiments indicate that immunological reactivity may frequently be used as a sensitive measure of conformational integrity.

Table. *Physical properties of sperm-whale metmyoglobin and apomyoglobin*

	Met-myoglobin	Apo-myoglobin	Reference
α-Helix (%)*	56—64	42—49	Breslow et al. (1965)
α-Helix (%)+	71	51—60	Harrison and Blout (1965)
$s^{\circ}_{20,w}$ ($\times 10^{-13}$ sec)	2.08	1.92	Crumpton and Polson (1965)
$D^{\circ}_{20,w}$ ($\times 10^{-7}$ cm^2/sec)	10.82	9.90	Crumpton and Polson (1965)
Intrinsic viscosity (ml/g)	3.33	4.16	Crumpton and Polson (1965)
Frictional ratio	1.14	1.23	Crumpton and Polson (1965)

* By optical rotatory dispersion and circular dichroism
\+ By optical rotatory dispersion

The dissociation at acid pH of myoglobin into apomyoglobin and haem is reversible. Fig. 3 shows that the addition of an equimolar amount of haematin to apomyoglobin at pH 7.0 results in an almost complete reformation of metmyoglobin as judged from the intensity of the Soret absorption band at 410 nm. In this

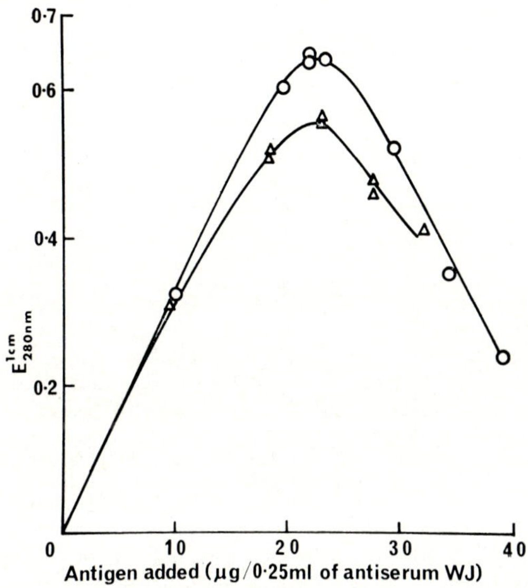

Fig. 2. Precipitin curves of sperm-whale metmyoglobin (△) and of apomyoglobin (○) with an anti-metmyoglobin serum. Increasing amounts of antigen in 0.25 ml of 0.9 % NaCl were incubated for 1 h at 37 °C and at 2 °C overnight with 0.25 ml of twofold-diluted antiserum WJ. The washed precipitates were dissolved in 0.6 ml of 0.1 M-NaOH and the extinction at 280 nm of the solution measured

particular experiment the end-point of the titration corresponded to 1.10 mole of haem/mole of apomyoglobin and to a 95 % yield of metmyoglobin; a more exact stochiometric relationship has been achieved by using borate buffer, pH 9.2, as the solvent [Breslow et al., 1965]. The reformation of metmyoglobin is associated with an almost complete recovery of the immunological [Reichlin et al.,

1963] and optical rotatory dispersion properties [BRESLOW et al., 1965; HARRISON and BLOUT, 1965] of the native protein. These results demonstrate that the conformational changes associated with the removal of haem are truly reversible and also emphasize that the conformation of metmyoglobin depends at least in part on the interaction of the protein with haem.

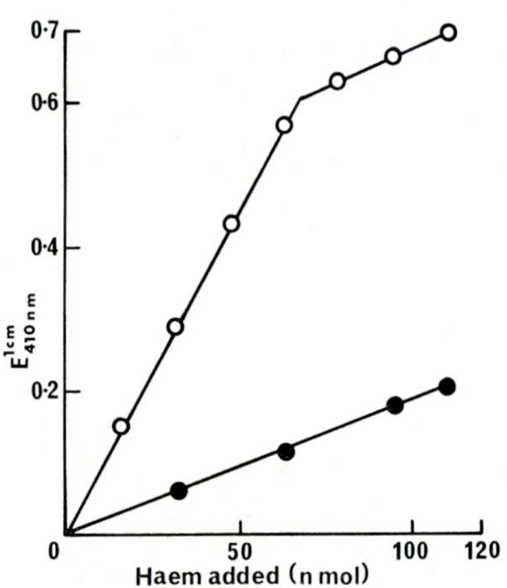

Fig. 3. Reformation of metmyoglobin. Increasing amounts of a solution of haematin (1.58 μmol/ml) were incubated for 22 h at 21 °C with either 61.3 nmol of apomyoglobin (○) in 1.07 ml of 0.1 M Na phosphate buffer, pH 7.0 or solvent alone (●). The solutions were diluted 15-fold prior to measuring the extinction at 410 nm

Ferrihaem in Antigen-Antibody Precipitates

As expected, antisera to metmyoglobin gave reddish-brown precipitates with metmyoglobin and white precipitates with apomyoglobin. The absorption spectrum of a solution of the metmyoglobin-antibody precipitate possessed a band with a maximum absorption at about 410 nm (the Soret band) characteristic of haem, whereas a solution of the apomyoglobin-antibody precipitate gave a negligible absorption at this wavelength (Fig. 4).

Antisera to apomyoglobin also precipitate both metmyoglobin and apomyoglobin [Fig. 5 and see Fig. 5; CRUMPTON, 1966] but, in contrast to the anti-metmyoglobin sera, the metmyoglobin-antibody precipitate was white and the spectrum of neither precipitate absorbed in the region of the Soret band (Fig. 6). All antisera collected from 12 rabbits at various times after immunization with apomyoglobin in complete Freund's adjuvant behaved in an identical manner. It is concluded that the precipitates formed by metmyoglobin with anti-apomyoglobin sera do not contain any ferrihaem.

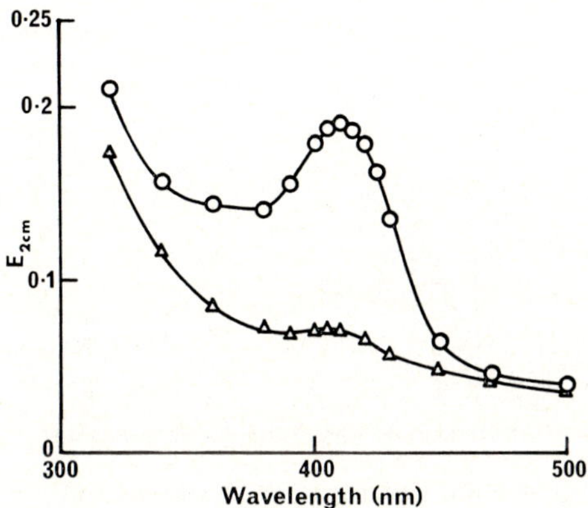

Fig. 4. Absorption spectra of precipitates formed by an anti-metmyoglobin serum. The precipitates produced by 0.5 ml of antiserum WC with 46 μg of metmyoglobin (○) and of apomyoglobin (△) were washed and dissolved in 1.2 ml of 0.1 M-NaOH 30 min prior to measuring the spectra

Fig. 5 also shows that myoglobin which had been treated with citraconic anhydride [Dixon and Perham, 1968] still precipitates with antisera to apomyoglobin, whereas no precipitate or negligible amounts only were formed with anti-metmyoglobin sera. Substitution of the amino groups of metmyoglobin by citraconyl groups caused a loss of the haem and an almost complete unfolding of the polypeptide chain as judged from the results of gel filtration, ultracentrifugal analyses and optical rotatory dispersion studies [M. J. Crumpton, unpublished observations]. Consequently, anti-apomyoglobin sera differed also from anti-metmyoglobin sera in possessing antibodies to the unfolded (denatured) protein. Furthermore, at least some of the antibodies to the unfolded protein combine with the native protein, since anti-apomyoglobin serum that had been adsorbed with citraconylated-apomyoglobin formed less precipitate with both apomyoglobin and metmyoglobin compared with the unadsorbed serum.

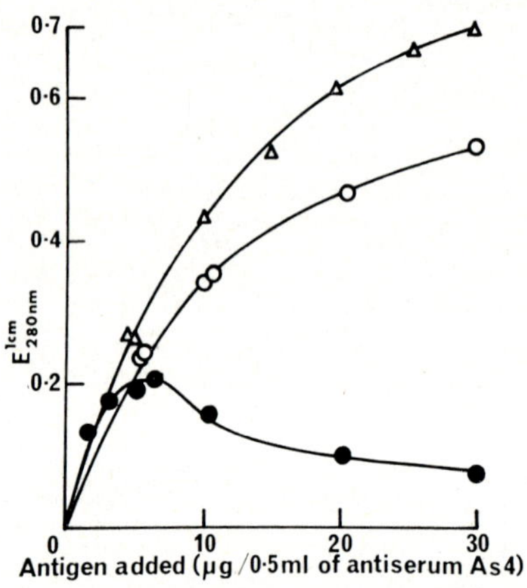

Fig. 5. Precipitin curves of sperm-whale metmyoglobin (○) and of apomyoglobin (△) with antiserum As4 to apomyoglobin. The amounts of precipitate formed by adding increasing amounts of citraconylated-apomyoglobin (●) to 0.5 ml of antiserum As4 are also shown. The washed precipitates were dissolved in 0.7 ml of 0.1 M-NaOH

Release of Ferrihaem from Metmyoglobin

The above results (Fig. 6) suggest that ferrihaem was released during the interaction of metmyoglobin with the antibodies to apomyoglobin. If this interpretation is correct then the 410 nm-absorption of metmyoglobin should be reduced by the apomyoglobin antibodies, as monomeric ferrihaem gives a sharp absorption band at 398 nm with ε of 1.22×10^5 [Inada and Shibata, 1962] whereas sperm-whale metmyoglobin absorbs maximally at 410 nm with ε of 1.66×10^5. Release of ferrihaem can, however, be followed spectrophotometrically only in the period prior to the precipitation of antigen-antibody complexes or by using monovalent antibody fragments, since precipitation increases the absorption due to light-scattering. Measurement of light scattering at 320 nm at various times after mixing 25 μg of metmyoglobin with an anti-apomyoglobin serum (0.8 ml of As3) indicated that precipitation did not occur until after 15 min at 22 °C. The absorption at 410 nm due to metmyoglobin, however, decreased by 85% within 1 min of adding

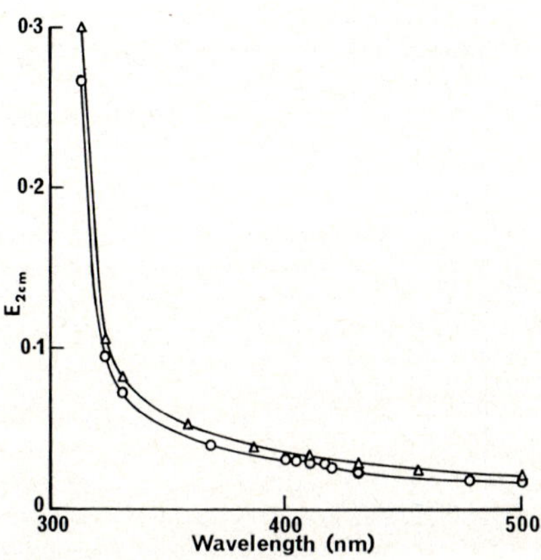

Fig. 6. Absorption spectra of precipitates formed by 60 μg of sperm-whale metmyoglobin (○) and of apomyoglobin (△) with 1.0 ml of antiserum As3 to apomyoglobin. The washed precipitates were dissolved in 0.9 ml of 0.1 M-NaOH 30 min prior to measuring the spectra

the apomyoglobin antibodies. In view of this very large decrease in absorption, it is concluded that the apomyoglobin antibodies caused a displacement of the ferrihaem from metmyoglobin.

As shown in Fig. 7 the monovalent Fab-fragments of the apomyoglobin antibodies also decreased the 410 nm-absorption of metmyoglobin, whereas no significant decreases were detected with either the Fab-fragments of normal rabbit immunoglobulin G or of purified metmyoglobin antibodies. The apomyoglobin-antibody fragments caused, however, a very much slower reduction in absorption (36% after 23 h; Fig. 7) than the whole antibodies. In this case, due to the smaller decrease in absorption, it is not certain that ferrihaem was actually released, although it seems reasonable to assume that the effects of the whole antibodies and their monovalent fragments are mediated by very similar mechanisms.

The reason for the differences between the effects of the anti-apomyoglobin serum and the preparation of Fab-fragments is not known, but the rate of decrease in absorption may be influenced by a variety of factors. For example, antibody activity may have

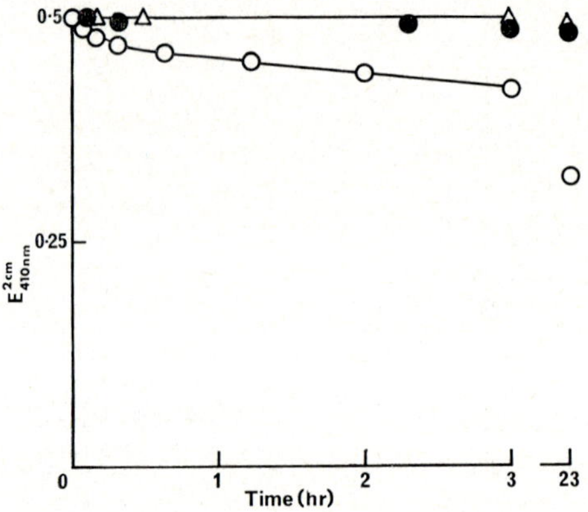

Fig. 7. Rate of decrease in 410 nm-absorption of 27 μg of metmyoglobin incubated at 22 °C with the Fab-fragments (equivalent to 0.8 ml of anti-serum As3) of the immunoglobulin G fraction of an anti-apomyoglobin serum (○). Controls contained equivalent amounts of the Fab-fragments of normal rabbit immunoglobulin G (●) or of purified metmyoglobin antibodies (△)

been lost during the isolation of the Fab-fragments. Also, in the presence of anti-apomyoglobin serum the decrease in absorption at 410 nm of free ferrihaem and possibly the release of ferrihaem may have been promoted by serum components, such as albumin and α- and β-globulins, that bind ferrihaem [Shinowara and Walters, 1963; Bunn and Jandl, 1968]. No promotion would, however, have occurred with the Fab-fragments due to the removal of these serum components during the preparation of the fragments.

Discussion

It has already been pointed out that the conformation of the protein moiety of metmyoglobin depends on the presence of ferrihaem and that removal of the ferrihaem causes a slight unfolding of the protein. As a result, it seems likely that the release of ferrihaem from metmyoglobin by the apomyoglobin antibodies is associated with an alteration in the conformation of the protein. In this case, metmyoglobin should not be reformed by adding haematin to the precipitate produced by metmyoglobin with an anti-apomyoglobin serum. Indeed, no binding of ferrihaem was detected on the addition of a two-fold molar excess of haematin to this precipitate, whereas complete reformation of metmyoglobin was achieved with the precipitate formed by apomyoglobin with an anti-metmyoglobin serum [Crumpton, 1966]. It was concluded from these results that the conformation of the protein moiety bound by apomyoglobin antibodies differs from that of native metmyoglobin. Similar conclusions have been invoked in order to explain the activation of defective enzymes and the enhancement of enzyme activity by antibody, and it seems likely that these effects and the release of ferrihaem from metmyoglobin are expressions of the same general phenomenon. Thus, enhancement of enzyme activity by antibody amounts to an induction of a higher affinity for substrate, resembles that of an allosteric activator and can be readily explained in terms of a conformational change of the enzyme. Furthermore, it would be very difficult if not impossible to explain the above phenomena in terms of rigid molecules whose conformations remain unaltered during combination with antibody.

The remodelling of the conformation of metmyoglobin by apomyoglobin antibodies may occur in three somewhat different but not mutually exclusive ways. First, the conformational change may be specifically induced by the antibody in that the antigenic determinants are forced to adopt a shape complementary to that of the antibody combining site. This explanation does not incorporate any pre-existing conformational equilibrium but presupposes that all conformational changes occur subsequent to the binding of the antibody. Although this model is consistent with the experimental data it is difficult to conceive of a specific antibody binding an antigenic determinant whose shape is not complementary to the antibody combining site unless interaction with a small portion of the determinant is sufficient to give a stable complex.

Second, although it is generally accepted that the ferrihaem of metmyoglobin is firmly bound to the protein moiety, dissociation of metmyoglobin is finite even at neutral pH and at room temperature [Banerjee, 1962]. Consequently, it is possible that the apomyoglobin antibodies react preferentially with the small amount of apomyoglobin that is in equilibrium with metmyoglobin and thereby displace the equilibrium in favour of dissociation. It was argued previously [Crumpton, 1966] that this mechanism is unlikely because the rate of decrease in the 410 nm-absorption of metmyoglobin by the Fab-fragments of apomyoglobin antibodies was very much faster than could be accounted for by the simple dissociation of metmyoglobin. This argument now, however, appears much less convincing due to two reasons. Firstly, the equilibrium constant for the dissociation of metmyoglobin may be higher (10^{-13} M) than that used previously (10^{-15} M at 25 °C and pH 7 [Banerjee, 1962] if, as seems likely, the free ferrihaem is initially monomeric [Inada and Shibata, 1962]. Secondly, although a decrease in 410 nm-absorption was previously equated with the release of ferrihaem, it now appears that this interpretation may not be justified for small decreases which may reflect a conformational change in the vicinity of the haem binding-site only. On the other hand, the above objections do not apply to the anti-apomyoglobin serum which apparently released ferrihaem very much faster than would have occurred if the antibodies had reacted with the apomyoglobin in equilibrium with metmyoglobin. Furthermore, although the promotion of dissociation of metmyoglobin by apomyoglobin antibodies represents a possible explanation for the observed release of ferrihaem, this model is obviously not applicable to all cases of antibody-induced changes in protein antigens.

The third explanation depends on the concept that the conformation of a protein is motile or flexible and undergoes small, rapid and reversible fluctuations. In this case, recognition and combination of a localized, transient shape with antibody would result in the freezing of this conformation. Furthermore, the process of stabilization of particular localized conformations by their specific antibodies would probably promote more extensive conformational changes elsewhere in the molecule. This explanation permits anti-metmyoglobin and anti-apomyoglobin sera to comprise populations of antibodies with different specificities that bind somewhat different conformations for the same portion of the metmyoglobin molecule. As a result, metmyoglobin bound to metmyoglobin antibodies can possess a different conformation from that of metmyoglobin bound to apomyoglobin antibodies. Direct evidence for the proposed variation in conformation, such as differences in the optical rotatory dispersion parameters of the soluble antigen-antibody complexes has not been obtained. However, the lack of reformation of metmyoglobin on the addition of haematin to apomyoglobin bound to apomyoglobin antibodies, referred to previously, is consistent with this proposal. As the release of ferrihaem from metmyoglobin is associated with an unfolding of the protein, the above explanation also predicts that anti-apomyoglobin sera would contain antibodies which select the more unfolded conformational states of the protein. This prediction appears to be verified by the observations that anti-apomyoglobin sera possess antibodies which precipitate apomyoglobin unfolded by treatment with citraconic anhydride, and that these antibodies also combine with metmyoglobin.

If this explanation is correct then combination of protein antigens with antibodies should impose a constraint upon the flexibility of the antigen molecules. Evidence in support of this view is provided by the decreased susceptibility of some enzyme-antibody complexes to heat and to variation in pH (e.g. penicillinase [ZYK and CITRI, 1968 (1), (2)]), and by the enhancement of the activities of certain enzymes by homologous antisera. For example, the increased affinity of ribonuclease S-protein for the S-peptide in the presence of antibodies to the native enzyme [CINADER et al., 1971] is most probably due to the stabilizing effect of the antibodies on the conformation of S-protein that binds the peptide most strongly.

Conclusions

It is concluded that antisera to apomyoglobin contain antibodies which induce alterations in the conformation of metmyoglobin leading to a release of the ferrihaem. The induction of the conformational changes is most probably mediated by the 'motility' [SCHECHTER et al., 1968] of the protein moiety which permits the selection by the apomyoglobin antibodies of a somewhat unfolded conformation that does not bind ferrihaem.

References

ASHMAN, R. F., METZGER, H.: Immunochemistry 8, 643 (1971).

— KAPLAN, A. P., METZGER, H.: Immunochemistry 8, 627 (1971).

Banerjee, R.: Biochim. biophys. Acta (Amst.) **64**, 368 (1962).
Breslow, E., Beychok, S., Hardman, K. D., Gurd, F. R. N.: J. biol. Chem. **240**, 304 (1965).
Bunn, H. F., Jandl, J. H.: J. biol. Chem. **243**, 465 (1968).
Chong, D. V., Goldner, M.: Canad. J. Microbiol. **16**, 647 (1970).
Cinader, B., Suzuki, T., Pelichová, H.: J. Immunol. **106**, 1381 (1971).
Crumpton, M. J.: Biochem. J. **100**, 223 (1966).
— Polson, A.: J. molec. Biol. **11**, 722 (1965).
— Wilkinson, J. M.: Biochem. J. **94**, 545 (1965).
Dickerson, R. E.: In: The Proteins, 2nd ed. 2, p. 634. (Neurath, H., Ed.). New York: Academic Press Inc. 1964.
Dixon, H. B. F., Perham, R. N.: Biochem. J. **109**, 312 (1968).
Edmundson, A. B.: Nature (Lond.) **205**, 883 (1965).
Feinstein, R. N., Jaroslow, B. N., Howard, J. B., Faulhaber, J. T.: J. Immunol. **106**, 1316 (1971).
Harrison, S. C., Blout, E. R.: J. biol. Chem. **240**, 299 (1965).
Hugli, T. E., Gurd, F. R. N.: (1) J. biol. Chem. **245**, 1930 (1970).
— — (2) J. biol. Chem. **245**, 1939 (1970).
Inada, Y., Shibata, K.: Biochem. biophys. Res. Commun. **9**, 323 (1962).
Kendrew, J. C., Watson, H. C., Strandberg, B. E., Dickerson, R. E., Phillips, D. C., Shore, V. C.: Nature (Lond.) **190**, 666 (1961).
Lehmann, F.-G.: Biochim. biophys. Acta (Amst.) **235**, 259 (1971).
Melchers, F., Messer, W.: Europ. J. Biochem. **17**, 267 (1970).
Metzger, H.: Ann. Rev. Biochem. **39**, 889 (1970).
Michaeli, D., Pinto, J. D., Benjamini, E.: Nature (Lond.) **213**, 77 (1967).
Pollock, M. R., Fleming, J., Petrie, S.: In: Antibodies to biologically active molecules, p. 139. (Cinader, B., Ed.). Oxford: Pergamon Press Ltd. 1967.
Reichlin, M., Hay, M., Levine, L.: Biochemistry **2**, 971 (1963).
— Bucci, E., Antonini, E., Wyman, J., Rossi-Fanelli, A.: J. molec. Biol. **9**, 785 (1964).
Rotman, M. B., Celada, F.: Proc. nat. Acad. Sci. (Wash.) **60**, 660 (1968).
Schechter, A. N., Morávek, L., Anfinsen, C. B.: Proc. nat. Acad. Sci. (Wash.) **61**, 1478 (1968).
Shinowara, G. Y., Walters, M. I.: Amer. J. clin. Path. **40**, 113 (1963).
Stryer, L.: Ann. Rev. Biochem. **37**, 25 (1968).
— Kendrew, J. C., Watson, H. C.: J. molec. Biol. **8**, 96 (1964).
Zyk, N., Citri, N.: (1) Biochim. biophys. Acta (Amst.) **159**, 317 (1968).
— — (2) Biochim. biophys. Acta (Amst.) **159**, 327 (1968).

Stabilization of Conformations of E. coli ß-Galactosidase by Specific Antibodies. Restrictions in Antigenic Determinants and Antibodies

F. Melchers and G. Köhler

Basel Institute for Immunology, Basel, Switzerland

and

W. Messer

Max-Planck-Institut für Molekulare Genetik, Berlin, Germany

With 7 Figures

Introduction

Enzymes as antigens are used for studies of protein-protein interactions as well as for immunological studies [1—8], since their catalytic properties allow for a specific and sensitive detection. The enzyme β-galactosidase from Escherichia coli in particular is useful for a number of reasons:

1. The enzyme protein is well immunogenic in different species of experimental animals. Most enzymes are inactivated by specific antibodies (see [5]). Antibodies against E. coli β-galactosidase, however, precipitate, but do not inactivate the enzyme [9]. Specific antibodies directed against the enzyme can therefore be detected by the catalytic activity of antibody-bound enzyme.

2. The enzyme is biochemically well characterized and easy to purify. A large number of different substrates are available (see [10, 11]). As little as one enzyme molecule can be detected [12].

3. The gene for β-galactosidase in Escherichia coli is one of the best-characterized structural genes. A large number of mutants have been mapped. They allow for easy localization of any new mutations within the gene (see [13]).

4. Nonsense, missense and deletion mutants exist, which produce inactive enzyme proteins in the range from relatively small polypeptides to molecules differing from the normal enzyme by only one amino acid [14, 15]. Such mutant proteins have been used to distinguish three classes of antigenic determinants on β-galactosidase: those of relatively short polypeptide chains, those of relatively long polypeptide chains, and those due to the polymeric structure of the enzyme [16].

Interactions between ß-Galactosidase and its Specific Antibodies

A high degree of heterogeneity must be expected in the interactions between the enzyme and its specific antibodies. Many antigenic determinants on the surface of the enzyme molecule interact with many structurally different specific antibodies. Different sequential as well as conformational determinants may exist on the enzyme molecule (for reviews see [17—19]). Within one animal any given antigenic determinant may or may not be immunogenic, e.g. may or may not evoke the response of specific antibody production. If one animal produces antibodies against a given determinant it may produce several antibodies with different structures, but with the same specificity. Furthermore, even in an inbred strain of mice, populations of antibodies will be produced which differ from animal to animal, but bind to the same determinant.

This heterogeneity both in antigenic determinants on an enzyme molecule and in specific antibodies produced against them complicates studies on enzyme-antibody interactions as a prototype of protein-protein interactions. It is therefore desirable to reduce this heterogeneity of both the antigenic determinants and of the specific antibodies.

We have attempted to reduce the heterogeneity of antigenic determinants by distinguishing out of all a restricted number of determinants to which specific antibodies bind which alter certain functions of the enzyme. A number of sites bind antibodies which stabilize the enzyme against heat denaturation. Other groups of sites are characterized by the binding of different antibodies which will activate a number of structurally different, enzymically inactive mutant enzymes to enzymic activity.

In all of these cases specific antibodies appear to stabilize conformations of the enzyme molecule. This stabilization of a conformation is measured as an increased stability of the active sites.

We have also attempted to reduce the heterogeneity of specific antibodies against a given determinant of the enzyme molecule. Clones of antibody-forming cells may be grown in sublethally-irradiated syngeneic mice, when limiting amounts of lymphoid cells from an enzyme-primed animal are injected in the presence of the enzyme. Such clones of cells produce populations of antibodies homogeneous both in their specificity for one site on the enzyme and in their polypeptide structure.

Results

Activation of Enzymically Inactive Mutant Proteins of β-Galactosidase to Enzyme Activity by Specific Antibodies

Antibodies can cause an increase of enzymic activity with a number of enzymes [3, 5]. The most drastic increase in activity has been found with a number of inactive mutant proteins of E. coli β-galactosidase [20, 21]. These defective proteins are the products of missense mutations at four sites within the structural gene for β-galactosidase ([22], genetic groups I, II, III, see Fig. 1). In the native form they are assembled in a four-subunit structure with a molecular weight for the subunits indistinguishable from that of the wild type enzyme [23].

Mutant β-galactosidase of genetic group I mutants (lac^{-}_{aba}-13, 71, 645, and 779) have been used to study the activation reaction with specific antibodies raised against the wild type enzyme.

Enzyme activity of mutant β-galactosidase in the absence of specific antibodies. The activity of mutant enzyme 645 in the absence of specific antibodies is around 1/2000 of the specific enzyme activity of the wild type enzyme between 5 and 35 °C. The mutant enzyme changes its activity with temperature as does the wild type enzyme, with an activation energy for the enzymic activity of 16.5 kcal/mole. This suggests that the active sites of wild type enzyme and of the residual activity in mutant enzyme may be similar.

Rate of activation of mutant enzymes by specific antibodies. Upon addition of specific antibodies enzymic activity in the population of mutant enzyme increases with time. The activation is a slow process: at 35 °C half-activation is reached in approximately 12 min.

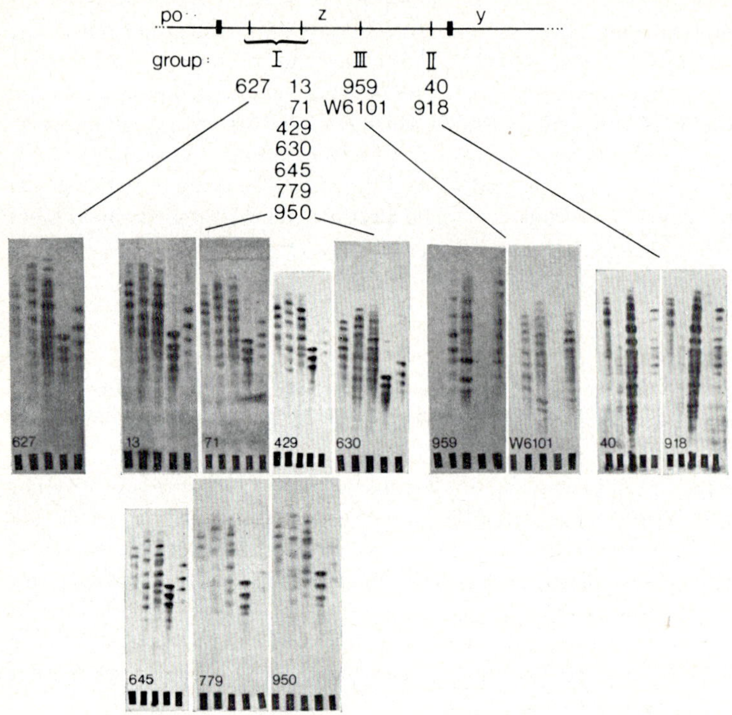

Fig. 1. Genetic map of lac^{-}_{aba}-mutants and isoelectric focusing spectra obtained with five antisera raised against the wild type enzyme in different rabbits, developed with the different lac^{-}_{aba}-mutant enzymes and the AMEF-producing mutant enzyme of ROTMAN and CELADA [20], W6101. Electrophoresis in 5% polyacrylamidegels was for 24 h at 4 °C under conditions described by AWDEH et al. [35]. Extracts of lac^{-}_{aba}-mutants (0.2 g bacteria/ml of buffer A containing 0.14 M NaCl, 10^{-3} M $MgCl_2$, 10^{-2} M potassium-phosphate, 10^{-5} M dithioerithrol, pH 7.0) were layered on top of the gels and incubated for 1 h at 37 °C in a moist chamber. Excess extract was washed off with buffer A. Then the gels were covered with a solution of 0.5% agarose containing 0.5 mg/ml 5-Bromo-4-chloro-3-indolyl-β-D-galactopyranoside, 1 mg/ml streptomycin, and 10% dimethylsulphoxide, kept at 45 °C. After the agarose layer had solidified the gels were incubated overnight at 37 °C in a moist chamber. Bands of activating antibodies are seen by the blue-green color of the indigo stain developed at the site of enzymic reaction

The rate of activation of mutant enzyme is measured as the slope of the curve obtained when the logarithm of the difference of activity at maximal activation (A_{max}) and of activity reached at a given time (A_t) is plotted against the time of exposure of mutant enzyme to specific antibodies. Different antisera activate one mutant enzyme population with rates, which differ at most 30% from each other (Fig. 2). We do not consider these differences in rates to be significant, hence we conclude that the rate of activation at a given temperature is independent of the four antisera used. (In the present experiments only antisera without demonstrable activation-inhibitory antibodies have been used.) Since log (A_{max}-A_t) is proportional to -t (Fig. 2) the rate-limiting step of the activation appears to be a first order reaction. CELADA, STROM and BODLUND [24] have reached the same conclusion for their mutant enzyme.

Dependence of rate of activation on concentrations of mutant enzyme and of specific antibodies. Concentrations of mutant enzyme and of specific antibodies were varied in the reaction mixture from a 50 fold excess of activating antibodies over mutant enzyme to a 10 fold excess of mutant enzyme over activating antibodies. The concentration of mutant enzyme was varied between 40 mg/ml and 0.04 mg/ml. Within these concentration limits the rates of activation were not significantly different at a given temperature. This indicates that the rate-limiting step in the activation reaction is a monomolecular reaction within the mutant enzyme molecules.

Dependence of rate of activation on temperature. Rates of activation were measured at seven different temperatures between 15 °C and 45 °C with four different antisera. Fig. 3 shows the Arrhenius plots. The regression of the logarithm of rate constants with the reciprocal absolute temperature (1/T) is linear over the temperature range between 15 and 35 °C. Above 40 °C the rate of reaction drops, due possibly to incipient denaturation of the enzyme molecules. An activation energy of 23.2 kcal/mole can be calculated for the activation reaction of the mutant enzymes for temperatures between 15 and 35 °C. This value is significantly higher than the maximum allowed for antigen-antibody binding reactions (14 kcal/mole [25]) and lower than the usual values for denaturation (e.g. 130 kcal/mole for ovalbumin [26]).

Activation of different mutant enzymes within one genetic group. Mutant enzymes 13, 71, 645, and 779 all show rates of activation at a given temperature, which are not significantly different from each other with two different antisera. We conclude that the activation reaction by specific antibodies is the same for all the mutant

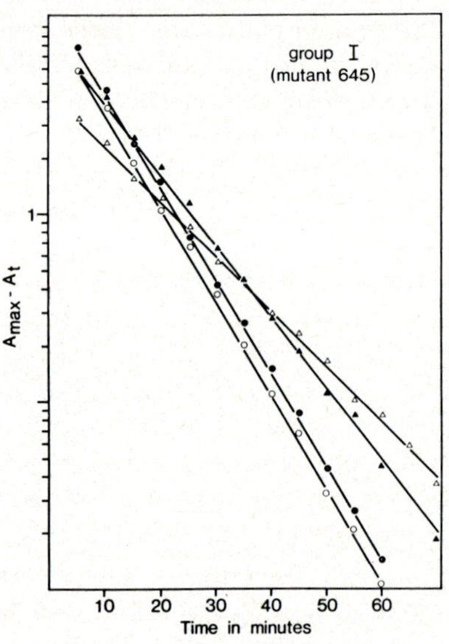

Fig. 2. Activation of mutant enzyme 645 as a function of time by different rabbit antisera raised against wild type β-galactosidase. (Antisera: Rabbit No. 23 △, No. 79 ▲, No. 97 ○, No. 99 ●). Mutant enzyme activation by antibody was done and enzyme activity determined as described previously [21]: 0.05 ml of mutant extract were mixed with 0.05 ml of antiserum and incubated for different periods of time at 30 °C. Then 0.25 ml p-nitrophenyl-β-D-galactoside, 3.3 mg/ml in 0.1 M phosphate-citrate buffer pH 6.3, was added and the mixture incubated for 5 min at 30 °C. The reaction was stopped by adding 3 ml of cold 0.1 M sodium-carbonate-buffer, pH 9.4. Absorption at 430 mμ was measured. All determinations of enzyme activity for a given time of incubation were done in quadruplicate

enzymes tested. It strengthens our earlier conclusions [21] that one group of mutant enzymes is correlated to one group of antigenic sites (see also below). Two conformations within a mutant enzyme population have also been inferred from immunological experiments [27].

The role of specific antibodies. Mutant enzyme was exposed to different rabbit antisera raised against the wild type enzyme. The activity of the fully activated mutant enzyme was measured in dependence of the concentration of specific antibodies. The activation curves have a central section with a slope of 1 (Fig. 4), suggesting that one bound antibody molecule is capable of activating one

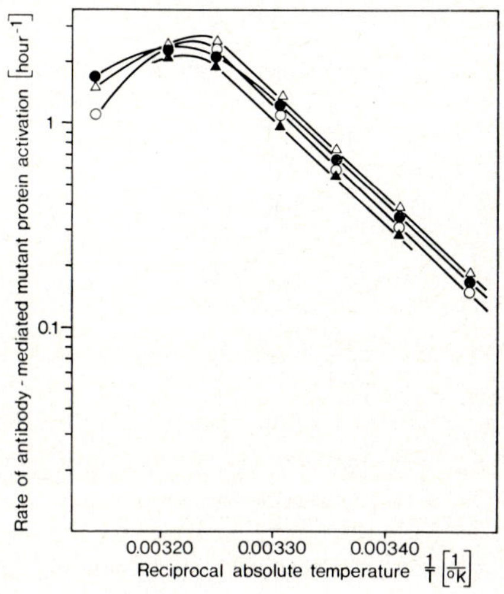

Fig. 3. Temperature-dependence of the rate of activation of mutant enzyme 645 with different antisera raised against the wild type enzyme. (Antisera: Rabbit No. 23 ▲, No. 97 △, No. 509 ○, No. 510 ●). Enzyme activities were determined at different temperatures as described in the legend of Fig. 2. The data are given in an Arrhenius plot

molecular unit of activity within the mutant enzyme. This has also been concluded by Rotman and Celada [20]. Rotman has also shown that one molecule of Fab-fragment of activating antibody is capable of activating the mutant enzyme [28]. Since it has been shown that one unit of activity, e.g. one active site, resides within one subunit of β-galactosidase [29] it is likely that one binding site of an antibody molecule activates one subunit within the mutant enzyme. We suggest therefore that the activation by antibodies is due to the stabilization of the active conformation of the subunit.

In antibody excess the mutant enzyme reaches an activity which is lower than that of the wild type enzyme (specific activity: 5×10^5 enzyme units/mg $\hat{=}$ 100%). Different antisera raised against the wild type enzyme activate the same mutant enzyme preparation to different levels of activity, expressed in percent values compared to the wild type enzyme in Fig. 4. This partial activation is thus

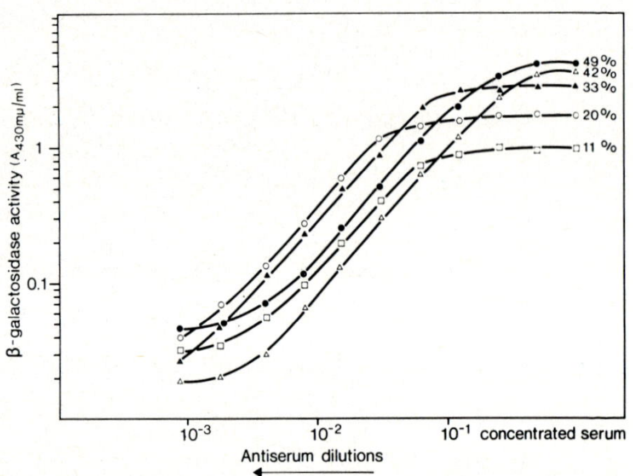

Fig. 4. Activation of mutant enzyme 645 by different concentrations of five rabbit antisera raised against the wild-type enzyme. (Antisera: Rabbit No. 23 □, No. 79 ▲, No. 97 △, No. 99 ●, No. 509 ○). Determinations of enzyme activity were done as described in the legend of Fig. 2. Incubation of mutant enzyme with antibody was for 45 min at 37 °C, followed by incubation with substrate for 30 min at 37 °C

due to the properties of the specific antibodies. At present we do not know whether it reflects the affinity of the antibodies, the presence of activation-inhibiting antibodies or other properties of antibodies.

Enhanced Stability against Heat Denaturation of Wild Type and Mutant ß-Galactosidase in the Presence of Specific Antibodies

Wild type β-galactosidase and group I and II mutant proteins exhibit an increased stability against heat denaturation when complexed with specific antibodies [30]. Group III mutant proteins, activated by specific antibodies to enzyme activity, do not show this increased stability against heat denaturation with certain selected antisera (Fig. 5). It suggests that stabilization against heat

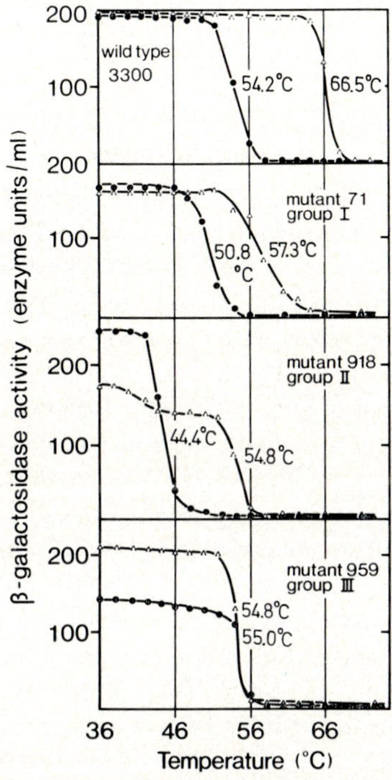

Fig. 5. Heat-denaturation of wild type and lac^{-}_{aba}-mutant β-galactosidase in the presence (— △ —) and in the absence (— ● —) of specific antibodies. Details of the experiment are described in [30]. Temperatures at which half of the enzyme activity is inactivated are indicated in the graphs

denaturation and activation of the mutant proteins to enzymic activity are effected by different populations binding to different antigenic determinants. It is interesting to note that group III mutant proteins seem to be examples of single mutations affecting two different functions of anti β-galactosidase antibodies and thus probably two different antigenic sites on the enzyme.

Other rabbit antisera stabilize the wild type and *all* mutant enzymes, e.g. also group III mutant enzymes, against heat denaturation. We conclude that some group III mutant enzyme-heat stabilizing antibodies are different from those stabilizing group I, II, or wild type enzyme. The findings illustrate again, that antibody production against a given determinant may vary from animal to animal.

The Activity of Single Active Sites in the Tetrameric Structure of ß-Galactosidase

The enzyme β-galactosidase of Escherichia coli is composed of four identical subunits (each of mol. wt. 135,000) and possesses four active sites (for a review see [10]). Hybrid enzyme molecules can be formed from mixtures of one of the antibody-activatable mutant enzymes (genetic group I) with the wild type enzyme in 8 M urea solutions [31], in which the enzyme molecules are believed to be disassembled and unfolded [32]. The formation of such hybrid enzyme molecules can be deduced from activity measurements of single hybrid enzyme molecules [29].

Although subunits are independent from one another in native enzyme preparations a cooperation between subunits exists during heat denaturation. In the absence of specific antibodies the heat-inactivated conformation of a subunit in hybrid enzyme molecules appears to force active subunits into an inactive form. In the presence of specific, heat-stabilizing antibodies wild type and mutant subunits in the hybrids behave as single, independent units of activity. The influence of neighbouring subunits, seen when the molecules are heat-inactivated without specific antibodies, is abolished by specific antibodies (Fig. 6). Thus heat-stabilizing antibodies stabilize the conformation of a subunit to be independent from neighbouring subunits. We do not know whether the total effect of heat-stabilization is due to this uncoupling.

The Heterogeneity of Antibody Molecules Directed against Different Determinants on ß-Galactosidase, which Bind Activating Antibodies

Earlier experiments suggested that antibody populations obtained from one rabbit activating group I and group II mutant proteins may be different [20]. We have developed a technique by

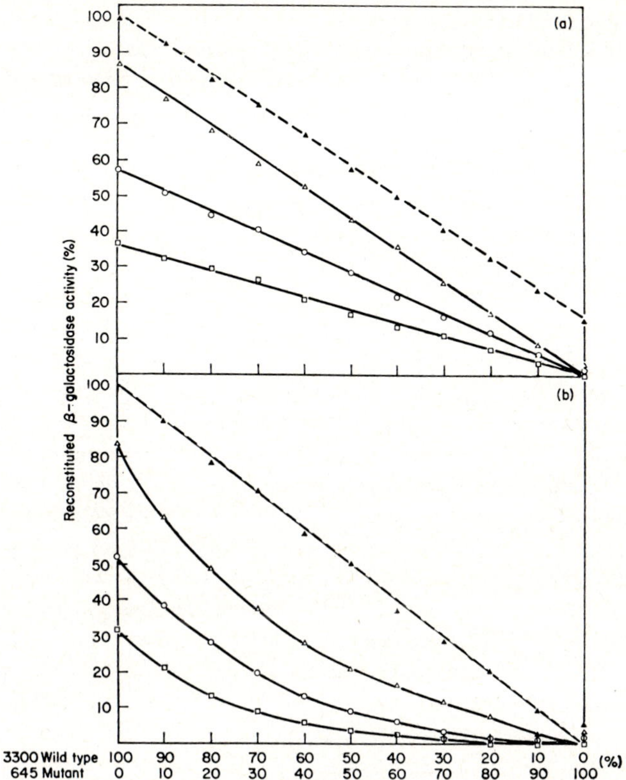

Fig. 6. a) Enzymic activity of hybrid β-galactosidase molecules in the presence of specific heat-stabilizing antibodies after heat inactivation at $T_m = 68$ °C. b) Enzymic activity of hybrid β-galactosidase in the absence of specific heat-stabilizing antibodies after heat inactivation at $T_m = 68$ °C. 0 min (-- ▲ -- ▲ --), 5 min (— △ — △ —), 10 min (— ○ — ○ —), and 20 min (-- □ -- □ --)

which isoelectric focusing spectra of antibodies with specificities against determinants on β-galactosidase can be detected in thin layers of polyacrylamide gels [33]. It is estimated that $\sim 10^{-2}$ μg ($\simeq 4 \times 10^{10}$ molecules) of 7 S antibody can be seen in one band on the gels. With this technique we have analyzed sera raised against the wild type enzyme in different rabbits for their capacity to activate the twelve antibody-activatable mutant enzymes (Fig. 1). In the serum of one animal different populations of specific antibodies activate the mutant enzymes. The twelve mutant enzymes can be arranged in groups according to the isoelectric focusing spectra of antibodies activating them. These groups correspond to the groups of mutants which have been defined by their position within the gene for β-galactosidase [22]. We conclude that three groups of antigenic determinants on the enzyme bind three populations of activating antibodies.

One population of antibodies activates mutant enzymes originating from mutations at two different sites within the gene (group I, Fig. 1). This demonstrates that the structure of an antigenic site may be altered by amino acid replacements some distance apart from each other in the polypeptide chain.

Different animals produce antibodies with different isoelectric focusing spectra. A restricted population of antibodies, e.g. the product of one to five clones of antibody-forming cells activate a group of mutant proteins. Some of the rabbits do not produce any antibodies activating group III mutant enzymes. This again demonstrates that only a restricted number of all determinants on the enzyme may be immunogenic in one given animal. The restricted heterogeneity of activating antibodies seen in some of the rabbit antisera may reflect a restricted heterogeneity within the structures of the antigenic sites to which activating antibodies can bind.

Homogeneous Activating Antibodies from Clones of Antibody-Forming Cells

When limiting numbers of lymphoid cells from mice, immunized once with antigen, are transferred into irradiated syngeneic mice together with the antigen, clones of cells forming antibodies with specificities for the antigen grow up [34]. The lymphoid cells of such host mice are believed to be inactivated by the irradiation procedure, leaving the body of the host mice as a passive environment for the growth and antibody production of the injected, antigen-primed lymphoid cells of the donor. We have assayed by the isoelectric focusing technique sera of host mice injected with β-galactosidase-primed lymphoid cells together with the enzyme, for the presence of antibodies activating the mutant enzymes (Fig. 7). The occurrence of activating antibodies in the host mice is depend-

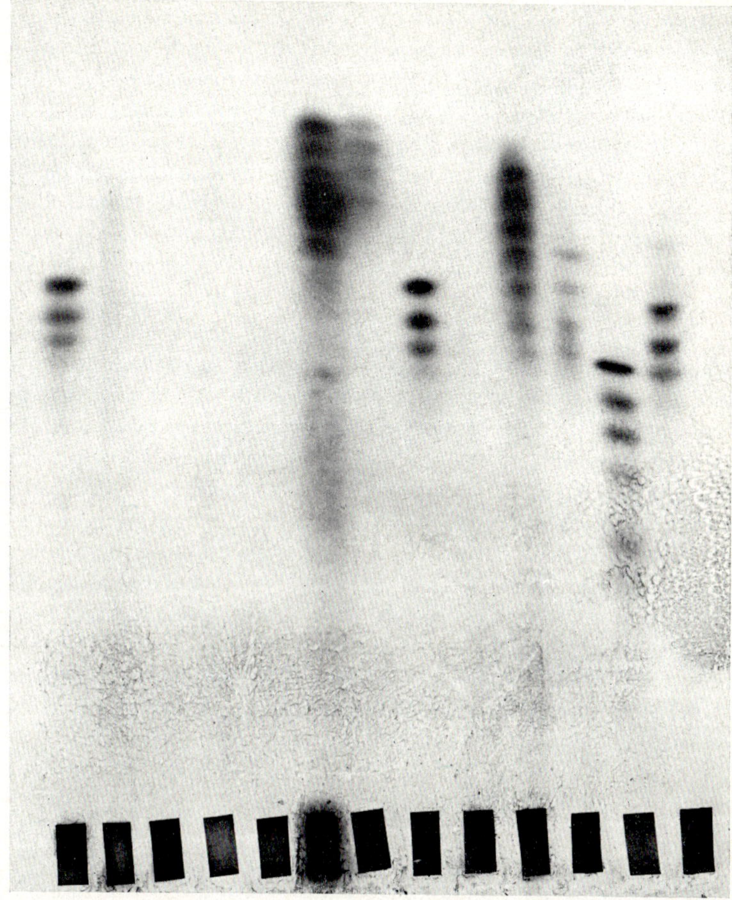

Fig. 7. Isoelectric focusing spectra of antibodies activating mutant enzyme 645 (group I) in sera of 13 X-ray irradiated host mice injected with 1×10^6 wild type enzyme-primed lymphoid cells together with 5 μg wild type β-galactosidase. Electrophoresis and development of the gels was done as described in the legend of Fig. 1. (Details of the cloning procedure (see also [34]) will be published elsewhere.) The product of a cell clone, an antibody with a given polypeptide structure, exhibits heterogeneity as detected by a set of different bands in the gels (see e.g. mice No. 1 and 8). This heterogeneity has been explained as single charge differences caused in the serum by loss of amide groups [35]

ent on the number of transferred cells. A clone of cells producing homogeneous antibody, detectable by its activation of group I mutant enzyme, appears in 7% of all host mice, when 2.5×10^5 lymphoid cells from a primed donor mouse are transferred. It is likely that one single cell is stimulated by antigen to proliferate and differentiate into a clone of cells producing homogeneous antibody. As more cells are transferred into a host, activating antibodies become more frequent, and at the same time also more heterogeneous in their isoelectric focusing spectra. Coincidence of two or more cells producing activating antibodies in the host increases.

Among the first cell clones producing activating antibodies which we have obtained in our laboratory, none activate group I *and* group II mutant enzymes simultaneously. However, some of the group I mutant enzyme-activating antibodies also activate group III mutant enzymes, while others activate only group I but not group III mutant enzymes. This indicates that the structures and sites on the enzyme of determinants, binding group I and group II activating antibodies, are completely different and apart from each other, while determinants binding group I and group III activating antibodies are in part structurally the same and thus partially overlapping in site on the enzyme. These results illustrate how a set of homogeneous antibodies may be used to determine structural similarities, and thus proximity of sites, of determinants on the surface of a macromolecule.

Summary

Antibodies against the enzyme β-galactosidase from Escherichia coli exert conformation-stabilizing effects upon the enzyme. A number of different functional tests, e.g. heat stabilization and mutant enzyme activation, indicate that different antigenic determinants, a restricted number out of all determinants on the enzyme molecule, are involved in the binding of different populations of specific antibodies. These antibodies show restricted heterogeneity in isoelectric focusing spectra. Homogeneous antibodies against a determinant are produced by clones of antibody-forming cells which can be obtained by injection of limiting numbers of immune competent lymphoid cells from an enzyme-primed donor mouse into a series of irradiated host mice.

The activation reaction of certain inactive mutant β-galactosidases to enzymic activity by wild type-specific antibodies has been studied in some detail. Measurements of the rates of activation of one group of mutant enzymes by specific antibodies suggest that the rate-limiting step of the activation reaction is a monomolecular reaction, with which a mutant enzyme population changes between two conformations. Activating antibodies bind to a wild type like conformation only, stabilize it, and thus change the equilibrium.

Acknowledgements. We thank S. Fazekas de St. Groth, N. Jerne, C. Spatz, C. Steinberg and T. Trautner for many stimulating discussions and for criticism; F. Celada and M. B. Rotman for a gift of their mutant strain W6101. The able technical assistance of D. Jablonski and I. Heueis is gratefully acknowledged.

References

1. Uriel, J.: Immunoelectrophoretic analysis of enzymes. In: Antibodies to biologically active molecules, p. 181. (Cinader, B., Ed.). Oxford, England: Pergamon Press 1967.
2. — Characterization of enzymes in specific immuneprecipitates. Ann. N. Y. Acad. Sci. **103**, 1956 (1963).
3. Pollock, M. R.: Penicillinase-antipenicillinase. Ann. N. Y. Acad. Sci. **103**, 989 (1963).
4. Suskind, S. R., Wickham, M. L., Carsiotis, M.: Antienzymes in immunogenic studies. Ann. N. Y. Acad. Sci. **103**, 1106 (1963).
5. Cinader, B.: Antibodies to enzymes — a discussion on the mechanisms of inhibition and activation. In: Antibodies to biologically active molecules, p. 85. Oxford: Pergamon Press 1967.
6. Avrameas, S., Bouteille, M.: Ultrastructural localisation of antibody by antigen labelled with peroxidase. Exp. Cell Res. **53**, 166 (1969).
7. Leduc, E. H., Scott, G. B., Avrameas, S.: Ultrastructural localization of intracellular immune globulins in plasma cells and lymphoblasts by enzyme-labelled antibodies. J. Histochem. Cytochem. **17**, 211 (1969).
8. Avrameas, S., Leduc, E. H.: Detection of simultaneous antibody synthesis in plasma cells and specialized lymphocytes in rabbit lymphnodes. J. exp. Med. **131**, 1137 (1970).
9. Cohn, M., Torriani, A. M.: Immunochemical studies with the β-galactosidase and structurally related protein of Escherischia coli. J. Immunol. **69**, 471 (1952).
10. Zabin, I., Fowler, A. V.: β-Galactosidase and thiogalactosidase transacetylase. In: The lactose operon, p. 27. (Beckwith, J. R., Zipser, D., Eds.). Cold Spr. Harb. Laboratory 1970.
11. Wallenfels, K., Weil, R.: β-galactosidase In: The enzymes, Vol. VII (Boyer, Ed.) (in press).
12. Rotman, B.: Measurement of activity of single molecules of β-D-galactosidase. Proc. nat. Acad. Sci. (Wash.) **47**, 1981 (1961).

13. Beckwith, J. R.: Lac: The genetic system. In: The lactose operon, p. 5. (Beckwith, J. R., Zipser, D., Eds.). Cold Spr. Harb. Laboratory 1970.
14. Newton, W. A., Beckwith, J. R., Zipser, D., Brenner, S.: Nonsense mutants and polarity in the lac operon of Escherichia coli. J. molec. Biol. **14**, 290 (1965).
15. Fowler, A. V., Zabin, I.: Colinearity of β-galactosidase with its gene by immunological detection of incomplete polypeptide chains. Science **154**, 1027 (1966).
16. — — β-galactosidase: Immunological studies of nonsense, missense and deletion mutants. J. molec. Biol. **33**, 35 (1968).
17. Sela, M.: Antigenicity. Some molecular aspects. Science **166**, 1365 (1969).
18. Crumpton, M. J.: The molecular basis of the serological specificity of proteins, with particular reference to sperm whale myoglobin. In: Antibodies against biologically active molecules, p. 61 (Cinader, B., Ed.). Oxford: Pergamon Press 1967.
19. Arnon, R.: Antibodies to enzymes. A tool in the study of antigenic specificity determinants. In: Current topics in microbiology and immunology, Vol. 54, p. 47. Berlin-Heidelberg-New York: Springer 1971.
20. Rotman, M. B., Celada, F.: Antibody-mediated activation of a defective β-D-galactosidase extracted from an Escherichia coli mutant. Proc. nat. Acad. Sci. (Wash.) **60**, 660 (1968).
21. Messer, W., Melchers, F.: The activation of mutant β-galactosidase by specific antibodies. In: The lactose operon, p. 305 (Beckwith, J. R., Zipser, D., Eds.). Cold Spr. Harb. Laboratory 1970.
22. — — Genetic analysis of mutants producing defective β-galactosidase which can be activated by specific antibodies. Molec. gen. Genet. **109**, 152 (1970).
23. Melchers, F., Messer, W.: The activation of mutant β-galactosidase by specific antibodies. Purification of eleven antibody-activatable mutant proteins and their subunits on sepharose immunosorbents. Determination of the molecular weights by sedimentation analysis and acrylamide gel electrophoresis. Europ. J. Biochem. **17**, 267 (1970).
24. Celada, F., Strom, R., Bodlund, K.: Antibody-mediated activation of a defective β-galactosidase (AMEF). Characteristics of binding and activation processes. In: The lactose operon, p. 891 (Beckwith, J. R., Zipser, D., Eds.). Cold Spr. Harb. Laboratory 1970.
25. Eisen, H. N., Karush, F.: Immune tolerance and an extracellular regulatory role for bivalent antibody. Nature (Lond.) **202**, 677 (1964).
26. Tanford, C.: In: Physical Chemistry of Macromolecules. New York and London: John Wiley and Sons 1961.
27. Celada, F., Ellis, J., Bodlund, K., Rotman, B.: Antibody-mediated activation of a defective β-D-galactosidase. Immunological relationship between the normal and the defective enzyme. J. exp. Med. **134**, 751 (1971).
28. Rotman, M. B.: Personal communication.
29. Melchers, F., Messer, W.: The activity of individual molecules of hybrid β-galactosidase reconstituted from the wild type and an inactive mutant enzyme, Submitted for publication (1972).

30. — — Enhanced stability against heat denaturation of E. coli wild type and mutant β-galactosidase in the presence of specific antibodies. Biochem. biophys. Res. Commun. **40**, 570 (1970).
31. — — Hybrid enzyme molecules reconstituted from mixtures of wild type and mutant Escherichia coli β-galactosidase. J. molec. Biol. **61**, 401 (1971).
32. ULLMAN, A., MONOD, J.: On the effect of divalent cations and protein concentration upon renaturation of β-galactosidase from E. coli. Biochem. biophys. Res. Commun. **35**, 35 (1969).
33. KÖHLER, G., MELCHERS, F.: Analysis of the heterogeneity of antibody molecules directed against different groups of determinants on the enzyme β-galactosidase from Escherichia coli by isoelectric focusing in gels. Manuscript in preparation.
34. ASKONAS, B. A., WILLIAMSON, A. R., WRIGHT, B. E. G.: Selection of a single antibody-forming cell clone and its propagation in syngeneic mice. Proc. nat. Acad. Sci. (Wash.) **67**, 1398 (1970).
35. AWDEH, Z. L., WILLIAMSON, A. R., ASKONAS, B. A.: Isoelectric focusing in polyacrylamide gel and its application to immunoglobulins. Nature (Lond.) **219**, 66 (1968).

Discussion

N. M. GREEN (London)—Chairman's remarks—: The results presented in the last two papers raise the interesting question whether the antibody (or the native β-galactosidase subunit) combines only with a small proportion of apomyoglobin (or of mutant subunits in a native conformation), thus shifting the conformation of the whole population by selection, or whether it plays a more active role by combining with the myoglobin (or the mutant conformation) and inducing a change to the more closely complementary apo (or native) conformation. I would like to ask Dr. CRUMPTON if it is possible to distinguish these alternatives in the case of the myoglobin, since one can make an approximate estimate of the amount of apomyoglobin present and plausible assumptions about its rate of reaction with antibody and compare this with the observed rate of loss of haem.

M. J. CRUMPTON (London): The equilibrium constant for the dissociation of 2 mole of metmyoglobin into apomyoglobin and dimeric haem has been reported to be 10^{-15} M at pH 7.5 and at 25 °C [BANERJEE, 1962]. The data of INADA and SHIBATA [1962] suggest that the association constant for the dimerization of haem is of the order of 10^{6} M^{-1}, in which case the dissociation constant of metmyoglobin is about 10^{-13} M. If the rate constant for the association of haem with apomyoglobin is assumed to be 10^{8} $M^{-1}sec^{-1}$, then the rate constant for dissociation will be 10^{-5} sec^{-1} and the half-time for dissociation will be about 7×10^{4} sec. (The rates of dissociation of haem from the α- and β-chains of human methaemoglobin A have been estimated to be 3.5×10^{-5} and 2.9×10^{-4} sec^{-1} respectively [BUNN and JANDL, 1968]).

The apomyoglobin antibodies reduced the 410 nm absorption of metmyoglobin to 15% of its initial value within 60 sec at 22 °C. Thus, the haem was apparently released very much faster than would have occurred if the antibodies had reacted with the apomyoglobin in equilibrium with metmyoglobin.

As a result, it appears most likely that the antibodies play an active role and promote a conformational change in the protein. In contrast to the whole antibodies, the monovalent Fab-fragments caused a much slower decrease in the 410 nm absorption (see Fig. 7) and, in this case, it is not possible to use the above kinetic argument to differentiate between the alternative mechanisms. Furthermore, due to the small overall decrease in absorption, there is some uncertainty as to whether the haem was actually displaced by the Fab-fragments. However, since the intensity of the 410 nm absorption is an indicator of the conformational integrity of myoglobin [e.g. ACAMPORA and HERMANS, 1967], the decrease caused by the Fab-fragments is at least consistent with a conformational change in the vicinity of the haem-binding site. Also, in spite of the inexplicable differences between the effects of the whole antibodies and their monovalent fragments, it seems very likely that these effects are mediated by the same basic mechanism.

ACAMPORA, G., HERMANS, J.: J. Amer. chem. Soc. **89**, 1543 (1967).
BANERJEE, R.: Biochim. biophys. Acta (Amst.) **64**, 368 (1962).
BUNN, H. F., JANDL, J. H.: J. biol. Chem. **243**, 465 (1968).
INADA, Y., SHIBATA, K.: Biochem. biophys. Res. Commun. **9**, 323 (1962).

E. HELMREICH (Würzburg): An intriguing possibility is, of course, to use the antibody to study the conformational change in the protein antigen. You have shown that antibody when it interacts with myoglobin squeezes the haem out of its pocket. Are there conditions which can affect the myoglobin in such a way that it becomes resistent to the conformational perturbation caused by the antibody-pH, temperature, ionic strength, chemical modification and so forth?

M. J. CRUMPTON (London): Although it would be of interest to determine conditions and/or treatments that prevent the release of haem by antibody, this approach is fraught with difficulties due to a variety of reasons. Firstly, antigen-antibody interaction is fairly sensitive to small changes in pH and ionic strength [e.g. HURWITZ et al., 1965] and secondly, high and low pH, increase in temperature and chemical modification such as citraconylation cause denaturation of myoglobin that is accompanied by displacement of the haem group. A more fruitful approach would probably be to use myoglobin that had been intramolecularly cross-linked by using various bifunctional reagents. I have not, however, explored this possibility.

HURWITZ, E., FUCHS, S., SELA, M.: Biochim. biophys. Acta (Amst.) **111**, 512 (1965).

E. Helmreich (Würzburg): A protein antigen has of course many antibody combining sites. How large is in fact the conformational change in myoglobin? What surface regions are involved?

M. J. Crumpton (London): The size of the conformational change accompanying the release of haem is not known. If a preparation of the pure antibodies was available then it should be possible to determine the magnitude of the change by optical rotatory dispersion or circular dichroism. Attempts to purify the antibodies by the dissociation of the precipitates formed by apomyoglobin with homologous antisera have, however, proved unsuccessful.

Unfortunately, I am also unable to answer your second question. Immunologically active peptides that represent at least some of the antigenic determinants of myoglobin have been isolated from digests of apomyoglobin. The selective adsorption of anti-apomyoglobin sera by these peptides should provide interesting information on whether the release of haem is mediated by a variety of determinants or a restricted portion of the surface. This approach has not, however, been pursued.

P. J. G. Butler (Cambridge): I have a question to Dr. Melchers. Without having the actual concentrations of mutant enzyme and antibody, it seems possible that a constant rate of activation at varying enzyme and antibody concentrations could be due to almost complete complex formation and then a uni-molecular rearrangement of this, rather than essentially being given by the prior uni-molecular re-arrangement of the enzyme alone.

Do the actual concentrations allow this or not?

F. Melchers (Basel): Concentrations of mutant enzyme and of specific antibodies were such that they ranged from a 10-fold excess of mutant enzyme over specific antibodies to a 50-fold excess of specific antibodies over mutant enzyme. We conclude that the reaction is monomolecular in these concentration ranges.

H. W. Hofer (Konstanz): In this context, it might be of interest that phosphofructokinase from rabbit skeletal muscle is also activated by low concentrations of specific antibodies. We explain this phenomenon by assuming that inactive subunits, formed by dissociation of the enzyme, undergo reassociation in the presence of antibodies. The activation was not observed when phosphofructokinase was preincubated under conditions which protect the enzyme against dissociation.

Specific antibodies exerted not only a marked influence with respect to optimum specific activity of phosphofructokinase but also altered the response of the enzyme to nonsaturating concentrations of the substrate fructose 6-phosphate. In the absence of antibodies, phosphofructokinase reached only 60% of its reference activity as determined at pH 7.6 and saturating concentrations of the substrates. The half-saturating concentration of fructose 6-phosphate was about 1.2 mM. Yet, the maximum velocity was 130% of the reference activity, and 0.8 mM of fructose 6-phosphate was sufficient for half-saturation when the enzyme had previously reacted with antibodies.

H. D. Söling (Göttingen): Dr. Melchers mentioned the possibility of using antigen-antibody reactions as a tool in enzymological studies. I want to contribute an example using the same enzyme as Dr. Hofer. We studied the effect of antibodies against purified phosphofructokinase from rat liver. The antigen-antibody reaction did not affect the catalytic activity of the enzyme, whereas the inhibitory effect of ATP was strongly inhibited. Preincubation of the enzyme with ATP on the other hand led to a significant inhibition of the antigen-antibody reaction. Apparently the antibody binds near or affects the conformation of the allosteric binding site for ATP.

H. W. Hofer (Konstanz): The antibodies against muscle phosphofructokinase seem to have different properties from those obtained by Dr. Söling against the liver enzyme. The rabbit muscle enzyme was completely inhibited by 10 mM ATP, even in the presence of antibodies. This inhibition could be reversed by 0.1 mM glucose 1,6-bisphosphate, thus demonstrating that neither the inhibitory site of ATP nor the activating site of the enzyme was blocked by the antibodies.

U. Henning (Tübingen): I think it really is not too surprising that you find *partial* activation of individual active sites. Partial inactivation (or, in fact, improvement) of single active sites is a well-known phenomenon in mutant (single-step) enzymes.

Did you say that your "break" in the Arrhenius function *is* due to denaturation, or could it be different kinetics?

F. Melchers (Basel): The break in the Arrhenius function could be either different kinetics or denaturation. It was a surprise to us that the active site of one mutant enzyme arising from a nonsense mutation at *one* site within the gene for β-galactosidase could be restored to different activities by different specific antibody populations.

Self-Assembly

TMV Protein Association and its Role in the Self-Assembly of the Virus

P. J. G. Butler

Medical Research Council's Laboratory of Molecular Biology, Hills Road, Cambridge CB2 2QH, Great Britain

With 6 Figures

Introduction

Tobacco mosaic virus (TMV) is a rod-shaped virus consisting of a single-stranded RNA molecule embedded in a helical groove between adjacent turns of the helically aggregated protein [for detailed reviews see Caspar, 1963 and Lauffer and Stevens, 1968]. The isolated RNA has been shown to be infective [Gierer and Schramm, 1956; Fraenkel-Conrat, 1956], but it is very susceptible to attack by nucleases and so is unstable on its own. However, the protein and the RNA have been shown to reassemble to give an infective particle which is indistinguishable from the virus, and in which the RNA is protected and very stable [Fraenkel-Conrat and Williams, 1955]. It is thus of interest to understand the nature and control of the protein association in view of the knowledge this may give of the self-assembly into the intact virus.

Aggregation Forms of the Protein

TMV protein occurs in many polymorphic forms which include the single helix, in which its structure and the bonding pattern between adjacent subunits is identical to that in the virus, and the A-protein, which consists of a mixture of small protein aggregates of which the detailed structure has only recently been described [Durham, Finch and Klug, 1971; Finch and Klug, 1971; Durham and Klug, 1971, 1972]. The aggregates in the A-protein, apart from the monomer, are believed to be two-layer aggregates, of which the smallest is the trimer. Larger two-layer aggregates can be formed from this by the addition of further subunits [Durham and Klug, 1971, 1972] and, because of the shape of the protein subunits, as these aggregates grow their ends curve round towards each other and finally meet to form a closed structure, the disk, with two rings each of 17 subunits. These disks can stack on top of each other either reversibly, into short stacks of disks, or else irreversibly [Durham, 1970; Carpenter, 1970] to give long stacked disk rods. This latter is the form in which disks were first recognised by Klug and Caspar [1960] from the electron micrographs of Nixon and Woods [1960].

Because of the large number of repeating units in these long rods, they are particularly suitable for structural analysis, and the structure of the disk, in the stacked disk rods, has been determined [Finch and Klug, 1971]. This shows a bonding pattern between the protein subunits which is essentially similar to that in the protein helix, but with an axial perturbation of the ends of the subunits, in the two rings of the disk, of about 5 Å towards each other. This perturbation of axially adjacent subunits from strictly equivalent positions is thought to occur in all the two-layer aggregates and it is this that renders the disk a closed structure in the axial, as well as the azimuthal, direction and so prevents the indefinite aggregation of the protein subunits which occurs in the protein helix, which is not such a closed structure.

Control of Mode of Aggregation

The extent of aggregation of TMV protein is known to be increased by decreasing pH, increasing ionic strength or increasing temperature [for review see Lauffer and Stevens, 1968]. However, the precise effect of these variables had not been determined until a recent investigation [Durham, 1970, 1972] in which the occurrence of the various aggregates over a range of conditions was studied. From this the conditions under which any given aggregate predominates were mapped [Durham et al., 1971] and such a map is shown in Fig. 1. The effect of temperature is merely to change the pH at which a particular boundary occurs — towards lower pH for a decreased temperature and towards higher pH for a raised temperature — thus increasing the degree of aggregation, but without any significant effect upon its mode. The effect of ionic strength is likewise to cause the formation of larger aggregates, but with little effect upon their type. The controlling variable for the mode of aggregation is thus the pH, with lower pH favouring the helical aggregates while higher pH favours the two-layer aggregates.

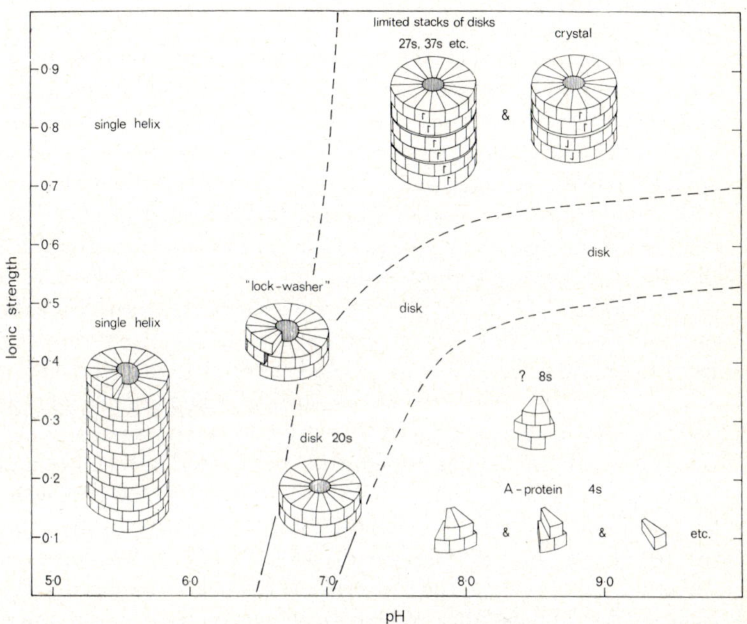

Fig. 1. Map showing ranges over which a particular polymeric species dominates the equilibrium [Durham et al., 1971]. The "lockwasher" is not well defined and represents a metastable and transitory state between disks and helix (see text for discussion). Boundaries are approximately correct for a protein concentration of 5 mg/ml at 20 °C

Since the protein from TMV does not contain any histidine and has an acetylated α-amino terminal, the observation that two protons are bound per subunit with a pK of about 7.5 in the virus [Caspar and Caspar, personal communication and quoted in Caspar, 1963] means that these groups must have their pK's altered by the local environment in the virus helix. From the effects of the binding of lead ions and also from the proton release upon dissociation of the virus helix by denaturing agents, Caspar [1963] deduced that the groups with the abnormal pK's were probably carboxyl-carboxylate pairs. Caspar and Caspar also titrated the protein helix and observed two groups titrating per subunit, but this time cooperatively as the protein helix disaggregates and too sharply for a simple titration.

This observation that the groups which titrate abnormally in the helix appear to titrate normally in the A-protein led us to the idea that the perturbed pK's might be a property of protein subunits in the helical mode of aggregation and would not be found in any of the two-layer aggregates. In order to test this hypothesis we wished to titrate the protein while it was locked in each of the aggregation modes, namely in the virus and the stacked disk rods. The titration curve for the virus was in complete agreement with that found by Caspar and Caspar, while the titration curve for the stacked disk rods showed very little proton binding between pH 6 and pH 10, with the difference curve between the titration curves of the protein in these two states corresponding to two groups titrating independently with pK's of 7.1 [A. C. H. Durham, A. Klug and P. J. G. Butler unpublished work and quoted in Durham and Klug, 1971]. To measure the effect of this proton binding upon the aggregation state of the protein, reversibly aggregating protein was titrated over a range of temperatures from 0 to 35 °C. The titration curves all fall between the two archetypal curves shown by the virus and the stacked disk rods, with the curves at higher temperature nearer to that for the virus and those at lower temperatures nearer to that for the stacked disk rods.

Since the critical difference between the archetypal curves results from the titration of two groups per subunit, the titration curves for the protein were analysed by subtracting a sigmoid curve of variable width and centre position, equivalent to two protons per subunit, from each protein titration curve while adjusting the width and centre position of the sigmoid curve until the resulting difference curve was similar to the titration curve for the stacked disk rods. The best sigmoid curve for each temperature thus approximates to the actual proton uptake of the two abnormal groups at that particular temperature, and these curves range from a very sharp curve for 0 °C to a curve very close to a normal titration curve at 35 °C. The best values for the centre positions and widths are shown in Fig. 2. At those temperatures where there was sufficient sedimentation data available, the positions of the centres were found to correspond to the pH below which an aggregate with a sedimentation coefficient greater than about 23 S is first found in the

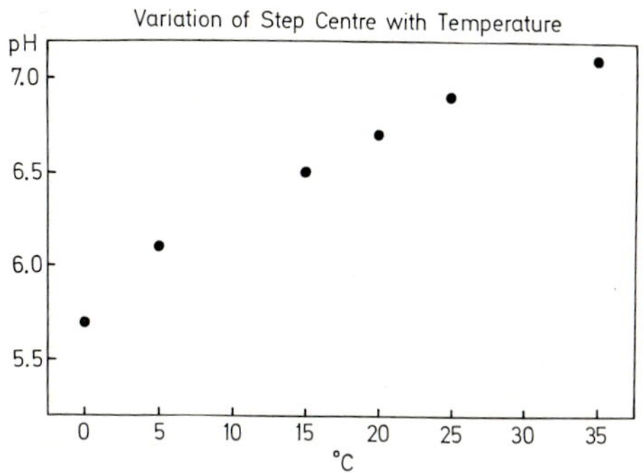

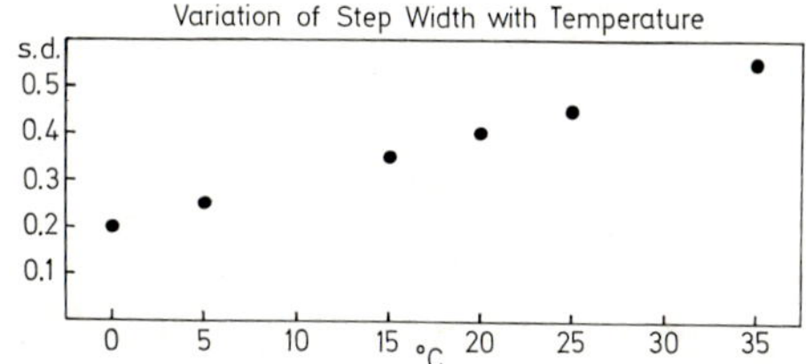

Fig. 2. Centre positions and widths (to include 90% of the effect) for sigmoid curves for uptake of protons by reversibly aggregating TMV protein between pH's 5 and 8. Upper limits indicate the values for a simple titration of a group with a pK of 7.1. See text for details of derivation

equilibrium mixture. This is fully compatible with the binding of the "Caspar" protons by the abnormally titrating groups having a controlling effect upon the aggregation mode of the protein.

Mechanism of Control by Protons

In order to understand this control more fully, we have formulated a model for the control of the mode of aggregation by the protonation of these Caspar groups in the helix [unpublished work of A. C. H. DURHAM, A. KLUG and P. J. G. BUTLER]. The formation of helix from the A-protein will involve the aggregation of monomers, or small aggregates, onto the growing helix, with an ensuing change in the pK of the carboxyl groups from a normal pK to about pK 7.1. For convenience of calculation, the free energy change per monomer aggregating into the helix can be factorized into two components — a free energy change due to the bonding interactions with the neighbouring subunits, without any protonation of the Caspar carboxyl groups, and a free energy change due to the protonation of these groups, which will relieve the unfavourable ionic interactions.

Because of the steepness of the proton binding curves, particularly at low temperatures, one obvious model for the transition was a cooperative aggregation, analogous to the transition between a gas and a liquid or solid. One such model is that for a "lattice gas", in which the free energy for the addition of each extra molecule to the condensed phase rises in proportion to the number of molecules which are already in that phase, thus giving a highly cooperative condensation. This model was applied to the equilibrium between the monomer and the helix and the bonding energy per subunit was adjusted until the transition between the monomer and the helix occurred at the pH of the centre of the proton binding curve. The values for the bonding energies for the unprotonated subunits range from about −1 kilocalorie at 0 °C to about −4 kilocalories at 35 °C and are linearly dependent upon the temperature. When allowance is made for the protonation of the helix, the values are similar to those found for the bonding of a subunit onto a growing A-protein aggregate at the same temperature [DURHAM and KLUG, 1972].

According to this model, the transition from the A-protein to the helix is extremely abrupt, more than 99% of the transition occurring within 0.1 pH unit, and no intermediate sized aggregates occur. In consequence, the proton binding curves for the Caspar groups show an abrupt jump at the pH where the transition occurs and, while this is a reasonable approximation to the experimetally determined curves at 0 °C and 5 °C, it does not agree with those found at higher temperatures. Furthermore, aggregates of a size intermediate between the small A-protein and the large helix are seen in the pH range where the transition is occurring [DURHAM,

1972]. We have therefore worked out an alternative model which includes the aggregation of the monomer into the A-protein, up to and including the disk, and then into helices from three turns upwards. The choice of three turns for the helical nucleus is somewhat arbitrary, but a variation in this size has only a minimal effect upon the overall picture. In this model, the mole fraction of any particular aggregate was calculated by determining its equilibrium constant for formation from monomer, from the free energy change for this reaction.

The protein subunits in a helix are considered to belong to two classes — those in the end turns of the helix (2×17 subunits) and those in the middle turns of the helix (all the other subunits). These two classes of subunits can have different bonding energies and also, while the non-terminal subunits are considered to have the two groups titrating with pK 7.1 (as measured in the virus helix), the end subunits are considered to have only a single abnormally titrating group, with a pK intermediate between 7.1 and 4.5 (as for a normal carboxylic acid residue). These differences between the end and the middle subunits were found to be necessary to fit the titration curves and also to allow the stable existence of intermediate sized aggregates in the transition region between the A-protein and the large helix. The bonding energies for subunits into the two-layer aggregates were taken directly from DURHAM and KLUG [1972].

By adjustment of the bonding energies for the unprotonated subunits inside the helix and at the ends of the helix and also the pK for the single abnormally titrating group on the end subunits, it was found possible to obtain a good fit to each of the proton binding curves from 15 to 35 °C, together with a pattern of aggregates which was compatible with the sedimentation patterns [DURHAM, 1972]. At lower temperatures the proton binding curves were not steep enough, as the model did not allow sufficiently cooperative aggregation. The actual bonding energies were even closer to those for the A-protein, under conditions where the two forms would be in equilibrium. At all temperatures the bonding energy for a subunit going onto the end of a helix is more favourable than that for a subunit going into the middle of the helix, while only a single pK is perturbed and that not as much as the two pK's which are perturbed per subunit in the middle of a helix. Since the perturbation of the pK is a measure of the unfavourable ionic interaction [see Caspar, 1963, for discussion] the observation that one of the groups titrates normally and the other one with a less perturbed pK in the end subunits, again agrees with their bonding being more favour-

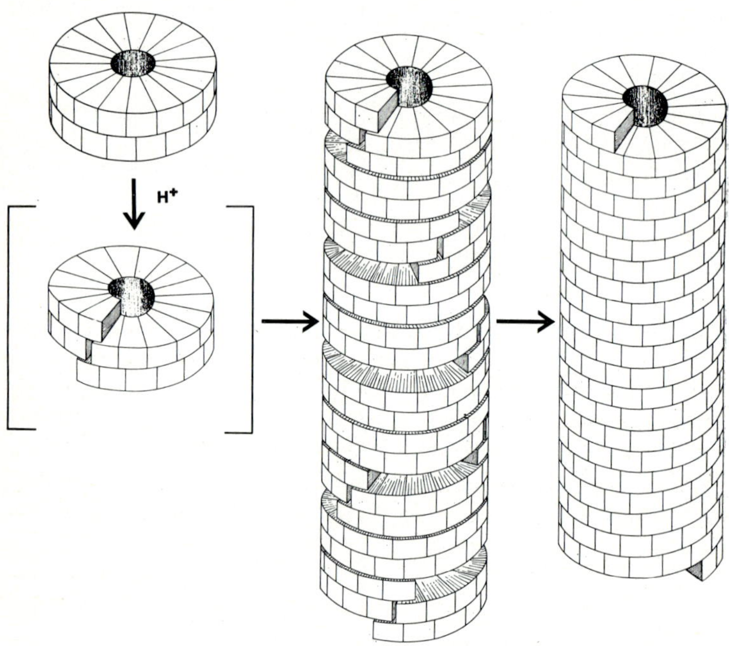

Fig. 3. Diagram of the effect of rapidly dropping the pH of a solution of TMV protein disks. The "lockwasher" (shown in brackets) has not been observed and is a hypothetical intermediate. The annealing of the imperfect helix to a perfect helix is slow and takes a period of hours [Durham et al., 1971]

able. This is presumably because of the greater freedom of subunits at the end of the helix to adapt their positions slightly with respect to their neighbours and thus to optimize their bonding. This enhanced stability of the end turns of a helix accounts for the stability of the small aggregates which would otherwise not be found in a system involving quasi-crystallization, in which the only stable aggregates tend to be the monomer and the infinite aggregates.

The key to the control of the aggregation mode lies in the protonation of the groups with the abnormal pK's. Without such protonation, the interaction which would change the pK of these groups in the helix is sufficiently unfavourable to prevent the formation of helix at all. These groups thus act as a "negative switch" to prevent the formation of the protein helix except at pH's below neutrality, where it can be overcome by protonation. In consequence, at or above neutral pH, the protein forms the two-layer aggregates. Since these are of limited size, unlike the helix which can grow in length indefinitely, this means that the protein is available in small aggregates when it has to interact with the RNA to coat and protect it. Clearly it would be a biological disaster for TMV if all its coat protein were in large aggregates and thus not available for this purpose.

It had previously been observed [FRAENKEL-CONRAT and SINGER, 1959, 1964] that the optimum conditions for the reassembly of TMV, from the isolated RNA and protein, were about pH 7.25 and at ionic strengths over 0.1. Under these conditions, the major protein aggregate is the disk and it was found that this could be switched into the helical mode of aggregation, without prior dissociation into smaller aggregates, simply by dropping the pH rapidly [DURHAM et al., 1971]. The reaction is shown schematically in Fig. 3 and involves the transformation of the disks into a hypothetical intermediate, the "lockwasher"; these lockwashers then stack end-to-end in imperfect register, to give nicked helices, which then anneal over a period of hours to the normal helix.

Role of Disks in Virus Assembly

A major problem in the assembly of the viral helix is the initiation of the helix growth. Until the first couple of turns of the helix are complete, some of the protein subunits will be bonded to only two neighbours and the growing nucleoprotein aggregate will therefore be relatively unstable. One possible solution to this problem would be for the disk to be formed and then to act as a preformed nucleus for the helix growth, being transformed into a short helix by the interaction with the RNA in a similar way to its transformation by the pH drop. A similar difficulty of formation of disks does not occur since they are two-layer aggregates and each layer is stabilized by its interaction with the other layer. After initiation with disks, the nucleoprotein helices could then continue to grow by the addition of protein subunits from the A-protein, either singly or a few at a time.

The effect of disks upon the rate of reassembly of TMV was investigated [Butler and Klug, 1971] and it was found that they did indeed increase the rate by more than an order of magnitude, but only when present at comparable concentrations to that of the A-protein. More surprisingly the omission of the A-protein does not reduce the rate of reassembly (Fig. 4) thus implying that the A-protein is not involved significantly in the reassembly. The effects upon the reaction rate of varying the RNA and protein concentrations [Butler, 1971 (1)] show that the initiation reaction is very rapid and takes place between a single RNA molecule and two protein disks, while subsequent growth is relatively slow and occurs by the addition of further disks onto the growing nucleoprotein helix. All of the thirty four protein subunits from the disk are added to the helix and the rearrangement reaction takes about $6^1/_2$ sec, giving a total time for the formation of a virus particle of just over 6 min, under optimal conditions.

TMV protein has been shown to form the normal nucleoprotein helix with poly A and poly I, as well as with its homologous RNA, but not with poly C, poly G, poly U, poly X or any of the other natural RNA's which have been tried [Fraenkel-Conrat and Singer, 1964]. When the rates of assembly were studied [Butler and Klug, 1971], only poly A and poly I gave any measurable reaction and even these were two orders of magnitude slower, on a molar basis, than the homologous TMV-RNA. Since there is no evidence for an unusual base composition in TMV-RNA, it is unlikely that the nucleotide sequence throughout the molecule will be the cause of this greater rate and so it is probable that it is due to an especially favourable nucleotide sequence on the RNA which gives the very rapid initiation. This idea was tested by digesting the TMV-RNA with exonucleases and following any change in the rates of reconstruction as the digestion proceeded. In this way, Butler and Klug [1971] found that digestion from the 5′-hydroxyl end of the RNA (with spleen phosphodiesterase) caused a steady loss of reconstitution activity, while similar digestion from the 3′-hydroxyl end (with snake venom phosphodiesterase)

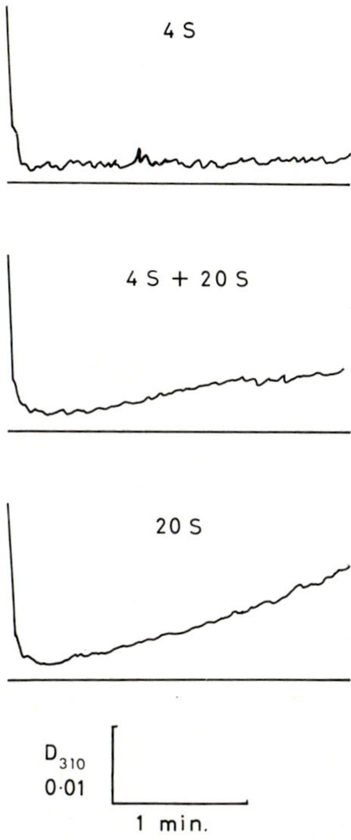

Fig. 4. Effect of protein aggregation state upon the rate of reassembly of TMV [Butler and Klug, 1971]. Reassembly was followed by the increase of turbidity at 310 nm and 25 °C. Reaction was in 0.1 M-sodium pyrophosphate, pH 7.3, with an RNA concentration of 0.05 mg/ml and both 4 S (A-protein) and disk concentrations of about 1 mg/ml

did not produce any effect (Fig. 5). From this they concluded that the rapid initiation was due to a special nucleotide sequence at the 5′-hydroxyl end of the TMV-RNA and they also estimated the length of this special sequence to be about 50 nucleotides.

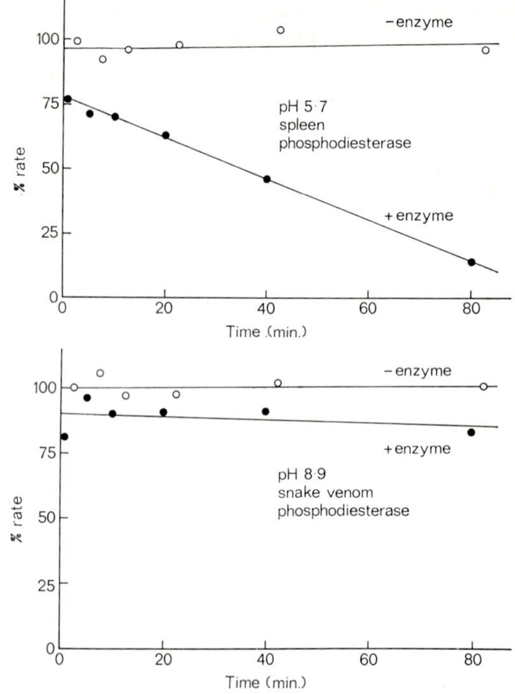

Fig. 5. Effect of exonuclease digestion of TMV RNA on the rate of reconstitution with protein disks [Butler and Klug, 1971]. Reconstitution conditions as in Fig. 4, and rates are expressed as percentage of initial rate before digestion. Digestions carried out at 25 °C with (a) spleen phosphodiesterase and (b) snake venom phosphodiesterase, while controls were incubated without enzyme. ● = with enzyme; ○ = without enzyme

Concluding Remarks

Although this use of a sub-assembly of the TMV protein for the assembly of the virus, as well as for its initiation (when the structural need is obvious), is somewhat surprising, in hindsight the advantages become more obvious. With the A-protein as the protein source, the subunits will be added singly or else a few at a time, and any unfavourable nucleotide sequence will be expected to slow the assembly down very badly. This effect has indeed been observed and such an unfavourable nucleotide sequence has been reported to affect the growth of the TMV particles when their lengths are about 70 nm [STUSSI, LEBEURIER and HIRTH, 1969]. With disks as the protein source, the interaction energy from the 34 protein subunits which will bind to the RNA at once might well be favourable enough to overcome any such local unfavourable regions on the RNA, thus explaining the lack of any obvious "preferred length" for partially coated particles. Thus not only is the rate of growth of the nucleoprotein helix over ten times faster from disks than from A-protein, even on a region of the RNA favourable for growth, but furthermore the rate from disks will not be affected by an adverse region of the RNA while that from A-protein will be reduced still further.

The need for some preformed protein aggregate to overcome the problem of the nucleation of the first turns of the helix has already been discussed. Besides filling this need, disks are also seen to confer the selectivity for its homologous RNA onto the TMV protein. At the stage of nucleation, two disks (with a total of 68 protein subunits) have to be transformed into "lockwashers" by interaction with a single turn of RNA and it is probably the need for an especially favourable interaction energy at this point that singles out the correct RNA for the virus formation. In this way the concerted interaction of the protein subunits of the disk with the RNA can at one time confer specificity, while subsequently reducing the dependence of the reaction rate on the RNA sequence.

One conceptual difficulty to the acceptance of such a model is the problem of forming a plausible picture of how the two turns of a disk can be incorporated into a growing helix. One possible picture [BUTLER, 1971 (1), (2)] is illustrated in Fig. 6. The initial interaction occurs between the special sequence at the 5′-hydroxyl end of the TMV-RNA and two protein disks (Fig. 6a). After the two disks have been converted into short helices, entrapping the first turn of the RNA (Fig. 6b), initiation is complete and growth can then ensue. Normally an incoming disk will interact with the growing nucleoprotein helix, which will have the RNA protruding from one end, and will "unroll" onto the top to enclose one turn of this RNA (Fig. 6c). At the end of this turn, a "ruck" will have been formed where the RNA protrudes sideways (Fig. 6d) and the RNA can then intercalate into this second turn of the protein helix through

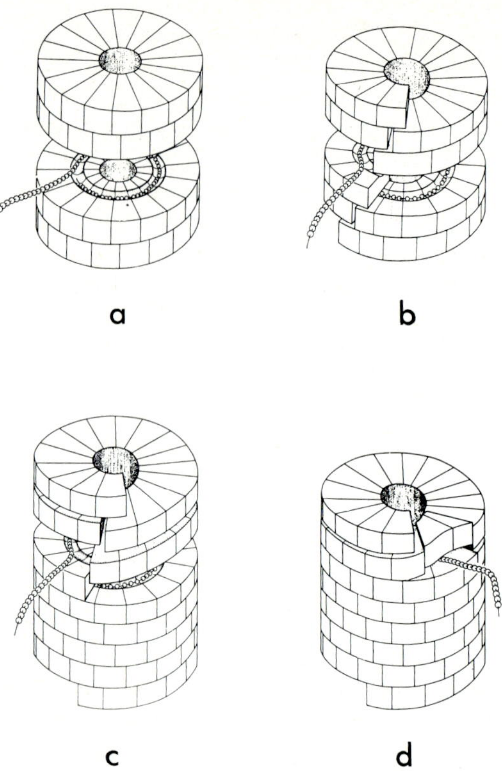

Fig. 6. Picture of assembly of TMV from disks and RNA [Butler, 1971 (1), (2)]. Initiation — a) 5′-hydroxyl end of TMV RNA interacts with two disks; b) disks dislocate to form short helices enclosing the end of the RNA. Growth — c) dislocated disk "unrolls" to entrap turn of RNA; d) second turn of RNA intercalates into helix through the "ruck" formed at the end of the first turn

this ruck. Although this mechanism may be inaccurate in detail, it does illustrate the possibilities of such a growth process.

Amongst the advantages of the use of a sub-assembly during the self-assembly of a complex structure is the probable reduction of the consequences of any errors in the assembly process [Crane, 1950]. TMV protein aggregation is not a process which is free from error (e.g. see Durham and Finch, [1972], for some of the aberrant and irreversible aggregates of the protein) and this effect may indeed be of some importance. What is more striking, however, is the way in which the disks are sub-assemblies which are not incorporated as such, but which are transformed during the final assembly by interaction with another component of the structure, in this case the viral RNA.

Although there are still further details of the molecular basis of the control to be elucidated, it is already clear that the association of the TMV protein and its subsequent assembly into the virus is an elegant example of the delicate balance of biological control.

References

Butler, P. J. G.: (1) Cold Spr. Harb. Symp. quant. Biol. **36**, 461 (1971).
— (2) Nature (Lond.) **233**, 25 (1971).
— Klug, A.: Nature (Lond.) New Biology **229**, 47 (1971).
Carpenter, J. M.: Virology **41**, 603 (1970).
Caspar, D. L. D.: Advanc. Protein Chem. **18**, 37 (1963).
Crane, H. R.: Sci. Monthly **70**, 376 (1950).
Durham, A. C. H.: Ph. D. Dissertation, Univ. of Cambridge, England (1970).
— J. molec. Biol. **67**, 289 (1972).
— Finch, J. T.: J. molec. Biol. **67**, 307 (1972).
— — Klug, A.: Nature (Lond.) New Biology **229**, 37 (1971).
— Klug, A.: Nature (Lond.) New Biology **229**, 42 (1971).
— — J. molec. Biol. **67**, 315 (1972).
Finch, J. T., Klug, A.: Phil. Trans. **B 261**, 211 (1971).
Fraenkel-Conrat, H.: J. Amer. chem. Soc. **78**, 882 (1956).
— Singer, B.: Biochem. biophys. Acta (Amst.) **33**, 359 (1959).
— — Virology **23**, 354 (1964).
— Williams, R. C.: Proc. nat. Acad. Sci. (Wash.) **41**, 690 (1955).
Gierer, A., Schramm, G.: Nature (Lond.) **177**, 702 (1956).
Klug, A., Caspar, D. L. D.: Advanc. Virus Res. **7**, 225 (1960).
Lauffer, M. A., Stevens, C. L.: Advanc. Virus Res. **13**, 1 (1968).
Nixon, H. J., Woods, R. D.: Virology **10**, 157 (1960).
Stussi, C., Lebeurier, G., Hirth, L.: Virology **38**, 16 (1969).

Intercellular Interactions

Surface Membrane Alterations and Relevance to Cell-Cell Interaction and Growth Control in Tissue Culture

M. M. Burger and K. D. Noonan

Dept. of Biochemical Sciences, Princeton University, Princeton, N. J. 08540, USA

With 13 Figures

It has been suggested that the cell surface may be a direct or indirect mediator of cellular growth control [1]. Many attempts have been made to correlate chemical changes of the surface membrane with alterations in cell growth [2, 3, 4, 5]. Recently, using plant agglutinins which bind to various carbohydrates on the cell surface, we have observed structural or configurational changes of the cell surface which may play a role in cell growth [6, 7, 8].

The agglutinin or lectin which we are working with was isolated from wheat germ or wheat germ lipase [9]. It is a glycoprotein with a molecular weigth of 23,500 Daltons which has recently been crystallized [10]. X-ray crystallography is presently being carried out by Dr. C. Wright in Dr. R. Langridges laboratory in Princeton.

Using wheat germ agglutinin and Concanavalin A (Con A), an agglutinin from Jack Beans [12], it has been shown that the surface configuration of transformed cells is different from that of normal cells [6, 13]. Transformed cells are agglutinable with several plant lectins; whereas the normal, parent cells are poorly or not agglutinable [9, 13, 14].

Three mechanism have been considered which might be responsible for the enhanced agglutinability of a tissue culture cell after treatment with an oncogenic virus, a chemical carcinogen or X-irradiation (Fig. 1).

The first mechanism is considered unlikely for a variety of reasons, the most compelling being that some oncogenic viruses contain as few as six genes thereby making it unlikely that the viral genome could code for all of the agglutinin specificities known. The distinction between the activation of receptor formation, i.e. mechanism II and an increase in agglutinability due to a rearrangement in the cell surface can only be made after a careful structural analysis of the isolated receptors from both transformed and parental surface membranes. Such detailed work has not yet been carried out.

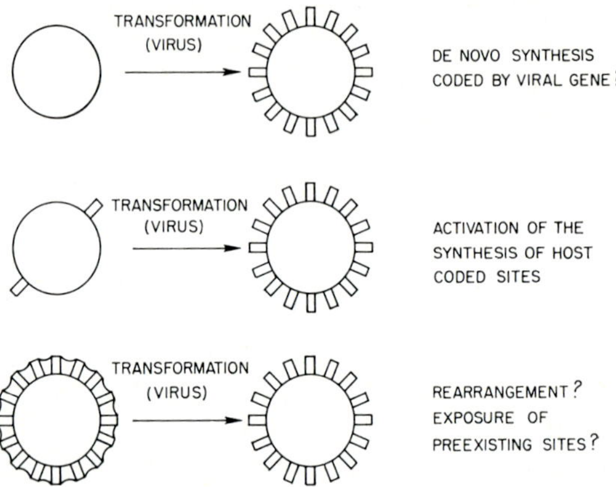

Fig. 1. Three possible mechanisms by which oncogenic transformation can render a cell agglutinable

Two experiments indicate that the formation of more receptor sites in transformed cells is not a likely explanation for the increased agglutinability. First we have compared the amount of partially purified surface receptor [15] (consisting of glycoprotein) isolated from transformed and parental cell surfaces. From preliminary studies the two surfaces seem to contain about the same number of agglutinin receptor sites [16]. Second, treatment of the parental cell surface with very small amounts of protease for a short period of time allows the normal cells to agglutinate to the same degree as the untreated, transformed cells [6] (Fig. 2). Since the proteases used were free of glycosidases and since very little, if any, glycoprotein was removed from the cell surface during the treatment, we suggested that the receptor site for lectins may be present in normal cells in a conformation which does not permit agglutination. However, conclusive evidence that the same receptor sites are present in the same quantities in normal and transformed cells is still missing.

The various types of cell surface changes which might occur after transformation include chemical changes as well as alterations in the macromolecular architecture of the surface [16] (Fig. 3).

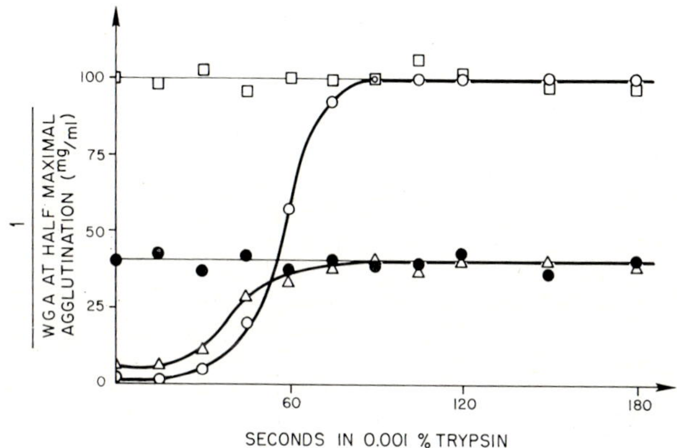

Fig. 2. At time 0, 3T3 cells or BHK cells were treated with 0.001 % Trypsin and agglutination with wheat germ agglutinin was monitored. △—△ BHK cells; ●—● Py-BHK cells; ○—○ 3T3 cells; □—□ Py3T3 cells

The cell surface alterations we suggested as possible explanations for increased agglutinability after transformation are summarized under topographical changes.

Two of the most popular explanations for increased agglutinability after mild protease treatment or transformation are summarized under tangential rearrangements of the surface membrane and includes exposure of some sites and clustering of other sites. At present it is not possible to choose between these alternative explanations although direct electronmicroscopic observation supports site clustering [17] while direct measurement of available receptor sites supports the exposure of sites. Below we briefly discuss recent results on the quantitation of lectin receptor sites.

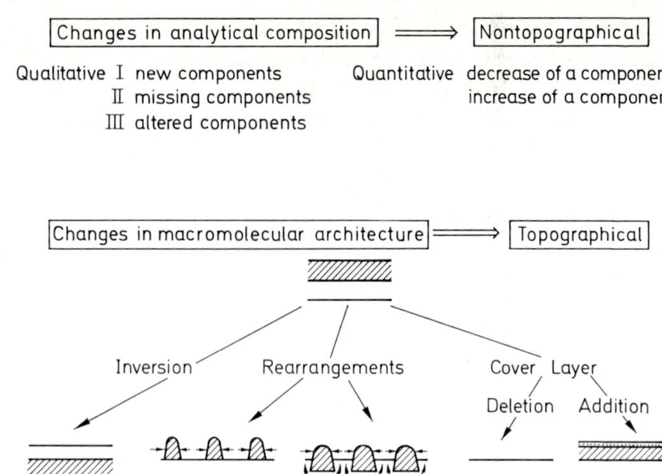

Fig. 3. Theoretically possible membrane alterations in the process of transformation

Agglutinin Binding

Using a ^{3}H-Con A binding assay we have found that transformed cells bind 3.5 to 5 times more ^{3}H-Con A than the normal cell [18].

In our assay system [19] normal and transformed cells are grown on Falcon Petri dishes (35 mm) to 80% confluency in Dulbecco modified Eagle's medium supplemented with 10% calf serum. The cells are then chilled for 5 min at 0 °C, washed two times with ice cold saline and incubated for 5 min with ^{3}H-Con A at 0 °C in phosphate buffered saline (pH 7.2) containing 1×10^{-3} M Mg^{2+}. The cells are then washed 5 times with ice cold saline and solubilized in 10% Triton X-100 at 37 °C and the extract is counted in a Triton-Toluene scintillation cocktail.

Fig. 4 demonstrates that mouse embryo fibroblasts transformed with Polyoma virus (Py3T3) bind 5 times more ^{3}H-Con A than the parental mouse embryo fibroblast line (3T3). Table 1 shows 95% of the ^{3}H-Con A can be inhibited from binding to the cells if the ^{3}H-Con A is preincubated with 10^{-3} M α-methyl mannopyranoside,

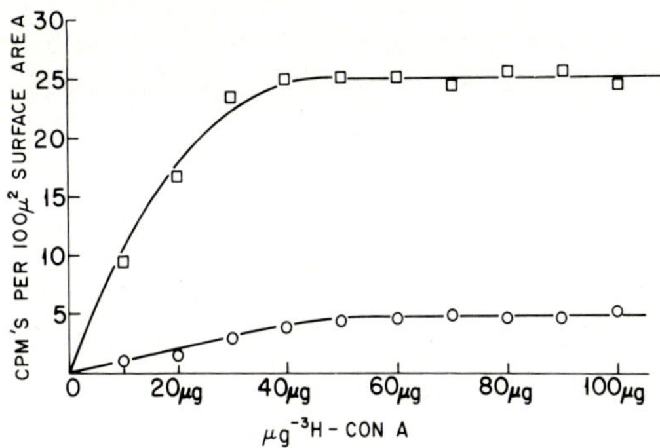

Fig. 4. ³H-Con A binding to 3T3 and Py3T3 cells. The assay was performed as described in the text. ○—○ 3T3 cells; □—□ Py3T3 cells

Table 1. *Hapten inhibition of ³H-Con A binding at 0 °C. CPM's/100μ^2 surface area*

Conc. Con A (µg/ml)	Py3T3		3T3	
	−α-MM	+α-MM	−α-MM	+α-MM
10	7.12	0.35	1.11	—
20	12.10	0.61	1.64	—
30	23.30	1.50	3.06	0.15
40	20.70	1.04	4.16	0.21
50	25.70	1.30	4.47	0.22
60	25.50	1.28	4.85	0.24
70	24.20	1.21	5.01	0.25
80	26.50	1.28	4.58	0.23
90	27.20	1.36	4.68	0.21
100	24.50	1.23	5.15	0.29

suggesting that the ³H-Con A is binding to the specific Con A receptor site on the cell surface.

Similar results have been obtained with other transformed cell lines including a Simian virus transformed 3T3 cell line, a minimal deviation hepatoma cell line and a temperature sensitive line which binds 3.5 times more ³H-Con A at the permissive temperature, where it grows as a transformed cell, than at the non-permissive temperature where it grows as a normal cell.

Fig. 5 demonstrates that synchronized 3T3 cells bind approximately 4.5 times more ³H-Con A at mitosis than during any other part of the cell cycle [18], an important observation for some of the points on mitotic cell surface alterations discussed at the end of this presentation.

Thus more agglutinin receptor sites appear to be present on the transformed cell surface than on the normal cell surface.

Care must be taken to perform the agglutinin binding at 0 °C. At any temperature above 15 °C interpretation of the binding assay becomes difficult due to specific and non-specific binding plus

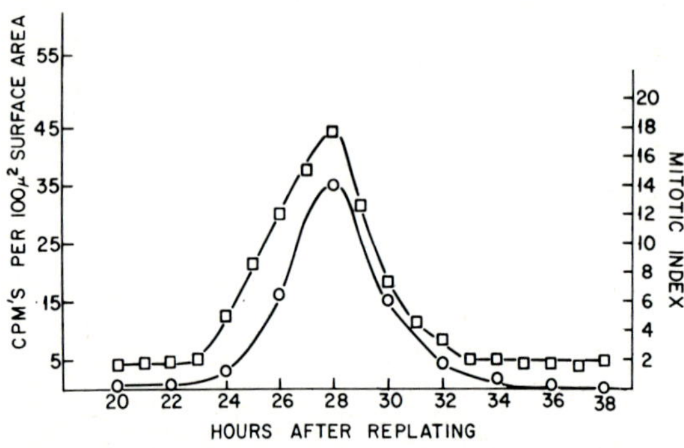

Fig. 5. ³H-Con A binding to 3T3 cells during mitosis. 3T3 cells were synchronized either by replating the cells from a monolayer or by feeding a confluent monolayer with 50% calf serum. ³H-Con A binding was performed as described in the text. The mitotic index was determined by counting the number of metaphase figures in a field using a phase microscope. □—□ ³H-Con A bound to 3T3 cells; ○—○ Mitotic Index

endocytosis of the radiolabelled agglutinin. The combination of these processes serves to obscure the true binding of the Con A molecule to its specific surface receptor, a consideration that applies particularly to two [20, 21] of the three publications that have appeared on the binding of labelled lectins to cell surfaces [22].

The Significance of the Increased Susceptibility of Transformed Cell Surfaces to Agglutination by Lectins

The following points suggest that surface changes detectable with lectins are closely related to viral transformation:

1. As we have indicated previously many transformed cells agglutinate better than do their parental cell line [13, 16, 23, 24]. Variant cells isolated from a transformed cell culture which have regained normal growth control, have been shown to agglutinate like a normal cell line.

2. Cells which were transformed with temperature sensitive oncogenic viruses under permissive conditions and then shifted to

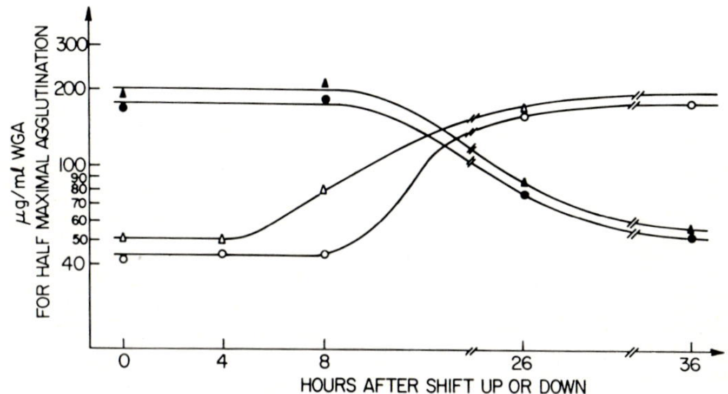

Fig. 6. Baby hamster kidney cells were transformed with a temperature sensitive mutant of polyoma virus (ts-3) under permissive conditions (32 °C). Two transformed clones were used. Clone 1 (△—△) and clone 7-C (○—○) were shifted from 32 °C to 39 °C at time 0. Such cells were later shifted back again from 39 °C to 32 °C at time 0 (clone 1 ●—● and clone 7-C ▲—▲)

a temperature non-permissive for the expression of the transformed state will behave like normal cells both in their growth properties and in their resistance to agglutination [27, 28] (Fig. 6).

Host range viral mutants cannot alter the surface if they are defective for infection and transformation [29].

3. Temperature sensitive host cell mutants [30] which loose the expression of the transformed state at the non-permissive temperature also loose their agglutinability [31].

4. The selection of poorly agglutinable cells from transformed cell cultures indicates that not only does the agglutinability of the cell line decrease but the cells also return to normal growth control [32, 33].

5. Mice immunized against the carbohydrate specific for WGA (di-N-acetylchitobiose) become resistant to challenge by tumor cells [34] (Fig. 7). This provides some preliminary evidence that the WGA site or nearby surface groups may be relevant for the tumor properties of the cell. Why such animals do not suffer from severe anemia is not yet clear although it would have been predicted since mouse erythrocytes agglutinate with WGA. This type of

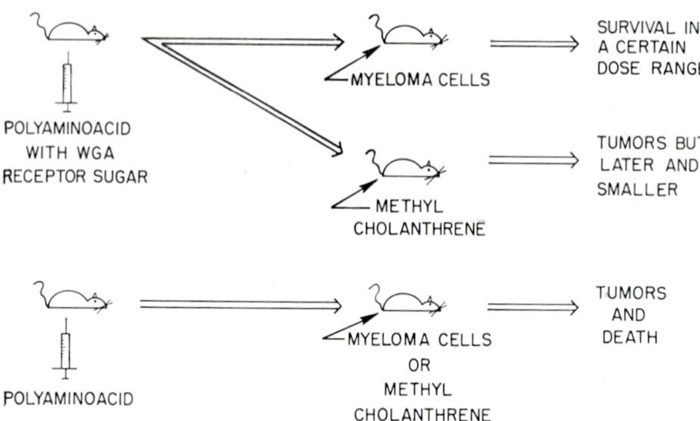

Fig. 7. The effect which the injection of a polyaminoacid containing the WGA receptor sugar has on protecting mice from a later challenge with myeloma cells

experiment has to be repeated with several other lectin sites and the exact mechanism of tumor cell killing *in vivo* has to be established before it can be used as evidence for a possible *in vivo* role of the lectin receptor sites on tumor cells.

The Agglutinin Receptor Site and Growth Control

1. Escape from Growth Control

Using tissue culture cells transformed with oncogenic viruses, the change in the surface membrane monitored with plant agglutinins has been correlated with the loss of contact inhibition [25]. Cell lines demonstrating low saturation densities agglutinate poorly whereas cell lines displaying the higher saturation densities agglutinate extremely well [16], suggesting that the availability of the agglutinin receptor site parallels the increase in saturation density and loss of growth control.

Two experiments have been done to test whether the surface alteration monitored with plant agglutinins is responsible for the observed growth patterns.

Since we had shown that the agglutinin receptor sites could be exposed by treatment of the normal cell surface with protease, we

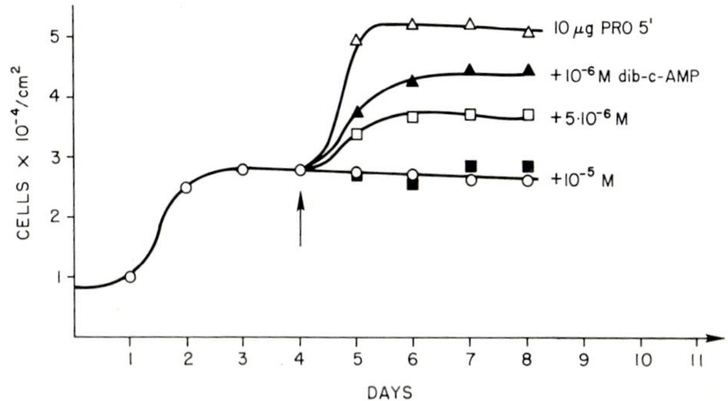

Fig. 8. Protease induced growth stimulation and inhibition of the stimulation by dibutryl-cyclic-AMP. At the arrow the medium was removed from 3.5 cm plastic dishes containing confluent 3T3 fibroblasts and 10 µg pronase in PBS added for 5 or 10 min. The dishes were rinsed twice thereafter with 1.5 ml PBS and refilled with the conditioned medium removed before treatment. Cells were counted daily. The average of six determinations is reported. To demonstrate inhibition, cells were incubated with varying concentrations of dib-c-AMP together with pronase and which was removed with the pronase

felt that one should treat normal mouse fibroblasts grown to confluency briefly with small amounts of protease to expose the agglutinin receptor sites. If a relationship existed between overgrowth of the monolayer and exposure of the agglutinin receptor sites, cell division should follow [7]. Fig. 8 demonstrates this to be the case.

Since the surface change lasts for only 4 to 10 h, after which the cells repair the protease induced alteration, normal cells escape contact inhibition for only one generation. However if these same cells, at their new saturation density, are again treated with protease they will divide and assume an even higher saturation density. Sefton and Rubin have demonstrated similar results with chick embryo fibroblasts [35].

We have shown that the protease does not activate inactive precursors in the serum. Furthermore we have demonstrated that

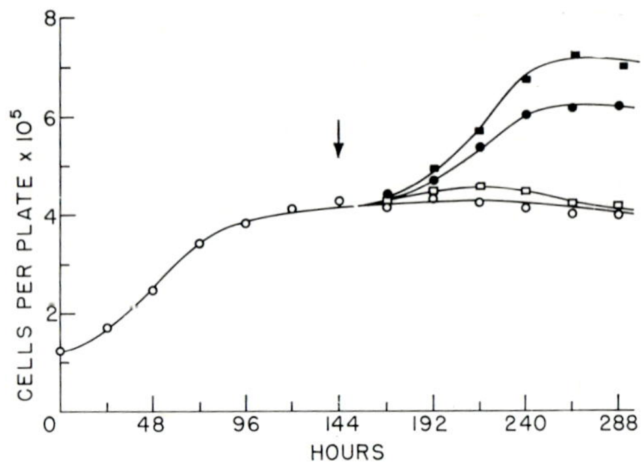

Fig. 9. Evaluation of the growth stimulatory effect of L1210 cells upon resting 3T3 mouse fibroplasts. ●—● 1 ml of peritoneal fluid withdrawn together with L1210 leukemia cells from a mouse carrying such cells gave a growth stimulatory effect which was probably due to the proteolytic enzymes in the ascites fluid. □—□ Medium from 3 days old L1210 leukemia cell cultures grown in tissue culture. Since no growth stimulation of 3T3 cells occurred it is not likely that L1210 cells secrete enough protease to stimulate 3T3 cell division. ■—■ Isolated L1210 plasma membranes added for 20 min or 3 h to the 3T3 cells and rinsed off with calcium, magnesium-free PBS gave as much of a growth stimulation as did proteases covalently attached to beads. ○—○ Controls

the protease acts on the cell surface since protease attached to beads which cannot enter the cell, are still active in inducing overgrowth.

We have previously suggested that surface proteases may bring about the agglutinable state in transformed cells and that a series of cascading proteases may be necessary to induce a growth response.

An earlier observation of ours has strengthened this working hypothesis. Transient addition of L1210 leukemia cells to confluent 3T3 fibroblasts induced the cells to divide in the same manner as did the addition of protease (Fig. 9). The normal parent cells of this leukemic cell (lymphocytes) was unable to trigger a similar response indicating that the active component was not found in the normal cell. Surface membrane components from L1210 cells initiated a similar response indicating that the active component was in the surface membrane [37]. The addition of the protease inhibitor tosyl-lysyl-chloromethyl-ketone (TLCK) with the L1210 surface membranes inhibited the growth response completely [37].

More recently in our laboratory (R. Remo) and in that of our former collaborator, H. P. Schnebli it has been found that several protease inhibitors inhibited the growth of transformed cells while not affecting the growth of normal cells. This suggests that normal cells may be less dependent on proteases to determine their pattern and rate of growth.

2. Reestablishment of Growth Control

In the converse experiment we have been able to reestablish density dependent inhibition of growth in transformed cell lines by covering the agglutinin receptor sites of transformed cells [8].

We have shown that treatment of a pure Con A fraction with trypsin or chymotrypsin produces a preparation which does not agglutinate transformed cells but which does bind to the agglutinin receptor site (Table 2). This suggested that we had produced a preparation which behaves like a monovalent Con A fraction [39].

Using this Con A molecule modified by trypsin or chymotrypsin treatment we have been able to reduce the final saturation of a number of transformed cell lines. Fig. 10 demonstrates that treatment of Py3T3 cells with different concentrations of the Con A preparation reduces the final saturation density of the cell line until it approximates that of the normal 3T3 cell line.

Fig. 11 demonstrates that this effect is due to the modified Con A attached to the cell surface. In this experiment Py3T3 cells + modified Con A are grown to their "saturation density". After one day at the monolayer, 10^{-3} M α-methyl mannopyranoside or α-methyl glucopyranoside is added and the growth pattern observed. Addition of these sugars removes the modified Con A from the cell surface. Within 24 h after the removal of the modified Con A from the cell surface, the cells begin dividing and eventually reach a saturation density of the untreated controls [8].

Table 2.

Cell type	Additions	Intact Con A necessary for half maximal agglutination (μg/ml)
3T3	—	1500
Py3T3	—	250
Py3T3	250 μg/ml T-Con A preincubated with cells for 20 min	1500
Py3T3	250 μg/ml T-Con A preincubated with 5×10^{-3} M α-MM; added to the cells, incubated for 20 min, washed and tested	250

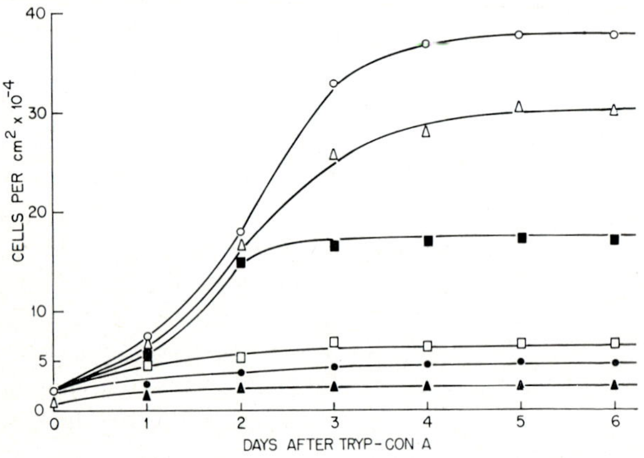

Fig. 10. Growth response of Py3T3 cells to varying concentrations of trypsinized Con A. ○—○ Py3T3, control, △—△ Py3T3 + 10 μg T-Con A, ■—■ Py3T3 + 25 μg T-Con A, □—□ Py3T3 + 50 μg T-Con A, ●—● Py3T3 + 75 μg T-Con A, ▲—▲ 3T3, control

From these experiments it may be suggested that the molecular structure and perhaps the configuration of the cell surface is important in *in vitro* growth control. The question which arises is how the "signal" for growth stimulation or cessation of growth gets from the membrane to the nucleus thereby controlling the division process.

It has recently been shown by a number of investigators [40, 41, 42] that the addition of dibutryl-cyclic AMP to cultures of transformed cells will significantly reduce the final saturation density of the particular cells. It has also been shown that in most cases tested the transformed cells contain less c-AMP than the normal cells [43]. Thus it has been hypothesized that a drop in c-AMP may be responsible for the loss of growth control exhibited by transformed cells.

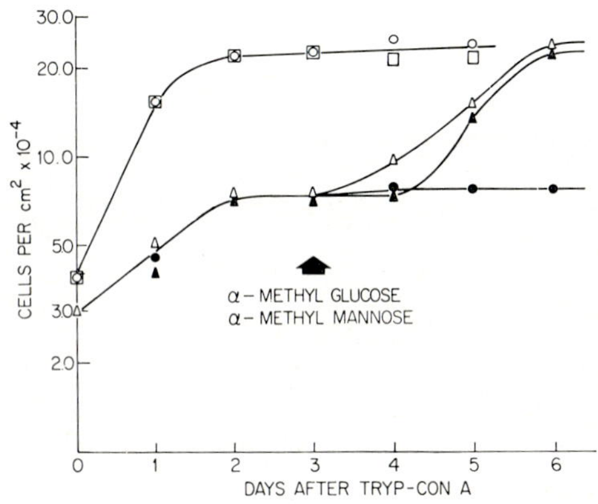

Fig. 11. Reestablishment of density dependent inhibition of growth in Polyoma virus transformed 3T3 mouse fibroblasts. ○—○ Py3T3, ●—● Py3T3 + T-Con A, □—□ T-Con A added together with 10^{-3} M α-methyl mannose from day 0, △—△ 10^{-2} M hapten added at day 3, ▲—▲ 10^{-3} M hapten added at day 3

Fig. 8 demonstrates that if dibutyryl-c-AMP is added immediately prior to or just after treatment of 3T3 cells with Pronase the subsequent division will be prevented, suggesting that protease treatment of the normal cells reduces the c-AMP levels of the cells and that this may play a role in inducing the cells to divide [44]. That 3T3 cells treated with protease do show a reduction in c-AMP levels has been demonstrated by SHEPPARD [43].

We have suggested earlier that cell division and growth control may be mediated in animal cells via the surface membrane [45], in

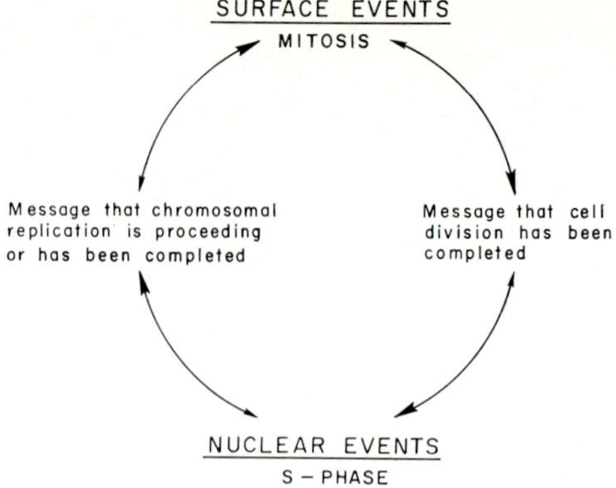

Fig. 12. Model for the interrelation of surface alterations in mitotic normal and transformed cells. For simplicity only the covered and uncovered state of the lectin receptor is considered but other mechanisms leading to the nonagglutinable and agglutinable state should not be excluded

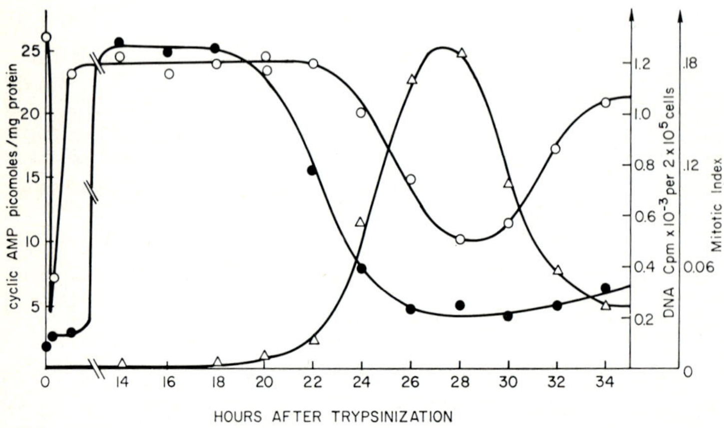

Fig. 13. 3T3 cells were synchronized by trypsinization of 4 days old monolayers and plated at 5 times lower densities with 20% calf serum in Dulbecco's modified Eagle's medium. The ratio of cells in mitosis to total cells (mitotic index) was determined with the phase microscope. All points are averages of three independent experiments. ○—○ cyclic-AMP level, ●—● DNA synthesis, △—△ mitotic index

a manner analogous to the scheme suggested by JACOB, BRENNER and CUZIN for bacteria (Fig. 12).

The hypothetical messenger triggering rounds of cell division could be a decline in intracellular c-AMP levels. Such a finding would corroborate the inhibition of growth stimulation by proteases described above. As can be seen in Fig. 13, we have found that the levels of c-AMP drop during mitosis [44]. We would like to suggest that this drastic decrease in intracellular c-AMP levels may be involved in the control process responsible for maintenance of the cell cycle although much work is needed to unequivocally support this hypothesis.

In summary we have shown that transformed cells have a different surface architecture than do normal cells and that investigation into the molecular basis of this alteration can now be taken up. We have also shown that this surface architecture may play a role in growth control. How the possible "message" is transported from the surface membrane to the nucleus is an open question, although c-AMP does appear to play some role in the process. Just what this role is must await further experimental approaches.

References

1. BURGER, M. M.: In: Current Topics in cellular regulation, Vol. 3 (HORECKER, B., STADTMAN, M., Eds.). New York: Academic Press 1972.
2. HAKAMORI, S., MURAKAMI, W. T.: Proc. nat. Acad. Sci. (Wash.) **59**, 254 (1968).
3. MEEZAN, E. W., BLACK, P. H., ROBBINS, P. W.: Biochemistry **8**, 2518 (1969).
4. BUCK, C. A., GLICK, M. C., WARREN, L.: Biochemistry **9**, 4567 (1970).
5. MORA, P. T., CUMAR, F. A., BRADY, R. O.: Virology **46**, 60 (1971).
6. BURGER, M. M.: Proc. nat. Acad. Sci. (Wash.) **62**, 494 (1969).
7. — Nature (Lond.) **227**, 170 (1970).
8. — NOONAN, K. D.: Nature (Lond.) **228**, 512 (1970).
9. — GOLDBERG, A. R.: Proc. nat. Acad. Sci. (Wash.) **57**, 359 (1967).
10. NAGATA, Y., BURGER, M. M.: J. biol. Chem. **247**, 2248 (1972).
11. WRIGHT, C., NAGATA, Y., LANGDRIDGE, R., BURGER, M. M.: FASEB Abstract, 1972.
12. AGRAWAL, B. B. L., GOLDSTEIN, I. J.: Biochem. J. **96**, 230 (1965).
13. INBAR, M., SACHS, L.: Proc. nat. Acad. Sci. (Wash.) **63**, 1418 (1969).
14. AUB, J. C., TIESLAU, C., LANKESTER, A.: Proc. nat. Acad. Sci. (Wash.) **50**, 613 (1963).
15. JANSONS, V. K., BURGER, M. M.: Submitted for publication.
16. BURGER, M. M.: In: Permeability and function of biological membranes (BOLIS, C., Ed.). Amsterdam: North Holland Publ. 1970.

17. Nicolson, G. L.: Nature (Lond.) New Biol. **233**, 244 (1971).
18. Noonan, K. D., Burger, M. M.: Submitted for publication.
19. — Thesis. Princeton University 1972.
20. Cline, M. J., Livingston, D. C.: Nature (Lond.) New Biol. **232**, 156 (1971).
21. Ozanne, B., Sambrook, J.: Nature (Lond.) New Biol. **232**, 156 (1971).
22. Arndt-Jovin, D., Berg, P.: J. Virol. **8**, 716 (1971).
23. Sela, B., Lis, H., Sharon, N., Sachs, L.: J. Membr. Biol. **3**, 267 (1970).
24. Uhlenbruck, G., Girlen, W., Pardoe, G. I.: Z. Krebsforsch. **74**, 171 (1970).
25. Pollack, R. E., Burger, M. M.: Proc. nat. Acad. Sci. (Wash.) **62**, 1074 (1969).
26. Inbar, M., Rabinowitz, M., Sachs, L.: Int. J. Cancer **4**, 690 (1969).
27. Eckhart, W., Dulbecco, R., Burger, M. M.: Proc. nat. Acad. Sci. (Wash.) **68**, 283 (1971).
28. Burger, M. M., Martin, G. S.: Nature (Lond.) New Biol. **237**, 9 (1972).
29. Benjamin, T. L., Burger, M. M.: Proc. nat. Acad. Sci. (Wash.) **67**, 929 (1970).
30. Renger, H. C., Basilico, C.: Proc. nat. Acad. Sci. (Wash.) **69**, 109 (1972).
31. Noonan, K. D., Renger, H. C., Basilico, C., Burger, M. M.: In preparation.
32. Ozanne, B., Sambrook, J.: In: 2nd Le Petit Colloquium on the biology of oncogenic viruses (Vrewey, E., Ed.). Amsterdam: North Holland Publ. 1971.
33. Culp, L. A., Black, P. H.: J. Virol. **9**, 611 (1972).
34. Shier, W. T.: Proc. nat. Acad. Sci. (Wash.) **68**, 2078 (1971).
35. Sefton, B. M., Rubin, H.: Nature (Lond.) **227**, 170 (1970).
36. Burger, M. M.: Ciba Symposium on growth control in cell cultures (Wolstenholme, G. E. W., Ed.). Boston: Little, Brown 1971.
37. — Current topics in developmental biology. (in press).
38. Schnebli, H. P., Burger, M. M.: Proc. nat. Acad. Sci. (Wash.) (in press).
39. Noonan, K. D., Burger, M. M.: Unpublished obersvation.
40. Hsie, A. W., Puck, T. T.: Proc. nat. Acad. Sci. (Wash.) **68**, 358 (1971).
41. Johnson, G. S., Friedman, R. M., Pastan, I.: Proc. nat. Acad. Sci. (Wash.) **68**, 425 (1971).
42. Sheppard, J. R.: Proc. nat. Acad. Sci. (Wash.) **68**, 1316 (1971).
43. — Nature (Lond.) **236**, 14 (1972).
44. Burger, M. M., Bombik, B., Breckenridge, B., Sheppard, J. R.: Nature (Lond.) (in press).
45. Fox, T. O., Sheppard, J. R., Burger, M. M.: Proc. nat. Acad. Sci. (Wash.) **68**, 244 (1971).

Discussion

M. J. Crumpton (London): Mac Pherson and others have reported that the glycolipids of normal cells differ from those of transformed cells. If these reports are correct then it seems likely that normal and transformed cells possess different glycosyl transferases and that the gene product associated with viral transformation is a glycosyl transferase. Do you agree with these views.

M. M. Burger (Princeton): I agree that this is an attractive hypothesis. Mora et al. [Proc. natl. Acad. Sci. (Wash.) **63**, 1290 (1969)] have shown that the activity of glycosyl transferases is reduced in transformed cells. Most recently Roth and Roseman have suggested that glycosyl transferases may play a role in cell—cell interactions.

M. J. Crumpton (London): What role, if any, do the glycolipids play in the interaction of the cell surfaces with plant lectins?

M. M. Burger (Princeton): At present this is an open question. Dr. Jansons and I have isolated a glycoprotein fraction which seems to be free of glycolipids which does inhibit WGA initiated agglutination. However this does not rule out the possibility that another glycolipid may also bind WGA [Hakamori: Biochem. biophys. Res. Commun. **33**, 563 (1968)].

M. J. Crumpton (London): You reported that Concanavalin A killed cells in culture. In contrast, although concanavalin also agglutinates lymphocytes, it induces these cells to enlarge and to undergo division. Is the reason for this apparent discrepancy known?

M. M. Burger (Princeton): Sachs [Nature (Lond.) **227**, 1244 (1970)] has independently demonstrated the same effect. Lymphocytes and fibroblasts must be entirely different cell types and must have entirely different regulatory systems, e.g. c-AMP activates division of lymphocytes while it inhibits division in fibroblasts. Furthermore if Con A is added to lymphocytes clustering and cap formation occurs while in fibroblasts we did not see such cap formation [Turner, R., Burger, M. M.: Unpublished observation]. (Comment on answer: Nicolson has reported that Con A also causes the collection of receptors of transformed cells.)

F. H. Bernhardt (Homburg/Saar): In your last scheme you indicated that serum participates in the regulation of cell growth in culture. Is there anything known about a serum factor which mediates this regulation?

M. M. Burger (Princeton): This has indeed been well studied, e.g. Kruse, P. F., Miedema, E.: J. Cell Biol. **27**, 273 (1965); Birierle, J. W.: Science **161**, 798 (1968) and much work from Holley's group, Proc. natl. Acad Sci. (Wash.) **68**, 2799 (1972).

R. Drzeniek (Hamburg): Some years ago we treated red blood cells with neuraminidase and observed the appearance of N-acetylglucosamine. May I suggest therefore the use of neuraminidase for the treatment of your cells to "unmask" N-acetylglucosamine-instead of using trypsin, pronase or other proteases.

M. M. Burger (Princeton): We have unexpectedly found neuraminidase to inhibit agglutination of cells with WGA. We believe that by cleaving the sialic acid from the WGA receptor site we may produce a conformational rearrangement of the receptor site thereby reducing agglutination.

Concluding Remarks

E. HELMREICH

Physiologisch-Chemisches Institut der Universität, Würzburg, Germany

I will not even try to give a comprehensive summary of the proceedings and discussions of this symposium. I will restrict myself to a few rather general remarks. In a recent fascinating paper, [Naturwissenschaften 58, 465, (1971)], MANFRED EIGEN develops mathematically his idea that the nonlinear coupling between complementary self-reproductive instruction inherent in nucleic acids and the enormous functional capabilities of the proteins, including specific recognition, catalysis and regulation, provides an explanation for the evolution of biological systems of ever-increasing complexity. This statement may serve both as an introduction and as a justification for a meeting on protein interactions. Protein-protein interactions mainly determine the phenomenological diversity of the structure and function of living matter. In particular, the conformational variability on the level of tertiary and quaternary structures enables the genetically instructed living systems to respond flexibly to environmental changes. However, generalizations from one protein to another are rather hazardous. In this Colloquium we have heard reports on several enzymes and non-catalytic proteins endowed with specific recognition consisting of identical or different subunits, among the latter being hemoglobin and antibodies. Among the enzymes discussed were aspartate transcarbamylase, glutamic dehydrogenase and phosphorylase and finally multienzyme complexes, i.e.: tryptophane synthetase, pyruvate dehydrogenase, fatty-acid synthetase, the respiratory chain, and the glycogen organelle of muscle. The last system impressively demonstrates how difficult it still is to decide how much of the regulatory potential of oligomeric enzymes is actually expressed under conditions closely resembling those in the living cell.

Progress in the field of protein-protein interactions, as in any other field of the natural sciences, relies on technical and methodical developments. We have seen impressive examples of the contributions made by the X-ray crystallographers and electron microscopists. Other important techniques involve the use of fluorescent and of other probes. The most impressive case where a biological function could be reduced to discrete chemical and physical events, is hemoglobin. There are very few other instances where it is profitable today to discuss a complex biological function in terms of discrete steps in the interaction of several different proteins. What comes to mind here is muscle contraction. Protein antigen-antibody interactions have interesting features because they show how heterologous interactions in general influence protein structure. One might speculate that even less specific interactions such as stabilization on surfaces, i.e. coupling proteins to a rigid matrix, may also alter considerably protein interactions. The scientific part of the meeting has just been concluded by Professor Max Burger who gave us an exciting glimpse of an area which is just beginning to yield to analysis at the molecular level, namely cell-cell interactions in tissue culture. Here we saw the first experimental approaches to the molecular basis of morphogenesis, with some of its very important biological and medical implications.

Of course, a quantitative analysis of protein interactions at all levels depends on our understanding of the physical and chemical forces involved. These forces are responsible for the propagation of conformational changes resulting from the binding of specific ligands, but they are also the driving force in the self-assembly of biological systems as exemplified by the tobacco mosaic virus. The present theory is treated in the important paper of Professor Shneior Lifson, who was unfortunately unable to deliver it in person, but kindly provided us with the manuscript.

Finally, I hope that all of you in the audience enjoyed listening to these interesting papers as much as did the organizers of this symposium, Rainer Jaenicke und myself. Therefore, we wish to thank our distinguished guests and speakers, who came from other European countries, from Israel, and from the United States, for giving us this splendid opportunity to discuss and obtain up-to-date information on diverse aspects of this fascinating field.